T0183746

Lecture Notes in Artificial Intelligence   11450

Subseries of Lecture Notes in Computer Science

More information about this series at http://www.springer.com/series/1244

Marija Slavkovik (Ed.)

Multi-Agent Systems

16th European Conference, EUMAS 2018
Bergen, Norway, December 6–7, 2018
Revised Selected Papers

 Springer

Editor
Marija Slavkovik ⓘ
University of Bergen
Bergen, Norway

ISSN 0302-9743   ISSN 1611-3349 (electronic)
Lecture Notes in Artificial Intelligence
ISBN 978-3-030-14173-8   ISBN 978-3-030-14174-5 (eBook)
https://doi.org/10.1007/978-3-030-14174-5

Library of Congress Control Number: 2019932172

LNCS Sublibrary: SL7 – Artificial Intelligence

This Springer imprint is published by the registered company Springer Nature Switzerland AG
The registered company address is: Gewerbestrasse 11, 6330 Cham, Switzerland

Preface

This volume contains revised versions of the papers presented at the 16th European Conference on Multi-Agent Systems (EUMAS 2018), which was held at the University of Bergen in Norway, during December 6–7, 2018. There were 34 submissions. Each submission was reviewed by three Program Committee members. The committee decided to accept 18 papers.

EUMAS 2018 followed the tradition of previous editions (Oxford 2003, Barcelona 2004, Brussels 2005, Lisbon 2006, Hammamet 2007, Bath 2008, Agia Napa 2009, Paris 2010, Maastricht 2011, Dublin 2012, Toulouse 2013, Prague 2014, Athens 2015, Valencia 2016, and Evry 2017) in aiming to provide a forum for presenting and discussing agents research as the annual designated event of the European Association of Multi-Agent Systems (EURAMAS). Both original submitted papers and already published work of relevance were presented at the conference. EUMAS 2018 was collocated with the 6th International Conference on Agreement Technologies (AT 2018). The proceedings of AT 2018 are published as a separate volume.

The year 2018 was marked by widespread public interest in artificial intelligence. Traditionally, EUMAS followed a spirit of providing a forum for discussion and an annual opportunity for primarily European researchers to meet and exchange ideas. For this reason, we encouraged submission of papers that report on both early and mature research. The peer-review processes carried out by both conferences put great emphasis on ensuring the high quality of accepted contributions.

The conference organizers would like to thank the administrative staff of the Department of Information Science and Media Studies at the University of Bergen that facilitated many aspects of the organization and without which the event would not have been feasible. I would also like to thank the University of Bergen, for making available the facilities in which the conference was held, and the student volunteers. We would like to acknowledge the financial support of the IKTPLUSS initiative of the Norwegian Research Council.

Last, but not least, we would like to express our appreciation for the EasyChair platform, which was used in the reviewing process and in the preparation of these proceedings.

January 2019

Marija Slavkovik

Organization

Program Committee

Stéphane Airiau	LAMSADE, Université Paris-Dauphine, France
Natasha Alechina	University of Nottingham, UK
Fred Amblard	IRIT, Université de Toulouse 1 Capitole, France
Merlinda Andoni	Heriot-Watt University, UK
Ana L. C. Bazzan	Universidade Federal do Rio Grande do Sul, Brazil
Olivier Boissier	Mines Saint-Etienne, Institut Henri Fayol, France
Thomas Bolander	Technical University of Denmark, Denmark
Vicent Botti	Universitat Politècnica de València, Spain
Ioana Boureanu	University of Surrey, UK
Cristiano Castelfranchi	Institute of Cognitive Sciences and Technologies, Italy
Sofia Ceppi	The University of Edinburgh, UK
Georgios Chalkiadakis	Technical University of Crete, Greece
Amit Chopra	Lancaster University, UK
Massimo Cossentino	National Research Council of Italy, Italy
Mehdi Dastani	Utrecht University, The Netherlands
Ronald de Haan	University of Amsterdam, The Netherlands
Marina De Vos	University of Bath, UK
Louise Dennis	University of Liverpool, UK
Catalin Dima	LACL, Université Paris Est - Créteil, France
Dragan Doder	IRIT - UPS, France
Sylvie Doutre	IRIT, Université de Toulouse 1, France
Amal El Fallah Seghrouchni	LIP6, Université Pierre et Marie Curie, France
Edith Elkind	University of Oxford, UK
Nicoletta Fornara	Università della Svizzera Italiana, Switzerland
Nicola Gatti	Politecnico di Milano, Italy
Malvin Gattinger	University of Groningen, The Netherlands
Benoit Gaudou	IRIT, Université de Toulouse, France
Nina Gierasimczuk	Technical University of Denmark, Denmark
Valentin Goranko	Stockholm University, Sweden
Davide Grossi	University of Groningen, The Netherlands
Andreas Herzig	CNRS, IRIT, Université de Toulouse, France
Magdalena Ivanovska	University of Oslo, Norway
Franziska Klügl	Örebro University, Sweden
Michał Knapik	ICS PAS, Poland
Dušan Knop	Charles University, Czech Republic
Joao Leite	Universidade NOVA de Lisboa, Portugal
Brian Logan	University of Nottingham, UK
Dominique Longin	IRIT-CNRS, France

Contents

Temporal Epistemic Gossip Problems

Martin C. Cooper$^{(\boxtimes)}$, Andreas Herzig, Frédéric Maris, and Julien Vianey

IRIT, CNRS, Univ. Toulouse, 31062 Toulouse Cedex 9, France
{martin.cooper,andreas.herzig,frederic.maris,julien.vianey}@irit.fr

Abstract. Gossip problems are planning problems where several agents have to share information ('secrets') by means of phone calls between two agents. In epistemic gossip problems the goal can be to achieve higher-order knowledge, i.e., knowledge about other agents' knowledge; to that end, in a call agents communicate not only secrets, but also agents' knowledge of secrets, agents' knowledge about other agents' knowledge about secrets, etc. Temporal epistemic gossip problems moreover impose constraints on the times of calls. These constraints are of two kinds: either they stipulate that a call between two agents must necessarily be made at some time point, or they stipulate that a call can be made within some possible (set of) interval(s). In the non-temporal version, calls between two agents are either always possible or always impossible. We investigate the complexity of the plan existence problem in this general setting. Concerning the upper bound, we prove that it is in NP in the general case, and that it is in P when the problem is non-temporal and the goal is a positive epistemic formula. As for the lower bound, we prove NP-completeness for two fragments: problems with possibly negative goals even in the non-temporal case, and problems with temporal constraints even if the goal is a set of positive atoms.

Keywords: Epistemic planning · Temporal planning · Gossip problem · Complexity · Epistemic logic

1 Introduction

The epistemic gossip problem defined in [11,12,21] is a problem in which n agents each have a secret. Agents communicate by calling other agents: during a call the two agents share all their knowledge, not only the secrets they have learned but also epistemic information concerning which agents know which information. The goal of this problem concerns agents' knowledge about other agents' secrets at various epistemic depths. For example, the goal may be shared knowledge of depth 2: all agents know that all agents know all secrets. Such goals can be described as logical formulas in Dynamic Epistemic Logic of Propositional Assignments and Observation DEL-PAO [17,18]. This setting generalises the well-known gossip problem [12,16] which has recently been analysed in the framework of dynamic epistemic logics [6]. We consider it to be an exemplary

M. Slavkovik (Ed.): EUMAS 2018, LNAI 11450, pp. 1–14, 2019.
https://doi.org/10.1007/978-3-030-14174-5_1

case of epistemic planning [7, 8, 20, 22] in which communication actions are used to spread knowledge among a network of agents.

Here, we enrich this setting by allowing to limit any communication to a set of instants (such as an interval during which the two agents involved in the call are both available). For an example, one can think of satellites on different orbits which can only communicate when they 'see' each other. In a more down-to-earth example, the interval during which a mobile communication is available is often limited by the charge capacity of the battery. Another variant we study is when certain calls must occur at given instants (for example for maintenance or security reasons).

What follows applies to either one-way or two-way communication and to either sequential or parallel communication. During a one-way call (such as a letter or email) information only passes in one direction, whereas during a two-way call (such as a telephone conversation) information passes in both directions. In the case of parallel communication, several calls between distinct pairs of agents may take place simultaneously, but an agent can only call one other agent at the same instant. The sequential version, in which only one call can take place at the same time, is of interest when the aim is to minimise the total number of calls.

We show that the temporal epistemic gossip problem is in NP even for a complex goal given in the form of a CNF and in the presence of constraints on the instants when calls can or must take place. This positive result, when compared to classical planning which is PSPACE-complete [9], follows from the reasonable assumption that knowledge is never destroyed. Moreover, we show that in the absence of temporal constraints and negative goals, temporal epistemic gossiping is in P. We then show maximality of this tractable subproblem in the sense that the problem becomes NP-complete in the following cases: in the presence of temporal constraints (even as weak as a simple upper bound on the execution time of a plan) and in the presence of negative goals (such as agent i should not learn the secret of agent j).

2   Definitions

First of all, we introduce a general framework for epistemic planning before focussing on the specific subproblem of epistemic gossiping.

Let *Prop* be a countable set of *propositional variables* and *Agt* a finite set of *agents*. A *knowledge operator* is of the form K_i with $i \in Agt$. An *atom of depth d* is any sequence of knowledge operators K_i of length d followed by a propositional variable. (So when the depth is 0 then the atom is just a propositional variable.) Atoms are noted α, α', etc. The atom $K_i p$ is of depth 1 and reads "agent i knows that p"; the atom $K_j K_i p$ is of depth 2 and reads "j knows that i knows that p"; and so on. The set of all atoms of depth at most d is noted $ATM^{\leq d}$. Observe that if the depth of atom $\alpha \in ATM^{\leq d}$ is strictly less than d then $K_i \alpha$ also belongs to $ATM^{\leq d}$. The set of atoms that are about the mutual knowledge of α by agents i and j up to depth d is:

$$ATM_{i,j}^{\leq d}(\alpha) = \{K_{k_1}\ldots K_{k_r}\alpha \mid k_1,\ldots,k_r \in \{i,j\} \text{ and } r + depth(\alpha) \leq d\}$$

Finally, the set of *boolean formulas* Fml_{bool} is comprised of formulas with the following grammar, where $\alpha \in ATM^{\leq d}$:

$$\varphi ::= \alpha \mid \neg\varphi \mid (\varphi \wedge \varphi)$$

A *state* is an assignment of truth values to all atoms in $ATM^{\leq d}$ and is represented by the set of atoms which are assigned the value true. Satisfaction of a boolean formula ϕ in a state s, noted $s \models \phi$, is defined in the usual way.

A *conditional action* is a pair $\mathsf{a} = \langle pre(\mathsf{a}), \mathit{eff}(\mathsf{a})\rangle$ where:

- $pre(\mathsf{a}) \in Fml_{bool}$ is a boolean formula: the *precondition* of a;
- $\mathit{eff}(\mathsf{a}) \subseteq Fml_{bool} \times 2^{ATM^{\leq d}} \times 2^{ATM^{\leq d}}$ is a set of triples ce of the form

$$\langle cnd(ce), \mathit{ceff}^+(ce), \mathit{ceff}^-(ce)\rangle,$$

the *conditional effects* of a, where $cnd(ce)$ is a boolean formula (the condition) and $\mathit{ceff}^+(ce)$ and $\mathit{ceff}^-(ce)$ are sets of atoms (added and deleted atoms respectively).

The result of executing action a in a state s is the state $(s \cup e^+) \setminus e^-$, where $e^+ = \bigcup_{ce\in \mathit{eff}(\mathsf{a}),s\models cnd(ce)} \mathit{ceff}^+(ce)$ and $e^- = \bigcup_{ce\in \mathit{eff}(\mathsf{a}),s\models cnd(ce)} \mathit{ceff}^-(ce)$.

In the case of the gossip problem, $Prop = \{s_{ij} \mid i,j \in Agt\}$ where s_{ij} reads "i knows j's secret". So $K_k s_{ij}$ means that agent k knows that i knows j's secret. The actions in the gossip problem are calls between two agents leading to an update of the two agents' knowledge. In one-way calls, after a call from agent i to agent j, agent j knows everything that agent i knew before the call and both know that they know these atoms. Indeed, they both know that they both know that they know these atoms, and so on up to the maximum epistemic depth d.[1] More formally, $call(i,j) = \langle pre(call(i,j)), \mathit{eff}(call(i,j))\rangle$ with $pre(call(i,j)) = \top$ and $\mathit{eff}(call(i,j)) = \{\langle K_i\alpha, ATM_{ij}^{\leq d}(\alpha), \emptyset\rangle \mid \alpha \in ATM^{\leq d}\} \cup \{\langle \top, ATM_{ij}^{\leq d}(s_{ji}), \emptyset\rangle\}$. Two-way calls have the same effect as two simultaneous one-way calls.

An instance of the depth-d temporal epistemic gossip problem (TEGP) is given by a tuple $\Pi = \langle Init, Goal, Agt, I_p, I_n\rangle$:

$$Init \subseteq ATM^{\leq d} \text{ such that } Init \text{ contains every } s_{ii}, \text{ for } i \in Agt$$
$$Goal \in Fml_{bool} \text{ is a conjunction of clauses}$$
$$I_p \subseteq \mathbb{N} \times (\mathbb{N} \cup \{\infty\}) \times Agt \times Agt$$
$$I_n \subseteq \mathbb{N} \times Agt \times Agt$$

where $Init$ is the initial state; $Goal$ is the goal we want to achieve in the form of a CNF formula (that we identify with a set of clauses); I_p is the set of intervals

---

[1] More generally, the caller's knowledge becomes common knowledge between i and j. We however have no common knowledge operator in our framework.

during which two agents can call each other and I_n is the set of instants when two agents must call each other. The set I_n of necessary calls may correspond to calls that have been programmed in the network for some other purpose. We suppose that I_n is included in I_p, in the sense that for every $\langle t, i, j \rangle \in I_n$ there is a $\langle t_1, t_2, i, j \rangle \in I_p$ such that $t_1 \leq t \leq t_2$. In this paper we always consider the initial state $Init = \{s_{ii} \mid i \in Agt\}$ in which all agents know their own secrets.

A set of calls A between agents induces a partial function between states, i.e. from $2^{ATM^{\leq d}}$ to $2^{ATM^{\leq d}}$. For a state $s \in 2^{ATM^{\leq d}}$:

$$
\mathsf{A}(s) = \begin{cases} \bot & \begin{aligned} &\text{if } \exists \mathsf{a} \in \mathsf{A} : s \nvDash pre(\mathsf{a}), \text{ or } \exists \mathsf{a}_1, \mathsf{a}_2 \in \mathsf{A} : \\ &\mathsf{a}_1 = call(i_1, j_1) \text{ and } \mathsf{a}_2 = call(i_2, j_2) \\ &\text{with } \{i_1, j_1\} \cap \{i_2, j_2\} \neq \emptyset \end{aligned} \\[2em] s \cup \bigcup_{\substack{\mathsf{a} \in \mathsf{A}, \\ ce \in eff(\mathsf{a}), \\ \text{and } s \vDash cnd(ce)}} ceff^+(ce) & \text{otherwise} \end{cases}
$$

Note that $\bot$ is not a state: a result $\bot$ represents a failure of the simultaneous execution of the set of actions A, either because a precondition does not hold or because one agent would be participating in two calls at the same time (which we assume to be impossible).

A *plan* is a relation $P \subseteq \mathbb{N} \times Agt \times Agt$. Given a plan P and a natural number t, the set of calls happening at instant t is $P(t) = \{(i,j) : (t,i,j) \in P\}$. We use $|P|$ to denote the number of distinct instants t for which $P(t) \neq \emptyset$. We use $T_P(k)$ to denote the k-th instant (in strictly increasing order of time) at which a call happens in P: i.e. $T_P(1) < \ldots < T_P(|P|)$ and $\forall t, P(t) \neq \emptyset \Leftrightarrow \exists k \in \{1, \ldots, |P|\}$, $T(k) = t$. Our modelling of time by the natural numbers implicitly imposes a fixed duration of one time unit for each call.

Given a TEGP $\Pi = \langle Init, Goal, Agt, I_p, I_n \rangle$, a plan P *satisfies the temporal constraints* of Π if and only if all the necessary calls are in P and every call in P is possible; formally: $I_n \subseteq P$ and for every $\langle t, i, j \rangle \in P$ there is a $\langle t_1, t_2, i, j \rangle \in I_p$ such that $t_1 \leq t \leq t_2$. Moreover, P *solves* the TEGP if and only if it satisfies the temporal constraints and there is a sequence of states $\langle s_0, \cdots, s_{|P|} \rangle$ such that

- $s_0 = Init$
- $s_{|P|} \vDash Goal$
- $s_{k+1} = P(T_P(k+1))(s_k)$ for every k with $0 \leq k < |P|$

where $P(T_p(k+1))$ is the set of actions at instant $T_p(k+1)$ and $P(T_p(k+1))(s_k)$ is the result of executing these actions in state s_k. By the definition above of $\mathsf{A}(s)$, the set of actions $P(T_p(k+1))$ at instant $T_p(k+1)$ cannot contain two calls involving the same agent. In the sequential version of the TEGP, a solution plan P must also satisfy $\forall t, card(P(t)) \leq 1$.

A TEGP defines in a natural way a call digraph G in which the vertices are the agents and the directed edges the possible calls. In the two-way version, G is a graph.

Example 1. Consider a network of five servers (which we call a, b, c, d and e) where each server can only communicate with a subset of the others. Note that all calls are assumed to be one-way in this example. As part of the maintenance program, a, b, c and d send a backup of their data to e every night and these backups can be sent to any server during the day (between 8:00 and 18:00). The others servers can communicate with each other at any moment if there is a communication link between them. The communication graph is depicted in Fig. 1.

There is some information on the server b that needs to be transferred to a, and c must know that the transfer is done. As the servers have different access rights, the information on server a should not be communicated to c. In the TEGP this can be represented by $Goal = K_c s_{ab} \land \neg s_{ca}$. There is a family of solution plans for this problem: $call(b, a)$ at instant t_1, $call(b, d)$ at instant t_2, $call(d, c)$ at instant t_3, where $t_1 < t_2 < t_3$ (together with the necessary calls $call(a, e)$ at instant 2, $call(b, e)$ at instant 4, $call(c, e)$ at instant 20 and $call(d, e)$ at instant 22).

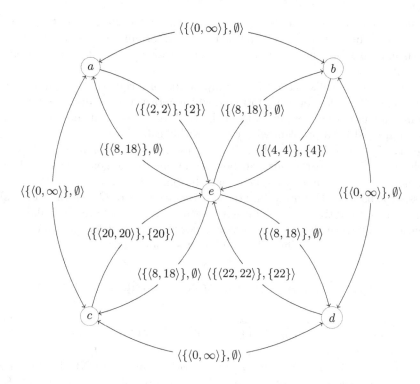

Fig. 1. Call graph for Example 1 involving necessary calls with e. A double-ended arrow represents two directed edges (i.e., the possibility of one-way calls in both directions).

On the same network, another question that we can ask is whether c can know a's data without a being aware of this. In this case, the goal is $s_{ca} \land \neg K_a s_{ca}$.

The answer is 'yes' since the following plan establishes the goal: the necessary $call(a, e)$ at instant 2, followed by $call(e, c)$ at an instant $t \in [8, 18]$ (together with the other necessary calls $call(b, e)$ at instant 4, $call(c, e)$ at instant 20 and $call(d, e)$ at instant 22).

3   Membership in NP

Proposition 1. *Let m be the number of clauses in the CNF of the goal. Let d be the depth of atoms for the problem. If a plan for an instance of a TEGP exists, then there is a plan with $md(n - 1) + |I_n|$ calls or less*

Proof. Let $\alpha \in ATM^{\leq d}$ be an atom. $K_i \alpha$ can only be true if there is a path in the graph G between i and some agent who knows α. Without loss of generality, we can assume that this path is cycle-free. In the worst case the length of this path is $n - 1$. Then, for any atom with an epistemic depth d, at most $d(n - 1)$ calls are needed for this atom to be true, by the concatenation of d paths of length $n - 1$.

The number of calls needed for a disjunction of formulas to be true is the maximum of the number of calls needed for each formula. In a CNF, there are only disjunctions over atoms, so the number of calls needed for a disjunction is $d(n - 1)$.

The number of calls needed for a conjunction of formulas is the sum of the number of calls needed for every formula, which here is at most $d(n-1)$. So, with m being the number of conjunctions in the CNF of the goal, at most $md(n - 1)$ calls are needed for a problem with only possible calls.

Thus, if a plan P exists, then P contains a subset Q of $md(n - 1)$ calls which are sufficient to establish all positive atoms in the goal. For a problem with necessary calls, it can happen that the plan Q does not contain all the necessary calls I_n of P; but adding these necessary calls to Q to form a plan Q' cannot destroy positive goals. All negative atoms in the goal are also valid after execution of Q' since all calls, and in particular the ones in P but not in Q', only establish positive atoms. Thus Q' is a plan of at most $md(n - 1) + |I_n|$ calls. $\square$

4   A Subproblem of the Temporal Gossip Problem in P

We say that a TEGP instance is *positive* if its goal is a CNF containing only positive atoms. A special case of TEGP is the class of positive non-temporally-constrained epistemic gossip problems $\Pi = \langle Init, Goal, Agt, I_p, I_n \rangle$ where $I_p = \{\langle 0, \infty, i, j \rangle : (i, j) \in E\}$, for some E, and $Goal$ is a positive CNF. In this case, E is the set of edges in the call digraph: if a call is possible (as specified by E), it is possible at any instant. On the other hand, there is no restriction on the set of necessary calls I_n.

Proposition 2. *The class of positive non-temporally-constrained epistemic gossip problems can be solved in polynomial time.*

Proof. There is a simple polynomial-time algorithm for positive non-temporally-constrained epistemic gossip problems: make all possible calls in some fixed order and repeat this operation $md(n-1)+|I_n|$ times. Call this sequential plan Q. By the proof of Proposition 1, if a solution plan exists, there is a sequential solution plan P of length at most $md(n-1)+|I_n|$. The actions of P necessarily appear as a subsequence of Q. Since the goal and preconditions of actions contain only positive atoms, the extra actions of Q cannot destroy any goals or preconditions. It follows that if a solution plan exists, then Q is also a solution plan. Thus this simple algorithm solves the class of positive non-temporally-constrained epistemic gossip problems in polynomial time.

Given an arbitrary instance of TEGP, we can construct a positive non-temporally-constrained instance by ignoring negative goals and temporal constraints (specified by I_p). This is a polynomial-time solvable relaxation of the original TEGP instance. This provides a relaxation which is inspired by the well-known delete-free relaxation of classical planning problems and is orthogonal to the relaxation of temporal planning problems based on establisher-uniqueness and monotonic fluents [13].

5   NP-completeness When Execution Time Is Bounded

The simplest temporal constraint is just a time limit on the execution of a plan. In the case of sequential plans this simply corresponds to placing a bound on plan length (which is equal to the number of calls) whereas in the parallel case execution time corresponds to the number of steps. We show in this section that this single constraint (a time limit on plan execution) is sufficient to render the epistemic gossip problem NP-complete. It is worth noting that the PSPACE complexity of classical planning is not affected by the possibility of placing an arbitrary limit on plan length, but the special case of delete-free planning passes from P to NP-hard when a bound is placed on plan length [9]. We show that this remains true for the specific case of gossiping problems.

We begin by studying the sequential case of TEGP.

Proposition 3. *The epistemic gossip problem with no temporal constraints but with a bound on the number of calls is NP-complete, even when the goal is a conjunction of positive atoms.*

Proof. We will exhibit a polynomial-time reduction from the well-known NP-complete problem SAT to the version of the epistemic gossip problem whose question is whether there is a sequential solution plan of length at most L. To do so, for a given set of clauses $\{C_1, \ldots, C_m\}$ we need the following agents:

- an agent S (the source),
- literal agents, i.e., agents for every variable and every negation of a variable (which we name, respectively, x^+ and x^-) for each variable x of the SAT instance,

– clause agents, i.e., agents for every clause (which we name C_i for the ith
 clause of the SAT instance).

Before performing this construction, we first add a dummy clause $(x \vee \neg x)$ for
each SAT variable x. This clearly does not change the semantics of the instance
but it does force us to specify the truth value of each variable in a solution of
the SAT instance.

The source agent S and clause agents can only communicate with literal
agents. The source agent S can communicate with every literal agent. A literal
agent can only communicate with S and those clauses it is a member of. The
graph G of communications is shown in Fig. 2 for a particular SAT instance. In
this example, $C_1 = (\neg x \vee y \vee z)$, $C_2 = (\neg y \vee z)$ and the clauses C_3, C_4, C_5 are
the dummy clauses $(x \vee \neg x)$, $(y \vee \neg y)$, $(z \vee \neg z)$.

A variable x is considered to be true (false) if S's secret passes through
x^+ (respectively, x^-) in the solution plan on its way to the agent representing
the dummy clause $(x \vee \neg x)$. The bound on the number of actions will prevent
the possibility of S's secret passing through both x^+ and x^-. So the choice of
whether S's secret passes through x^+ or x^- determines an assignment to the
variable x in the SAT problem.

The goal of this instance of TEGP is that every clause agent knows the secret
of S ($Goal = \bigwedge_{C_i} s_{C_i} S$). Now set the bound on plan length to be $L = 2n + m$,
where n is the number of variables in the SAT instance and m the number of
clauses in the original instance. With the new dummy clauses, the total number
of clauses is $n + m$.

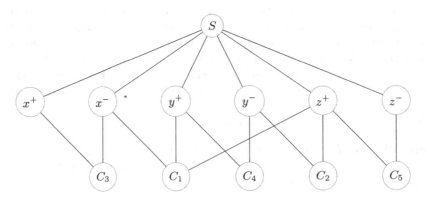

Fig. 2. Representation of the formula $(\neg x \vee y \vee z) \wedge (\neg y \vee z)$ as a temporal epistemic
gossip problem in which the question is whether there is a plan using no more than 8
calls.

In a solution plan P we require at least n calls, one to either x^+ or x^-,
for the n SAT variables x, in order for S's secret to be able to reach the agent
corresponding to the dummy clause $x \vee \neg x$. P must also contain at least $n + m$
calls to the clause agents C_i (including the dummy clauses) to establish the

goals $s_{C_i,S}$. A solution plan of length precisely $2n + m$ corresponds to a solution of the corresponding SAT instance since such a plan defines a unique assignment to all variables that satisfies all clauses. For example, the solution $x = false$, $y = false$, $z = true$ to the SAT instance of Fig. 2 corresponds to the following solution plan of length 8: S calls x^-; S calls y^-; S calls z^+; x^- calls C_1; z^+ calls C_2; x^- calls C_3; y^- calls C_4; z^+ calls C_5. This reduction from SAT is clearly polynomial.

Proposition 1 proves the existence of a polynomial-length certificate for positive instances of the decision version of TEGP. Such certificates (solutions) can be verified in polynomial time. Thus TEGP $\in$ NP. Since the epistemic gossip problem with no temporal constraints but with a bound on the number of calls is clearly still in NP, this completes the proof of NP-completeness.

The proof of Proposition 3 was given for the case of two-way communications. It is trivial to adapt it to the case of one-way communications (for example, by only allowing calls from S to literal agents and from literal agents to clause agents).

We now consider the parallel version of the TEGP. Recall that in the parallel version of the TEGP, several calls may take place at each step, provided no agent is concerned by more than one call at each step.

Proposition 4. *The parallel version of the epistemic gossip problem with no temporal constraints except for a bound on the number of steps is NP-complete even when the goal is a conjunction of positive atoms.*

Proof. By the same argument as in the proof of Proposition 1, the problem is in NP. We complete the proof by exhibiting a polynomial reduction from 3SAT which is well known to be NP-complete. Given an instance I_{3SAT} of 3SAT, by introducing sufficiently many new variables x' which are copies of old variables x (together with the clauses $x \lor \neg x'$, $\neg x \lor x'$ to impose equality of x and x') we can transform I_{3SAT} into an equivalent instance in which each literal does not occur in more than three clauses. This is a polynomial reduction since we need to introduce at most one copy of each variable x per clause in which it occurs in I_{3SAT}. Therefore, from now on, we suppose that each literal occurs in at most two clauses in I_{3SAT}.

We construct an instance I of the epistemic gossip problem which has a parallel solution plan of length $2p$ if and only if I_{3SAT} is satisfiable. We choose the value of p to be strictly greater than $n+3$, where n is the number of variables in I_{3SAT}. To be concrete, we can choose $p = n + 4$. We add to I_{3SAT} $p - n$ new dummy variables $x_{n+1}, \ldots, x_p$ none of which occur in the clauses of I_{3SAT}. In I there is an agent S (the source), literal agents x_i^+, x_i^- for each variable x_i ($i = 1, \ldots, p$), and a clause agent C_j for each of the clauses C_j ($j = 1, \ldots, m$) of I_{3SAT}. For each variable x_i ($i = 1, \ldots, p$), we also add a dummy-clause agent D_i which we can consider as representing the dummy clause $x_i \lor \neg x_i$. Instead of linking these basic agents directly, we place paths of new agents between these basic agents. Between agent S and agent x_i^+ we add a path of length $p + 1 - i$. Similarly, we add a new path of the same length between S and agent x_i^-. For

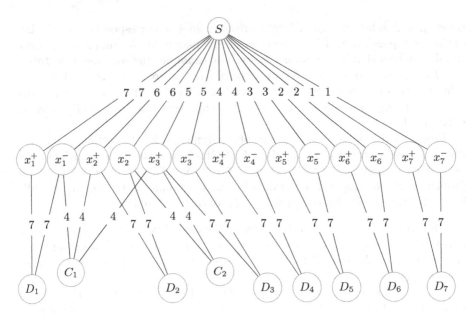

Fig. 3. Representation of the formula $(\neg x \lor y \lor z) \land (\neg y \lor z)$ as a temporal epistemic gossip problem in which the question is whether there is a parallel plan using no more than 14 steps.

$i = 1, \ldots, p$, we add two new paths both of length p between the literal agents x_i^+ and x_i^- and the dummy-clause agent D_i. For each clause C_j of I_{3SAT}, we also add three new paths of length q from the agents corresponding to the literals of C_j to the agent C_j, where $q = p - 2 = n + 1$. The resulting network is shown in Fig. 3 for an example instance. The numbers on edges in this figure represent the length of the corresponding path. For example, there are 6 intermediate agents (not shown so as not to clutter up the figure) between the agents S and x_1^+. The goal of I is

$$\left(\bigwedge_{i=1}^{p} s_{D_i S} \right) \land \left(\bigwedge_{j=1}^{m} s_{C_j S} \right)$$

In order to establish the goal $s_{D_i S}$, the secret of S has to follow a path from S to D_i. The shortest paths from S to D_1 are of length $2p$ and pass through either x_1^+ or x_1^-. Recall that our aim is to find a plan whose execution requires at most $2p$ steps. Thus, to establish $s_{D_1 S}$ in $2p$ steps, during the first step, S must call the first agent on the path to x_1^+ or the first agent on the path to x_1^-. The shortest path from S to D_2 is of length $2p - 1$, so during the second step, S must call the first agent either on the path to x_2^+ or the first agent on the path to x_2^-. By a simple inductive argument, we can see that during step i $(i = 1, \ldots, p)$, S must call the first agent on the path to x_i^+ or x_i^-. We can consider that the choice of whether S's secret passes through x_i^+ or x_i^- determines an assignment to the variables x_i. Due to the diminishing lengths of these paths as i increases, S's secret arrives simultaneously at the literal agents, either x_i^+ or x_i^-, for $i = 1, \ldots, p$. Another p steps are then required to send in parallel this secret to the dummy-clause agents D_j, for a total number of steps of $2p$. Almost simultaneously (within two time units), S' secret arrives at the clause agents C_j, provided it has passed through one of the agents corresponding to the literals of C_j. The length of paths from literal agents (x_i^+ or x_i^-) to clause agents C_j is $q = p - 3$ which is slightly less than p to allow for the fact that a literal agent, say x_i^+, may have to send S's secret along at most four paths: first towards D_i, then towards the (at most) three clauses in which x_i occurs.

It is important to note that S is necessarily occupied during the first p steps, as described above, so if S were to try to send its secret both to x_i^+ and x_i^- the secret could not arrive via the second of these paths at a clause agent C_j in less than $2p - n + q = 3p - n - 3$ steps which is greater than the upper bound of $2p$ steps (since $p = n + 4$). By our construction, the goal $s_{C_j}S$ is established only if the assignment to the variables x_i determined by the solution plan satisfies the clause C_j. Hence, parallel solution plans of length $2p$ steps correspond precisely to solutions of I_{3SAT}. We have therefore demonstrated a polynomial reduction from 3SAT to the parallel version of the epistemic gossip problem with a bound on the number of steps.

The following corollary follows from the fact that we can place an upper bound L on the number of steps in a plan by simply imposing via I_p an interval of possible instants $[1, L]$ for all calls.

Corollary 1. *TEGP is NP-complete.*

6 NP-completeness of Gossiping with Negative Goals

We show in this section that even without temporal constraints or a bound on plan length, when we allow negative goals the problem of deciding the existence of a solution plan is NP-complete.

Proposition 5. *The epistemic gossip problem with possibly negative goals is NP-complete even in the absence of any temporal constraints or bound on plan length.*

Proof. The same argument as in the proof of Proposition 1 shows that the problem belongs to NP since it is a subproblem of TEGP.

To complete the proof, it suffices to give a polynomial reduction from SAT. Let I_{SAT} be an instance of SAT. We will construct a call graph G and a set of goals such that the corresponding instance I_{Gossip} of the epistemic gossip problem is equivalent to I_{SAT}. Recall that the nodes of the call graph G are the agents and the edges of G the communication links between agents.

For each propositional variable x in I_{SAT}, we add four nodes x^+, x^-, b_x, d_x to G joined by the edges shown in Fig. 4(b). There is a source node S in G and edges (S, x^+), (S, x^-) for each variable x in I_{SAT}. For each clause C_j in I_{SAT}, we add a node C_j joined to the nodes corresponding to the literals of C_j. This is illustrated in Fig. 4(a) for the clause $C_j = \neg x \vee y \vee z$. The solution plan to I_{Gossip} will make S's secret transit through x^+ (on its way from S to some clause node C_j) if and only if $x = true$ in the corresponding solution to I_{SAT}.

For each clause C_j in I_{SAT}, G contains a clause gadget as illustrated in Fig. 4(a) for the clause $\neg x \vee y \vee z$. We also add $s_{C_j s}$ to the set of goals. Clearly, S's secret must transit through one of the nodes corresponding to the literals of C_j (x^-, y^+ or z^+ in the example of Fig. 4) to achieve the goal $s_{C_j s}$.

To complete the reduction, it only remains to impose the constraint that a's secret transits through at most one of the nodes x^+, x^-, for each variable x of I_{SAT}. This is achieved by the negation gadget shown in Fig. 4(b) for each variable x. We add the goals $s_{d_x b_x}$ and $\neg s_{d_x s}$ for each variable x, and the goal $\neg s_{C_j b_x}$ for each variable x and each clause C_j (containing the literal x or $\neg x$). The goal $s_{d_x b_x}$ ensures that b_x's secret transits through x or $\neg x$. Now, recall that we assume that during a call, agents communicate all their knowledge. Suppose that b_x's secret transits through x^+: then S's secret cannot transit through x^+ before b_x's secret (because of the negative goal $\neg s_{d_x s}$) and cannot transit through x^+ after b_x's secret (because of the negative goal $\neg s_{C_j b_x}$). By a similar argument, if b_x's secret transits through x^-, then S's secret cannot transit through x^+. Thus, this gadget imposes that S's secret transits through exactly one of the nodes x^+, x^-.

We have shown that I_{SAT} has a solution if and only if I_{Gossip} has a solution. Since the reduction is clearly polynomial, this completes the proof.

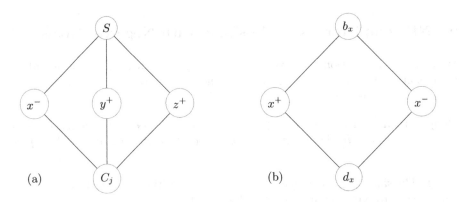

(a)   (b)

Fig. 4. (a) Gadget imposing the clause $C_j = \neg x \vee y \vee z$; (b) Gadget imposing the choice between x and $\neg x$.

The NP-completeness shown in the proof of Proposition 5 for two-way communication, would not be affected by a restriction to one-way communication. Similarly NP-completeness holds for both the sequential and parallel versions of the gossip problem.

7   Discussion and Conclusion

We have defined temporal epistemic gossip problems and have investigated their complexity. Our results are in line with previous results concerning epistemic planning: it is possible to add an epistemic dimension to planning, thus increasing expressibility, without increasing complexity [10].

In our approach agents are not introspective: s_{ij} does not imply $K_i s_{ij}$, and $K_k s_{ij}$ does not imply $K_k K_k s_{ij}$. This only concerns positive introspection: negative introspection cannot be expressed. Positive introspection can however be enforced by adding axioms $s_{ij} \rightarrow K_i s_{ij}$, and $K_k s_{ij} \rightarrow K_k K_k s_{ij}$. We however did not do so in order to simplify presentation.

We have assumed a centralized approach in which a centralized planner decides the actions of all agents. Several other researchers have recently studied distributed versions of the classical gossip problem where the agents have to decide themselves whom to call, based on the knowledge (and ignorance) they have [1–5,14,15]. An interesting avenue of future research would be to consider the epistemic gossip problem in this framework.

Several other variants of our centralized model could also be investigated, including the precondition that i has to know the telephone number of j in order to call j and telephone numbers are communicated in the same way as secrets. In another variant, the secrets can be passwords which are no longer constants since each agent i can change their own password [12].

References

1. Apt, K.R., Grossi, D., van der Hoek, W.: Epistemic protocols for distributed gossiping. In: Ramanujam, R. (ed.) Proceedings Fifteenth Conference on Theoretical Aspects of Rationality and Knowledge, TARK 2015. EPTCS, vol. 215, pp. 51–66 (2015). https://doi.org/10.4204/EPTCS.215.5
2. Apt, K.R., Grossi, D., van der Hoek, W.: When are two gossips the same? Types of communication in epistemic gossip protocols. CoRR abs/1807.05283 (2018). http://arxiv.org/abs/1807.05283
3. Apt, K.R., Kopczynski, E., Wojtczak, D.: On the computational complexity of gossip protocols. In: Sierra, C. (ed.) Proceedings of the Twenty-Sixth International Joint Conference on Artificial Intelligence, IJCAI 2017, pp. 765–771. ijcai.org (2017). https://doi.org/10.24963/ijcai.2017/106
4. Apt, K.R., Wojtczak, D.: Common knowledge in a logic of gossips. In: Lang, J. (ed.) Proceedings Sixteenth Conference on Theoretical Aspects of Rationality and Knowledge, TARK 2017. EPTCS, vol. 251, pp. 10–27 (2017). https://doi.org/10.4204/EPTCS.251.2
5. Apt, K.R., Wojtczak, D.: Decidability of fair termination of gossip protocols. In: Eiter, T., Sands, D., Sutcliffe, G., Voronkov, A. (eds.) IWIL@LPAR 2017 Workshop and LPAR-21 Short Presentations, vol. 1. Kalpa Publications in Computing, EasyChair (2017). http://www.easychair.org/publications/paper/342983
6. Attamah, M., van Ditmarsch, H., Grossi, D., van der Hoek, W.: A framework for epistemic gossip protocols. In: Bulling, N. (ed.) EUMAS 2014. LNCS (LNAI), vol. 8953, pp. 193–209. Springer, Cham (2015). https://doi.org/10.1007/978-3-319-17130-2_13

7. Aucher, G., Bolander, T.: Undecidability in epistemic planning. In: Rossi, F. (ed.) Proceedings of the 23rd International Joint Conference on Artificial Intelligence, IJCAI 2013, 3–9 August 2013, Beijing, China, pp. 27–33. IJCAI/AAAI (2013). http://www.aaai.org/ocs/index.php/IJCAI/IJCAI13/paper/view/6903
8. Bolander, T., Andersen, M.B.: Epistemic planning for single and multi-agent systems. J. Appl. Non-Classical Logics 21(1), 9–34 (2011). https://doi.org/10.3166/jancl.21.9-34
9. Bylander, T.: The computational complexity of propositional STRIPS planning. Artif. Intell. 69(1–2), 165–204 (1994). https://doi.org/10.1016/0004-3702(94)90081-7
10. Cooper, M.C., Herzig, A., Maffre, F., Maris, F., Régnier, P.: A simple account of multi-agent epistemic planning. In: Kaminka et al. [19], pp. 193–201. https://doi.org/10.3233/978-1-61499-672-9-193
11. Cooper, M.C., Herzig, A., Maffre, F., Maris, F., Régnier, P.: Simple epistemic planning: generalised gossiping. In: Kaminka et al. [19], pp. 1563–1564. https://doi.org/10.3233/978-1-61499-672-9-1563
12. Cooper, M.C., Herzig, A., Maffre, F., Maris, F., Régnier, P.: Simple epistemic planning: generalised gossiping. CoRR abs/1606.03244 (2016). http://arxiv.org/abs/1606.03244
13. Cooper, M.C., Maris, F., Régnier, P.: Monotone temporal planning: tractability, extensions and applications. J. Artif. Intell. Res. 50, 447–485 (2014). https://doi.org/10.1613/jair.4358
14. van Ditmarsch, H., van Eijck, J., Pardo, P., Ramezanian, R., Schwarzentruber, F.: Epistemic protocols for dynamic gossip. J. Appl. Logic 20, 1–31 (2017). https://doi.org/10.1016/j.jal.2016.12.001
15. van Ditmarsch, H., Grossi, D., Herzig, A., van der Hoek, W., Kuijer, L.B.: Parameters for epistemic gossip problems. In: Proceedings of LOFT 2016 (2016)
16. Hedetniemi, S.M., Hedetniemi, S.T., Liestman, A.L.: A survey of gossiping and broadcasting in communication networks. Networks 18(4), 319–349 (1988). https://doi.org/10.1002/net.3230180406
17. Herzig, A., Lorini, E., Maffre, F.: A poor man's epistemic logic based on propositional assignment and higher-order observation. In: van der Hoek, W., Holliday, W.H., Wang, W. (eds.) LORI 2015. LNCS, vol. 9394, pp. 156–168. Springer, Heidelberg (2015). https://doi.org/10.1007/978-3-662-48561-3_13
18. Herzig, A., Maffre, F.: How to share knowledge by gossiping. AI Commun. 30(1), 1–17 (2017). https://doi.org/10.3233/AIC-170723
19. Kaminka, G.A., et al. (eds.): 22nd European Conference on Artificial Intelligence, Frontiers in Artificial Intelligence and Applications, ECAI 2016, vol. 285. IOS Press (2016)
20. Löwe, B., Pacuit, E., Witzel, A.: DEL planning and some tractable cases. In: van Ditmarsch, H., Lang, J., Ju, S. (eds.) LORI 2011. LNCS (LNAI), vol. 6953, pp. 179–192. Springer, Heidelberg (2011). https://doi.org/10.1007/978-3-642-24130-7_13
21. Maffre, F.: Ignorance is bliss: observability-based dynamic epistemic logics and their applications. Ph.D. thesis, Paul Sabatier University, Toulouse, France (2016). https://tel.archives-ouvertes.fr/tel-01488408
22. Muise, C.J., et al.: Planning over multi-agent epistemic states: a classical planning approach. In: Bonet, B., Koenig, S. (eds.) Proceedings of the Twenty-Ninth AAAI Conference on Artificial Intelligence, 25–30 January 2015, Austin, Texas, USA, pp. 3327–3334. AAAI Press (2015). http://www.aaai.org/ocs/index.php/AAAI/AAAI15/paper/view/9974

Partial and Full Goal Satisfaction in the MUSA Middleware

Massimo Cossentino, Luca Sabatucci[(✉)], and Salvatore Lopes

National Research Council,
Istituto di Calcolo e Reti ad Alte Prestazioni (ICAR-CNR), Palermo, Italy
{massimo.cossentino,luca.sabatucci,salvatore.lopes}@icar.cnr.it

Abstract. Classical goal-based reasoning frameworks for agents suppose goals are either achieved fully or not achieved at all: unless achieved completely, the agents have failed to address them. This behavior is different from how people do and therefore is far from real-world scenarios: in every moment a goal has reached a certain level of satisfaction.

This work proposes to extend the classical boolean definition of goal achievement by adopting a novel approach, the Distance to Goal Satisfaction, a metric to measure the distance to the full satisfaction of a logic formula.

In this paper we defined and implemented this metric; subsequently, we extended MUSA, a self-adaptive middleware used to engineer a heterogeneous range of applications. This extension allows solving real situations in which the full achievement represented a limitation.

Keywords: Partial goal satisfaction · Metric · Multi-agent system

1 Introduction

Exploring alternative options is at the heart of rational and self-adaptive behavior. In our applications, agents always try to find plans to fully achieve their goals. However, this is different from how people do. A person often adopts plans for partially achieving her/his goals.

Recently, a research direction claims agents should be able of reaching a better approximation of scenarios arising from the real world. Indeed, Zhou and Chen argue that reasoning with a partial satisfaction of goals is an essential issue for achieving this result [19]. For instance, GoalMorph [18] is a framework for context-aware goal transformation to facilitate fault-tolerant service composition. It achieves planning with partial goal satisfaction by reformulating failed goals into problems that can be solved by the planner.

Two main trends exist in dealing with partial goal satisfaction. The former studies partial goal satisfaction by looking at the goal model as a whole (whether it is a goal hierarchy or a goal graph). Conversely, the second trend focuses on the single goal entity, by entering into the details of how it is represented.

© Springer Nature Switzerland AG 2019
M. Slavkovik (Ed.): EUMAS 2018, LNAI 11450, pp. 15–29, 2019.
https://doi.org/10.1007/978-3-030-14174-5_2

Concerning the first direction, Letier and Van Lamsweerde [6] present a technique for quantifying the impact of alternative system designs on the degree of goal satisfaction. The approach consists of enriching goal models with a probabilistic layer for reasoning about partial satisfaction. Within such models, non-functional goals are specified in a precise, probabilistic way; their specification is interpreted in terms of application-specific measures; the impact of alternative goal refinements is evaluated in terms of refinement equations over random variables involved in the system's functional goals.

This paper focuses on the second research direction, i.e., studying the partial satisfaction of a single goal.

An interesting contribution comes from Zhou et al. [20]. They define the partial implication to capture partial satisfaction relationship between two propositional formulas. According to their propositional language, a formula like $x \wedge z$ partially implies $x \wedge y$. Indeed, x, which can be considered as a part of $x \wedge y$, is a logical consequence of $x \wedge z$.

Van Riemsdijk and Yorke-Smith [17] propose a higher-level framework based on metric functions that represent the progress towards achieving a goal. Progress appraisal is the capability of an agent to assess which part of a goal it has already achieved. However, the framework is abstract: authors do not further detail how a partial ordering function may be implemented. According to the authors, it can be based on a wide range of either domain-independent parameters, such as time, utility, number of subgoals, and domain-dependent metrics.

Thangarajah et al. [15,16] adopt an approach base on resource analysis to provide a BDI agent with a quantified measure of effort with respect to the amount of resources consumed in executing a goal with respect to the total resource required. This kind of approach is strictly domain-dependent, and it may be applied only when there is a clear link between goals and resources.

This paper presents a novel approach for enabling agents to reason on partial satisfaction of single goals (where a goal condition is given in predicate logic). The approach is based on the definition of a Distance to Goal Satisfaction that measures the distance to the full satisfaction of a logic formula, given the current state of the world. To practically implement this metric we adopt an analogy between the logic formulas specifying goal satisfaction and an equivalent electric circuit.

The approach has been modeled to be consistent with the classical logic semantics. It is essentially domain-independent but easily extensible for considering other domain-dependent parameters. An interesting side effect of this approach is that the metric allows to calculate the 'cost for the full satisfaction', thus providing a powerful tool for selecting among alternative strategies.

We applied this metric in the context of a self-adaptive middleware (MUSA) in a heterogeneous range of application contexts (from IoT to maritime IT applications).

The remaining of the paper is structured as follows: Sect. 2 presents the middleware for self-adaptive systems and how we encountered the need for dealing with different degrees of goal satisfaction. Section 3 provides definitions for the core concepts of the proposed approach. Section 4 provides technical details about how MUSA has been extended for supporting the new metric. Section 5 discusses some of the key aspects of the approach, and finally, Sect. 6 summarizes the approach and reports some future works.

2   Motivation

MUSA (Middleware for User-driven Self-Adaptation) [11,13] is a middleware for composing and orchestrating distributed services according to unanticipated and dynamic user needs. It is a platform in which (1) virtual enterprises can deploy capabilities that wrap real services, completing them with a semantic layer for their automatic composition; (2) analysts and users can inject their goals for requesting a specific outcome. Under the hypothesis that both goals and capabilities refer to the same ontology, agents of the system can compose available services into plans for addressing user's requests.

The enablers of the MUSA core vision are: (i) representing *what* and *how* as run-time artifacts the system may reason on (respectively goals and capabilities); (ii) a reasoning system for connecting capabilities to goals; (iii) finally a common grounding semantic, represented with some formalism.

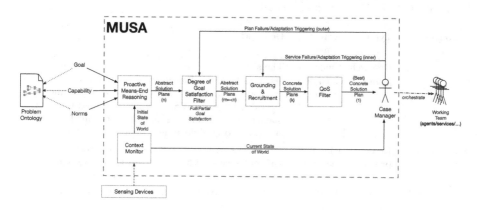

Fig. 1. Main components of MUSA.

The main abstractions in MUSA are that of *Capability*, which represents the knowledge about what the system can do, and that of *Goals*, representing the user's expectations. In MUSA, a user can describe the desired functionality via a high-level goal-oriented description language [14]. At run-time, once a goal model is specified, it may be injected in the system.

The concept of capability comes from the AI (planning actions [3]) and service-oriented architecture (micro-services [7]). MUSA presents this dual nature by adopting a separation of the abstract capability – a description of the effect of an action that can be performed – and the concrete capability – a small, independent, composable unit of computation that produces some concrete service.

Figure 1 illustrates the logical schema of MUSA with its macro components. The proactive means-end reasoning [11] is the main component, which aggregates abstract capabilities for deducing possible solutions, each one that can guarantee the desired goal state. The set of solutions are, at this stage still abstract (i.e., not bounded to real services), but this allows evaluating to which extent they will satisfy the goals. This way, MUSA can apply a filter to the valid solutions in order to select the best ones of those above a prescribed threshold (according to the application context in which it operates).

After discarding the non-satisfying solutions, MUSA creates a temporary link between abstract capabilities and real services via the Grounding and Recruitment module (the link is temporary because it can change at run-time for self-adaptation purposes), thus generating concrete solutions. These can be finally evaluated from other (non-functional) perspectives, for instance evaluating the fitness toward a QoS value.

MUSA has been employed in a heterogeneous set of application contexts. A selection of these is shortly illustrated in Table 1.

Table 1. Summary of research projects and case studies where the MUSA middleware has been employed between 2013 and 2018.

Acronym	Type	App. name	Description
IDS	Research project	Innovative document sharing	The aim has been to realize a prototype of a new generation of a digital document solution that overcomes current operating limits of the common market solutions. MUSA has been adopted for managing and balancing human operations for enacting a digital document solution in a SME
Smart grid	Research project	Cloud mashup	MUSA has been adopted as mashup engine, i.e., for self-configuring ad-hoc orchestration of existing services in order to address run-time business
Smart travel	Case study	Travel agency system	MUSA provides the planning engine that creates a travel-pack as the composition of several heterogeneous travel services. Traveler's goals drive the planning activity
UPA4SAR	Research project	IoT and robotic home assistance	MUSA is the core for implementing the behavior of smart devices deployed in a simulated environment
SPS reconfiguration	Case study	Electric management	MUSA is the core for the algorithm for run-time reconfiguration of the shipboard power system in case of malfunctioning

During its usage designers and developers have pointed out the need for extending the flexibility of MUSA goal definition in order to deal with many degrees of goal satisfaction. To clarify this concept, we report some examples.

Example 1: in the Smart Travel application [10], the tourist is able to configure its travel request by injecting a set of user's goals. For instance (s)he can desire "to book a 5-star hotel in the center of *TownX* with a pool". Sometimes too constrained goals produced zero results in the travel configuration so that the tourist has to manually change its goals with no information about what part of the goal was unsatisfiable. To solve this problem, the system should include a mechanism to relax the condition, thus finding solutions that partially solve the problem (for instance either a hotel with a pool but out of town, or in downtown with no pool).

Example 2: the Shipboard Power System (SPS) reconfiguration problem consists in acting on electric switches for changing power flows and ensuring ship's components are powered on when necessary. In the MUSA application [12], the different ship's missions are coded by using a set of goals (with different priorities). This proved to be an interesting case study for system self-adaptation [1]. An example of a goal is "the navigation system, the communication system, and the radar should be powered". In case of severe malfunctioning, when the main generators are off, the auxiliary generators are switched on. However, they can not produce enough power for the three components work properly. For such a reason MUSA should produce plans for partially restoring the energy on board, even if not all goals are fully satisfied.

Example 3: the UPA4SAR project will deliver an IoT based robotic home assistance to help elderly patients to remain at home, thus avoiding hospitalization or admission to long-term care institutions. In MUSA application [8] this high-level goal is decomposed in a set of subgoals for daily monitoring of patient activities and providing services for helping individuals to live with greater independence and promoting the optimal level of well-being. One of the subgoals of the system is something like "in the afternoon, propose some entertainment activity, as long as there is time". Actually, the goal is often unfeasible because there are (frequent) situations in which an elderly patient changes its mood in the meanwhile the system is proposing its activities. In these situations, the system cannot complete its goals, (because the activity does not terminate). In a boolean goal satisfaction approach, the system will declare a failure, whereas reasoning with partial satisfaction would be more appropriate.

Currently, the ability of MUSA to deal with goal satisfaction is inspired to the classical Kripke model, thus allowing associating a logical condition to a true-false interpretation. If W is the current state of the world and $W \models c$, then we can assert c is true; otherwise, it is false. This model is not suitable for calculating a partial degree of satisfaction when a formula is neither fully addressed nor totally unachieved. The next section introduces the background to let agents the ability to evaluate and measure partial goal satisfaction. The implementation and technical details are provided in Sect. 4.

3   Full and Partial Goal Satisfaction

In this section, we will provide some definitions of the most relevant concepts in the proposed approach. Let us draw a border to define our system; this is one of the most sensitive operations in many different problems. The system should include all the parts of the world that significantly participate in defining the behavior of the system itself. In the following, we will refer to discrete-time systems, and we will address successive time slots as successive steps in time.

Let be 'k' an arbitrary moment in time. Let us suppose our system is controlled by a set of inputs $U[k]$, where U is a vector of n elements, and k defines the time step where we consider the values of its elements. This system is supposed to be, at k, in a state $X[k]$ where X is a vector of m elements; the output of the system is designated as $Y[k]$, again a vector, of p elements.

We suppose there is a relationship among these vectors as follows:

$Y(k) = f(X(k), U(k))$

It is relevant to say we suppose to be able to observe the elements of the state vector X and the output vector Y (for instance by measuring them). An interesting consideration comes from Jackson, in fact, he studies the impact of environmental phenomena on the system [4], and he states that "All environment phenomena mentioned in the requirement are shared with the machine" [5]. For this reason, we can suppose that the environment is providing a part of the input vector while the remaining part is the result of some software computation.

Referring to such a model of our system, we can define the concept of goal:

Def. 1 - Goal: A goal specifies a desired condition for the output of the system, more specifically, a goal is a condition on the values of a subset of the output vector elements y_i.

This means a goal may aggregate values from different elements of the output vector by means of logic operators or mathematical formulas in order to specify the desired condition on the output of the system.

In order to better clarify that we will refer to two examples related to different systems:

1. An electric system composed of 3 loads (bulb lights), two generators providing input power to the system and three on/off switches to control the flow of power reaching the loads. In this case, the input vector is composed of the two values of voltage for the generators, the state vector of the on/off condition of each switch and the output vector of the current values in the three loads.
2. A production plant composed of two buildings: workshop and office. This plant operates 24/24 h all days of the week (except for holidays) for the workshop part while the office is open only on working days (Monday to Friday). The system is an advanced ambient intelligence system that can control (and personalize by specific needs and preferences) ambient temperature, humidity, light level (illumination) in the different places of the plant also according to some specific tasks the painting room requires higher temperatures to dry painting quickly. The system can control air conditioning, artificial light and rolling shutters in a continuous (analog) way.

Referring to the first example, goals may specify conditions like:

- $g_1 = [I_1 > 0]$ that means the first load is powered;
- $g_2 = [I_1 > 0 \text{ AND } I_2 = 0]$ that means the first load is powered while the second one is not;
- $g_3 = [I_3 = 4 \text{ Ampere}]$ that is a more detailed specification of the expected outcome allowing for the detection of anomalies in the system such as failures or unexpected negative influences by the environment (such as a flood causing current dispersion);

Looking at the second example, some goals may be:

- $g_4 = [L_{WRoom01} = 0]$ meaning that light in the workshop room 01 has to be switched off;
- $g_5 = [L_{Office101} = 500]$ meaning that illumination in the office 101 has to be 500 Lux;
- $g_7 = [L_{WRoom01} = 0 \text{ AND } L_{Office101} = 0]$ meaning that office 101 and workshop room 01 lights are switched off.

Being goals applied to the values of the output vector elements, it is interesting to note that more than one instances of this vector (i.e., the combination of different values of the output elements) may verify the same goal. Such a situation is represented in Fig. 2 where three different value instances of the output vector are shown (instance values of the vector are represented as upper-signed). We can suppose some goal of our system may be verified by the first two instances $(\bar{Y}', \bar{Y}'')$. This may happen for instance because the goal requires the element y_1 be greater than zero and the element y_2 be lesser than zero. Now let us suppose for the instance vector $\bar{Y}'''$ only the first one of these two conditions is verified. The goal is not satisfied by this set of output values but it is also obvious that the system is some way near to the desired condition.

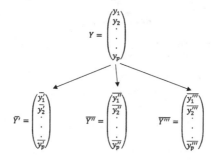

Fig. 2. Instances of the output vector with different element values.

In order to support the evaluation of such situations, we introduce the concept of distance to goal satisfaction as reported in the definition below.

Def. 2 - Distance to Goal Satisfaction: The Distance to Goal Satisfaction (DGS) is the distance of the current output vector $\bar{Y}$ from the nearest one that satisfies goal conditions.

Many different ways may be conceived to measure this distance, we decided to measure that by considering an electric analogy, and we introduced the measure of Resistance to Satisfaction (R2S) as a metric for DGS. We consider each condition specified in the goal as a piece of an electric circuit, more specifically as a resistor. The value of this resistor will be 0 (short circuit) if the condition is true (it means the corresponding value of the output vector verifies one of the goal conditions) while the value will be R_{max} if the condition is false where is a great value of resistance representing the open circuit condition. Let us consider the Y''' vector that does not verify goal conditions. For instance, let us suppose the goal specifies one unique condition: $y_1 > 0$. If the value of $y_1''' > 0$ then the value of the corresponding R2S $(y_1''') = 0$ otherwise it will be R2S $(y_1''') = R_{max}$ If the goal condition implies the evaluation of more than one element of the output vector we compose them as a growing complexity circuit where the AND condition is represented as a series of the resistors representing the two AND variables and the OR condition is represented as a parallel of the resistors representing the two OR variables. The NOT operator is simply dealt by inverting the measure values: if the result is true, R2S $= R_{max}$, otherwise R2S $= 0$. A description of the formulas and equivalent circuits is reported in Fig. 3.

Goal	R2S	Equivalent circuit
$g = \varphi\,(y_i)$	If $\begin{cases} \varphi\,(y_i) = True \Rightarrow R2S\,(\varphi\,(y_i)) = 0 \\ \varphi\,(y_i) = False \Rightarrow R2S\,(\varphi\,(y_i)) = R_{max} \end{cases}$	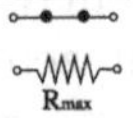
$g = y_i \wedge y_j$	$R2S = R_i + R_j$ where: If $\begin{cases} y_k = True \Rightarrow R_k = 0 \\ y_k = False \Rightarrow R_k = R_{max} \end{cases}$ with $k=i,j$	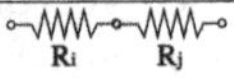
$g = y_i \vee y_j$	$\dfrac{1}{R2S} = \dfrac{1}{R_i} + \dfrac{1}{R_j}$ where: If $\begin{cases} y_k = True \Rightarrow R_k = 0 \\ y_k = False \Rightarrow R_k = R_{max} \end{cases}$ with $k=i,j$	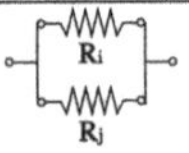
$g = !y_i$	If $\begin{cases} \varphi\,(y_i) = True \Rightarrow R2S\,(\varphi\,(y_i)) = R_{max} \\ \varphi\,(y_i) = False \Rightarrow R2S\,(\varphi\,(y_i)) = 0 \end{cases}$	

Fig. 3. Measures of the distance to goal satisfaction for logic formulas.

Of course, if the distance to goal satisfaction for a specific situation is 0, then the goal is fully satisfied.

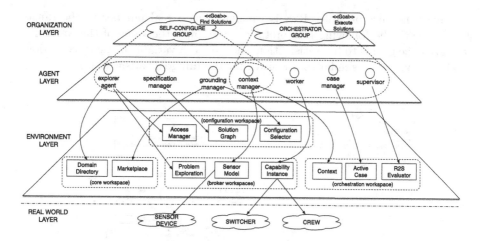

Fig. 4. The agents and artifacts view of the MUSA architecture.

4 The Agent and Artifact Architecture

This section presents the current MUSA architecture derived by that illustrated in [13] in 2017. It has been realized as a multi-agent system implemented in the Jason language [2] where the environment is modeled by an independent computation layer that encapsulates functionalities the agents can access and manipulate. These entities are called artifacts and are programmed via CArtAgO [9] as Java classes.

The new architecture is shown in Fig. 4 and it is decomposed in three layers (organization, agent and environment) whereas the main functionalities are organized in four workspaces: (i) core, (ii) configuration, (iii) broker and (iv) orchestration. The *core* workspace provides general facilities for the registration of agents and services. The *configuration* workspace is responsible for the self-configuration ability, providing a distributed algorithm for the discovery of solutions. The *broker* workspace is responsible for interfacing the real services available for creating and executing solutions. The orchestration workspace is responsible for granting the self-adaptive orchestration of the services that are contained in a solution. For more details about all the agents and artifacts of the solution, please refer to [13].

In the following, we provide more details about the supervisor agent of the orchestration workspace, who is responsible for checking at run-time the degree of satisfaction of goals.

4.1 Implementing the Metric for Partial Satisfaction

In order to provide MUSA with this feature, we will refer to the definitions provided in Sect. 3 where the *ResistanceToFullSatisfaction* (R2S) metric has been introduced.

Given the current state, R2S is calculated as a real number in the range $[0 - \infty]$, with the following interpretation: the lower the value of this variable, the closer the goal is to full satisfaction. Indeed, each variable in a goal condition produces something similar to a resistor for the full goal satisfaction.

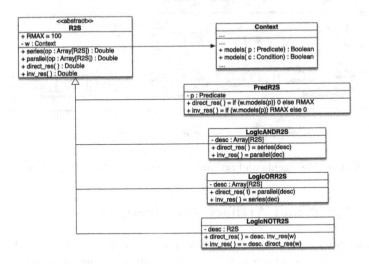

Fig. 5. Class diagram for implementing the ResistanceToFullSatisfaction metric. For reasons of conciseness we revealed the body of some of the methods with an inline notation (example: direct_res(w) = parallel(desc)).

The class diagram for implementing this metric is shown in Fig. 5 where R2S is the main abstract class that contains R_{Max} and a reference to the current Context object, and two public methods for calculating series and parallel of resistances.

To describe the algorithm, we start focusing on the NOT operator that is not directly translated into a formula. Indeed it changes the way how R2S is calculated for its operand:

- the negation of a predicate condition p implies, $R2S_{!p} = R_{Max}$ when $W \models p$ otherwise it is 0;
- the negation of an AND condition between a couple of predicates is calculated as a parallel of resistances: $R2S_{!AND} = \frac{1}{\sum \frac{1}{R2S_i}}$;
- whereas the negation of a OR condition is a series of resistances: $R2S_{!OR} = \sum R2S_i$.

In other words, these formulas are the opposite of those described in Fig. 3. To generalize this behavior, we define two modalities for calculating R2S: direct (the normal way - Fig. 3), and inverse (the negated way). We extend the notation with $R2S^!$ for indicating the latter modality.

Consequently, we can set: $R2S_{!\phi} = R2S_\phi^!$

For coherence, the abstract class (and all its descendants) have a couple of public methods: *direct_res* uses the direct modality, whereas *inv_red* uses the inverse modality.

The whole procedure for calculating R2S is recursive, and it is summarized in the following schema:

Formula	Class	$R2S(\phi)$	$R2S^!(\phi)$
$\phi = p$ (a predicate)	PredR2S	$0 \mid R_{Max}$	$R_{Max} \mid 0$
$\phi = \psi_1 \wedge \psi_2$	LogicANDR2S	$R2S_{\psi_1} + R2S_{\psi_2}$	$\dfrac{1}{\frac{1}{R2S_{\psi_1}} + \frac{1}{R2S_{\psi_2}}}$
$\phi = \psi_1 \vee \psi_2$	LogicORR2S	$\dfrac{1}{\frac{1}{R2S_{\psi_1}} + \frac{1}{R2S_{\psi_2}}}$	$R2S_{\psi_1} + R2S_{\psi_2}$
$\phi = !\psi$	LogicNOTR2S	$R2S^!(\psi)$	$R2S(\psi)$

5   Discussion

To discuss the introduced metrics, we will now propose a few examples. Let us consider the following goals:
$g_1 = (A \vee B) \wedge C$
$g_2 = A \vee (B \wedge C)$
and their corresponding equivalent circuits and resistances to satisfaction as reported in Fig. 6.

Goal	$g_1 = (A \vee B) \wedge C$	$g_2 = A \vee (B \wedge C)$
Equivalent circuit	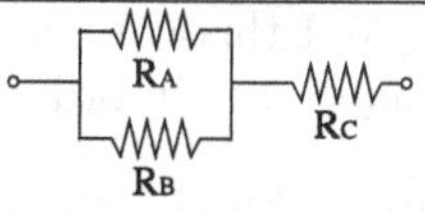	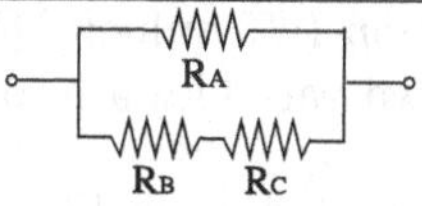
Resistance to satisfaction	$R2S_1 = \dfrac{R_A \cdot R_B}{R_A + R_B} + R_C$	$R2S_2 = \dfrac{R_A \cdot (R_B + R_C)}{R_A + R_B + R_C}$

Fig. 6. Examples of goals with related equivalent circuits and distance to satisfaction metrics.

Considering the different combinations of true/false values for the three variables, we can calculate the results as reported in Fig. 7 under the hypothesis: $R_A = R_B = R_C = R_{max}$. As it can be seen, regarding goal g_1, we can note three different conditions for full goal satisfaction (combinations # 3, 5, 7) and five different partial satisfaction conditions (combinations # 0, 1, 2, 4, 6). This situations are worth some considerations:

#	A	B	C	$R2S_1$
0	0	0	0	$\frac{3}{2}R_{max}$
1	0	0	1	$\frac{R_{max}}{2}$
2	0	1	0	R_{max}
3	0	1	1	0
4	1	0	0	R_{max}
5	1	0	1	0
6	1	1	0	R_{max}
7	1	1	1	0

#	A	B	C	$R2S_2$
0	0	0	0	$\frac{2}{3}R_{max}$
1	0	0	1	$\frac{R_{max}}{2}$
2	0	1	0	$\frac{R_{max}}{2}$
3	0	1	1	0
4	1	0	0	0
5	1	0	1	0
6	1	1	0	0
7	1	1	1	0

Fig. 7. Results for the resistance to satisfaction in the two example goals according to the different combinations true/false (1/0) of the variables (by hypothesis: $R_A = R_B = R_C = R_{max}$).

1. The first consideration is straightforward: if we associate some kind of cost evaluation to the satisfaction of each of the logical variables in the goal formula, and we assume each of them has the same cost, we can see that dealing with full goal satisfaction, conditions # 3, 5 are less costly than condition # 7. Of course, this could not be the case if the satisfaction of each variable is assumed to have a different cost. This allows for the selection of the most economical plan if more than one is available for goal satisfaction.
2. We may have three different levels of partial satisfaction for goal g_1 according to the different values of R2S. The smallest one (condition #1, value: $\frac{R_{max}}{2}$) represents the situation where there are two alternative paths towards full goal satisfaction (that is making either A or B true). Conditions #2, 4, 6 report a higher value for $R2S_1 = R_{max}$ and this is an intuitive representation of the fact that there is only one optional path towards full goal satisfaction, making C true. The highest value of the distance to goal satisfaction is represented by condition (#0 value: $\frac{3}{2}R_{max}$) and this correctly represents the fact that at least two variables are to be made true in order to achieve the goal.

Similar considerations may be drawn for the second goal of Fig. 6 and corresponding results of Fig. 7.

Another advantage of the proposed approach is that it allows the runtime evaluation of the progression towards goal satisfaction. This feature is relevant in the solution space exploration performed by MUSA. In fact, by continuously measuring the distance to goal satisfaction the system can monitor the successful execution of the selected strategy and to estimate the distance from the optimal path even if multiple goals are pursued at the same time.

Besides, the approach proves useful in monitoring goals where some proposition is to be maintained over time (maintenance goals) rather than reached once (achievement goals). While maintaining a goal, the immediate identification of a more than zero distance to satisfaction may trigger some corrective action that could also depend on the measured distance. In fact, in some conditions, this distance may indicate the system is in a stable or unstable situation, and different actions need to be performed. Maintenance goals rise other issues related to the role of time in their conditions. These issues may only partially be solved with the proposed approach: an extension is required, with some temporal logic, to explicitly manage (and measure) clauses where, for instance, the goal is to be maintained for (or after/before) a specific amount of time.

Finally, a very interesting improvement to the proposed approach could come from the introduction of different values of resistance for each of the variables in the goal proposition. This could represent a finer-grained representation of the user preferences in terms of the conditions inside a goal if it could not be fully satisfied. A clear example of that could be provided by considering the goal of finding a five-star category hotel, with pool, in the city center (already proposed in Sect. 2). Supposing in a specific city there is not such a hotel; the user would prefer a five-star hotel in the city center without a pool or a five-star hotel with a pool far from the city center? The proposed approach, theoretically speaking, fully supports this feature but this is not currently implemented in the MUSA middleware. We are working to introduce that in the next sub-release.

6   Conclusions

This work proposes to extend the classical definition of goal achievement that may represent a limit to the way agents reason and adapt themselves. The presented approach allows reasoning with partial satisfaction of goals. We presented the Distance to Goal Satisfaction, and a metric to measure the distance to the full satisfaction of a logic formula representing the goal.

We applied this metric in the context of a self-adaptive middleware (MUSA) in a heterogeneous range of application contexts (from IoT to maritime IT applications).

We are now working to extend the MUSA implementation with a specific support for different degrees of preference in the satisfaction of specific portions of the goal as discussed in Sect. 5.

Finally, we are also working on a major improvement of the approach that includes the adoption of Linear Temporal Logic in order to support temporal specifications in goals.

References

1. Agnello, L., Cossentino, M., De Simone, G., Sabatucci, L.: Shipboard power systems reconfiguration: a compared analysis of state-of-the-art approaches. In: Smart Ships Technology 2017, pp. 1–9. Royal Institution of Naval Architects (RINA) (2017)
2. Bordini, R., Hübner, J., Wooldridge, M.: Programming Multi-Agent Systems in AgentSpeak Using Jason, vol. 8. Wiley-Interscience, Chichester (2007)
3. Gelfond, M., Lifschitz, V.: Action languages. Comput. Inf. Sci. 3(16), 1–16 (1998)
4. Jackson, M.: Problem Frames: Analysing and Structuring Software Development Problems. Addison Wisley, Boston (2001)

5. Jackson, M., Zave, P.: Deriving specifications from requirements: an example. In: Proceedings of the 17th International Conference on Software Engineering, ICSE 1995, pp. 15–24. ACM, New York (1995). https://doi.org/10.1145/225014.225016

6. Letier, E., Van Lamsweerde, A.: Reasoning about partial goal satisfaction for requirements and design engineering. In: ACM SIGSOFT Software Engineering Notes, vol. 29, pp. 53–62. ACM (2004)

7. Namiot, D., Sneps-Sneppe, M.: On micro-services architecture. Int. J. Open Inf. Technol. 2(9), 24–27 (2014)

8. Napoli, C.D., Valentino, M., Sabatucci, L., Cossentino, M.: Adaptive workflows of home-care services. In: Proceedings of 27th IEEE International Conference on Enabling Technologies: Infrastructure for Collaborative Enterprises, WETICE 2018 (2018)

9. Ricci, A., Viroli, M., Omicini, A.: CArtAgO: a framework for prototyping artifact-based environments in MAS. In: Weyns, D., Parunak, H.V.D., Michel, F. (eds.) E4MAS 2006. LNCS (LNAI), vol. 4389, pp. 67–86. Springer, Heidelberg (2007). https://doi.org/10.1007/978-3-540-71103-2_4

10. Sabatucci, L., Cavaleri, A., Cossentino, M.: Adopting a middleware for self-adaptation in the development of a smart travel system. In: De Pietro, G., Gallo, L., Howlett, R.J., Jain, L.C. (eds.) Intelligent Interactive Multimedia Systems and Services 2016. SIST, vol. 55, pp. 671–681. Springer, Cham (2016). https://doi.org/10.1007/978-3-319-39345-2_60

11. Sabatucci, L., Cossentino, M.: From means-end analysis to proactive means-end reasoning. In: Proceedings of the 10th International Symposium on Software Engineering for Adaptive and Self-Managing Systems, pp. 2–12. IEEE Press (2015)

12. Sabatucci, L., Cossentino, M., Simone, G.D., Lopes, S.: Self-reconfiguration of shipboard power systems. In: Proceedings of the 3rd eCAS Workshop on Engineering Collective Adaptive Systems (2018)

13. Sabatucci, L., Lopes, S., Cossentino, M.: MUSA 2.0: a distributed and scalable middleware for user-driven service adaptation. In: De Pietro, G., Gallo, L., Howlett, R.J., Jain, L.C. (eds.) KES-IIMSS 2017. SIST, vol. 76, pp. 492–501. Springer, Cham (2018). https://doi.org/10.1007/978-3-319-59480-4_49

14. Sabatucci, L., Ribino, P., Lodato, C., Lopes, S., Cossentino, M.: GoalSPEC: a goal specification language supporting adaptivity and evolution. In: Cossentino, M., El Fallah Seghrouchni, A., Winikoff, M. (eds.) EMAS 2013. LNCS (LNAI), vol. 8245, pp. 235–254. Springer, Heidelberg (2013). https://doi.org/10.1007/978-3-642-45343-4_13

15. Thangarajah, J., Harland, J., Morley, D.N., Yorke-Smith, N.: Quantifying the completeness of goals in BDI agent systems. In: Proceedings of the Twenty-First European Conference on Artificial Intelligence, pp. 879–884. IOS Press (2014)

16. Thangarajah, J., Harland, J., Yorke-Smith, N.: Estimating the progress of maintenance goals. In: Proceedings of the 2015 International Conference on Autonomous Agents and Multiagent Systems, pp. 1645–1646. International Foundation for Autonomous Agents and Multiagent Systems (2015)

17. van Riemsdijk, M.B., Yorke-Smith, N.: Towards reasoning with partial goal satisfaction in intelligent agents. In: Collier, R., Dix, J., Novák, P. (eds.) ProMAS 2010. LNCS (LNAI), vol. 6599, pp. 41–59. Springer, Heidelberg (2012). https://doi.org/10.1007/978-3-642-28939-2_3

18. Vukovic, M., Robinson, P.: GoalMorph: partial goal satisfaction for flexible service composition. In: 2005 International Conference on Next Generation Web Services Practices, NWeSP 2005, pp. 6-pp. IEEE (2005)

19. Zhou, Y., Chen, X.: Partial implication semantics for desirable propositions. In: KR, pp. 606–612 (2004)
20. Zhou, Y., Van Der Torre, L., Zhang, Y.: Partial goal satisfaction and goal change: weak and strong partial implication, logical properties, complexity. In: Proceedings of the 7th International Joint Conference on Autonomous Agents and Multiagent Systems, vol. 1, pp. 413–420. International Foundation for Autonomous Agents and Multiagent Systems (2008)

Generalising the Dining Philosophers Problem: Competitive Dynamic Resource Allocation in Multi-agent Systems

Riccardo De Masellis[1(✉)], Valentin Goranko[1,2], Stefan Gruner[3], and Nils Timm[3]

[1] Stockholm University, Stockholm, Sweden
{riccardo.demasellis,valentin.goranko}@philosophy.su.se
[2] University of Johannesburg, Johannesburg, South Africa
[3] University of Pretoria, Pretoria, South Africa
{sgruner,ntimm}@cs.up.ac.za

Abstract. We consider a new generalisation of the Dining Philosophers problem with a set of agents and a set of resource units which can be accessed by them according to a fixed graph of accessibility between agents and resources. Each agent needs to accumulate a certain (fixed for the agent) number of accessible resource units to accomplish its task, and once it is accomplished the agent releases all resources and starts accumulating them again. All this happens in succession of discrete 'rounds' and yields a concurrent game model of 'dynamic resource allocation'. We use the Alternating time Temporal Logic (ATL) to specify important properties, such as goal achievability, fairness, deadlock, starvation, etc. These can be formally verified using the efficient model checking algorithm for ATL. However, the sizes of the resulting explicit concurrent game models are generally exponential both in the number of resources and the number of agents, which makes the ATL model checking procedure generally intractable on such models, especially when the number of resources is large. That is why we also develop an abstract representation of the dynamic resource allocation models and develop a symbolic version of the model checking procedure for ATL. That symbolic procedure reduces the time complexity of model checking to polynomial in the number of resources, though it can take a worst-case double exponential time in the number of agents.

Keywords: Dining philosophers games · Dynamic resource allocation · Alternating time temporal logic ATL · Symbolic model checking

1 Introduction and Related Work

The *dining philosophers problem* [11] is a well-established example for illustrating the problems of resource allocation in distributed computing [7]. In its original

---

[2]V. Goranko—Visiting professorship.

© Springer Nature Switzerland AG 2019
M. Slavkovik (Ed.): EUMAS 2018, LNAI 11450, pp. 30–47, 2019.
https://doi.org/10.1007/978-3-030-14174-5_3

version it involved 5 'philosophers' sitting at a round table, where 1 fork is placed between each pair of neighboured philosophers and they share it. The description of this scenario is well-known [11] and we do not need to repeat it here. However we should point out that we assume in this paper that philosophers' actions, thinking or eating, are instantaneous. The problem is to design a distributed protocol for picking up forks that ensures that each philosopher will get to eat repeatedly. The relevant properties here are *liveness*, as well as *deadlock*- and *starvation-freedom*. Technically the 'philosophers' represent processes of a distributed system, 'forks' are shared resources, and 'eating' is performing a computational task.

Since the introduction of the original problem, several generalisations have been published, for example the *drinking philosophers* [6], where each resource is still shared between two philosophers, but the resource distribution can now be arbitrary. Accordingly, a philosopher may have access to more than two resources and may also have more than two neighbours. The solution of [8] also uses communication between neighbours in order to determine on what each philosopher can do next. The *generalised* dining philosophers problem [12] permits that one resource may be shared between more than two philosophers. Still each philosopher has access to exactly two resources and needs these two resources in order to eat. The solution of [12] implements each philosopher as a random-based algorithm whereby all these algorithms run asynchronously. The randomised solution guarantees deadlock-freedom. The original problem has also been generalised by allowing mobility. *Mobile philosophers* [14] are able to move around the table, which results in a resource accessibility relation that changes over time. In [9] a solution to the mobile philosophers problem is presented that ensures mutual exclusion, liveness and self-stabilisation. The solution requires that the philosophers follow a certain access pattern that determines the orders of requests and the direction of moving around. All these problems fall under the broader category of *resource allocation problems*.

In this paper we present a *new generalisation* of the dining philosophers problem, involving a set of agents and a set of resource units which can be accessed by them according to a fixed bipartite graph of accessibility between agents and resources. Each agent needs to accumulate a certain (fixed for the agent) number of accessible resource units to accomplish its task. Once it is accomplished the agent releases all resources and starts accumulating them again. Thus, all agents compete for resources and attempt simultaneously to acquire them in a distributed way, in a discrete succession of rounds. In contrast to the drinking philosophers problem and the so far proposed solutions for it, we assume in our approach that the total resource demand of each philosopher remains the same in each round. However, we do not assume the restriction that a resource may be shared only between two philosophers. Moreover, our agent-based scenario does not involve communication between philosophers (other than possibly coordinating on their joint strategy when acting as a team). Thus, our problem formulation is a further generalisation of the one in [12] in the sense that we allow for arbitrary demands and arbitrary access topologies.

The scenario described above can be naturally modelled as a *concurrent game model* [4], [10, Chap. 9] of 'dynamic resource allocation'. For such models we use a version of the *alternating time temporal logic* ATL [4] to specify and verify their important properties, such as goal achievability, fairness, etc. An important feature of ATL is that its model checking has complexity which is linear in both the length of the formula and the size of the model. However, the sizes of the resulting explicit concurrent game models are generally exponential both in the number of resources and the number of agents, which makes the explicit model checking problem generally intractable. This is particularly bad when the number of resources is large. To avoid the resource-based exponential blow-up we also develop an *abstract representation* of the dynamic resource allocation models and develop a symbolic version of the model checking procedure for ATL. Working in such abstractions reduces the time complexity of model checking to polynomial in the number of resources. However, that is done at the cost of worst-case double exponential time complexity bound in the number of agents, so the symbolic algorithm is only guaranteed to outperform the explicit one when the number of resources is much larger than the number of agents.

In addition to these technical results, to our best knowledge our work is the first that presents an agent-based solution to a *generalisation* of the dining philosophers problem, even though a multi-agent approach to the *classical* problem was used in [5]. Besides modelling each philosopher as an agent, the solution in [5] uses an additional 'manager' agent (scheduler) who grants permission to acquire and release resources. With the scheduler as a central agent this approach can result in reduced parallelism in comparison to a decentralised solution. By contrast, our solution is fully distributed, without any central authority.

A similar definition of the generalised dining philosophers problem can be found in [15]. There, philosophers and resources are nodes of a bipartite graph that characterises the accessibility of resources. In contrast to our approach, however, a philosopher that requests a resource is blocked until he is eventually able to acquire it. The solution of [15] is based on (NP-complete) graph-colouring and guarantees robustness, deadlock- and starvation-freedom.

Lastly, some works on *resource-bounded reasoning* [1–3] are conceptually related to the approach presented here, although they are quite different in the framework and proposed solution. Indeed, while our reasoning tasks focus on how to obtain resources, resource-bounded reasoning abstracts this aspect away and is about which properties can be guaranteed given a set of resources and assuming that actions have costs.

The structure of the paper is as follows. In Sect. 2 we introduce our generalisation of the dining philosophers problem, viz dining philosophers game (GDP). In Sect. 3 we propose a variant of the alternating time temporal logic ATL for specifying and verifying properties of dining philosophers games. Section 4 develops an abstraction which represents sets of configurations by means of symbolic expressions, used for the symbolic model-checking algorithm developed in Sect. 5. We end with a brief concluding Sect. 6. The proof of our main theorem is placed in the Appendix A.

2 Generalisation of the Dining Philosophers Problem

Definition 1 (Generalised dining philosophers game (GDP)). *A GDP game is a tuple:*

$$\mathcal{G} = (Agt, Res, d, Acc) \ \ where:$$

- *$Agt = \{a_1, \ldots, a_n\}$ is a non-empty set of **agents**;*
- *$Res = \{r_1, \ldots, r_m\}$ is a non-empty set of **resource units** (of the same type);*
- *$d : Agt \to \mathbb{N}^+$ is a **demand function** defining the number of resources that each agent needs in order to carry out its tasks;*
- *$Acc \subseteq Agt \times Res$ is an **accessibility relation** denoting which resources agents can access. The set of resources that are accessible to an agent a is $Acc(a) = \{r \in Res \mid Acc(a,r)\}$, and we always assume:*
  - *$|Acc(a)| \geq d(a)$ for each $a \in Agt$ and*
  - *$\forall r \in Res, \exists a \in Agt.Acc(a,r)$.*

The above definition statically describes a game that is played in turns by agents, as explained later. The scenario is similar in spirit to the dining philosophers problem, where agents are philosophers and resources are forks. However here: *(i)* each resource can be shared by any set of agents (not only by two adjacent philosophers) as specified by *Acc* relation and *ii* each agent needs a generic (fixed for the agent) number of resources (not specifically two) in order to carry out its abstract task as described by the demand function *d*.

Example 1. The graph describes agents a_1, a_2, a_3, the resources $r_1 - r_6$, and the accessibility relation of a GDP game $\mathcal{G}$.

The game is fully specified once the demand function *d* is defined, e.g. $d(a_i) = 2$ for each $i = 1, 2, 3$.

Intuitively, the objective of each agent a_i is to acquire, gradually over time, the number of resources it needs by means of 'request' actions. Actually, each agent can perform several types of actions, as formalised below.

Definition 2 (Actions). *Given a GDP game $\mathcal{G}$, the **set of actions** Act is the union of the following types of actions:*

- ***request actions:*** *$\{req_r^a \mid a \in Agt, r \in Acc(a)\}$*
- ***release actions:*** *$\{rel_r^a \mid a \in Agt, r \in Acc(a)\}$*
- ***release-all actions:*** *$\{rel_{all}^a \mid a \in Agt\}$*
- ***idle actions:*** *$\{idle^a \mid a \in Agt\}$.*

The game is played in rounds, each of which consists of a tuple of (simultaneously executed) actions, one for each agent. Before any round, each agent holds a certain number of resources, and it can: request an accessible resource; release a resource that it holds; carry out its task, and then release all its resources at the same round; or idle.

Disallowing requests for multiple resources at the same time (or, release of multiple resources but not all of them) is a purely conceptual choice to keep the framework simple and essential: however, the results presented in the paper carry over without essential complications when the restrictions above are dropped.

The dynamics of the game is thus given in terms of a system of possible transitions between *configurations* over time. Configurations describe which resources each agent currently possesses.

Definition 3 (Configurations). *Given a GDP game $\mathcal{G}$, a **configuration** in $\mathcal{G}$ is a function $c : Res \rightarrow Agt^+$, where $Agt^+ = Agt \cup \{null\}$. If $c(r) = a$ then resource r is assigned in c to agent a. If $c(r) = null$ then r is unassigned in configuration c. We denote by c_0 the **initial configuration**, where $c_0(r) = null$ for each $r \in Res$, and by Conf the set of all possible configurations in $\mathcal{G}$.*

Example 2. Consider $\mathcal{G}$ as in Example 1. Here is a possible configuration c in $\mathcal{G}$: $c(r_1) = null$; $c(r_2) = a_1$; $c(r_3) = null$; $c(r_4) = a_2$; $c(r_5) = a_2$ and $c(r_6) = null$.

In each configuration, only a subset of all actions is executable by each agent.

Definition 4 (Actions' Availability). *The **availability of actions to agents** at configurations is a function $av : Conf \times Agt \rightarrow 2^{Act}$ defined component-wise as follows, for each $c \in Conf$ and $a \in Agt$:*

1. *if $|c^{-1}(a)| \geq d(a)$ then $av(c, a) = \{rel_{all}^a\}$;*
2. *otherwise:*
    (a) *$rel_{all}^a \notin av(c, a)$;*
    (b) *$req_r^a \in av(c, a)$ iff $c(r) = null$;*
    (c) *$rel_r^a \in av(c, a)$ iff $c(r) = a$;*
    (d) *$idle^a \in av(c, a)$.*

Intuitively: 1 and 2a say that when, and only when, agent a holds all the number of resources that it needs for achieving its goal, a *must* release them all; 2b says that a can request resource r iff r is accessible by a and is currently available; 2c states that a can release r iff it currently has it; and 2d says that a can always idle, unless it must release its resources.

Example 3. Consider the configuration c defined in Example 2. Then $av(c, a_1) = \{req_{r_1}^{a_1}, rel_{r_2}^{a_1}, req_{r_3}^{a_1}, idle^{a_1}\}$; $av(c, a_2) = \{rel_{all}^{a_2}\}$ and $av(c, a_3) = \{req_{r_6}^{a_3}, idle^{a_3}\}$.

Note that agents may request resources only if they are *currently* available in the configuration (no waiting queues). This has two implications: *(i)* agents cannot yet request resources that are about to be released by another agent and *(ii)* an agent that has just reached its goal and has just released all of its resources can request again any of these resources in the next turn, i.e., as soon as they are available again to everyone. We assume here full knowledge/observability by all agents of both the game and the current configuration: they know the other agents, their demand function, the accessibility relation as well as the current configuration. However, they cannot observe the actions taken by the others at any given round, until that round is completed.

Definition 5 (Action Profile). *Given a game $\mathcal{G}$ an **action profile** in $\mathcal{G}$ is a mapping $ap : Agt \to Act$. We denote with AP the set of all action profiles. Moreover, we say that ap is **executable** at $c \in Conf$ when for each $a \in Agt$ we have $ap(a) \in av(c, a)$.*

Given a configuration c and an action profile ap, we define the respective successor configuration c' and a **game step** (c, ap, c') component-wise, as follows. Firstly, in order for (c, ap, c') to be a legitimate game step, ap must be *executable* at c. Then, an agent a will keep holding in c' a resource r that it has in c unless it releases it with its action in ap, resulting in r being unassigned in c'. Lastly, an agent a will acquire a requested resource r in that step if and only if r is available and a is the only one requesting r; otherwise, i.e., whenever there is a request conflict for r, it remains *unassigned* for the sake of a *fully deterministic transition function* (see below). This choice is again to keep the framework simple: having nondeterministic evolutions does not affect the abstraction presented later, which only depends on configurations and the logical language.

Definition 6 (Game Steps and Game Transition Function). *Given a GDP game $\mathcal{G}$, a **game step** in $\mathcal{G}$ is a triple (c, ap, c') denoted by $c \xrightarrow{ap} c'$, where $c, c' \in Conf$ and:*

(i) ap is an executable action profile at c in $\mathcal{G}$, and
(ii) c' is such that for each $r \in Res$:

1. *if $c(r) = null$, then:*
   (a) if $((\exists a.ap(a) = \mathsf{req}_r^a \wedge \forall a'.a' \neq a \to ap(a') \neq \mathsf{req}_r^{a'})$ then $c'(r) = a)$;
   (b) otherwise $c'(r) = c(r) = null$;
2. *otherwise, let $c(r) = a$ for some (unique) agent a; then:*
   (a) if $(ap(a) = \mathsf{rel}_r^a \vee ap(a) = \mathsf{rel}_{all}^a)$ then $c'(r) = null$;
   (b) otherwise $c'(r) = c(r) = a$.

*The **game transition function** of $\mathcal{G}$ is the set $\rho(\mathcal{G})$ of all game steps in $\mathcal{G}$.*

Example 4. Consider the following action profiles in Example 2:
$$ap'(a_1) = \mathsf{idle}^{a_1}; \; ap'(a_2) = \mathsf{rel}_{all}^{a_2}; \; ap'(a_3) = \mathsf{req}_{r_6}^{a_3}$$
$$ap''(a_1) = \mathsf{req}_{r_3}^{a_1}; \; ap''(a_2) = \mathsf{req}_{r_3}^{a_2}; \; ap''(a_3) = \mathsf{req}_{r_5}^{a_3}$$
The respective resulting configurations from performing ap' and then ap'' at configuration c are:
$c'(r_1) = null; \; c'(r_2) = a_1; \; c'(r_3) = null; \; c'(r_4) = null; \; c(r_5) = null; \; c'(r_6) = a_3$.
$c''(r_1) = null; \; c''(r_2) = a_1; \; c'(r_3) = null; \; c'(r_4) = null; \; c(r_5) = a_3; \; c'(r_6) = a_3$.

Plays in a GDP game $\mathcal{G}$ are (infinite) sequences of game steps in $\mathcal{G}$, defined by means of the transition system $\mathfrak{G} = (Conf, \rho(\mathcal{G}))$, which we call **configuration graph** of $\mathcal{G}$. We also define the **local configuration graph of $\mathcal{G}$ generated by c_0**, that is the restriction $(\mathfrak{G}, c_0) = (Conf(c_0), \rho(\mathcal{G}))$ of $\mathfrak{G}$, where $Conf(c_0)$ is the set of only those configurations in $Conf$ which are reachable from the initial configuration c_0 by $\rho(\mathcal{G})$.

Proposition 1. *The size of* $(\mathfrak{G}, c_0)$ *(hence, also of* $\mathfrak{G}$*) is, in general, exponential in the number of resources in* $\mathcal{G}$*.*

Proof. Take $\mathcal{G}$ where $\forall a, r. Acc(a, r)$ and $\forall a. d(a) = |Res|$. Then each assignment of resources to agents is a reachable configuration, thus $|Conf| = |Agt|^{|Res|}$.

3   Logic for Specification and Verification of GDP Games

Agents in GDP games may, but need not, cooperate in pursuing their goals, which may be *positive*, i.e. eventually acquiring the necessary number of resources to achieve their individual goals, or *negative*, i.e. preventing others from achieve their goals, or combined and more complex. Thus, our choice of language and formalism $\mathcal{L}_{GDP}$ for specification and verification of GDP games is naturally a (slight) variation of the *alternating time temporal logic* ATL [4], which allows to express temporal properties φ parameterised by a set of agents A in multi-agent games. The main construction in ATL is $\langle\!\langle A \rangle\!\rangle\, \varphi$, the intuitive meaning of which is that the coalition of agents in A has a joint strategy to collaborate in order to achieve φ, no matter how the opponent agents in $Agt \setminus A$ may counter-act. As customary when reasoning at this level of abstraction, we do not focus on how agents in the same coalition should coordinate to achieve the objectives, but we assume such a coordination mechanism is already in place.

Definition 7 (Syntax). *The formulae of* $\mathcal{L}_{GDP}$ *are defined as follows:*

$$\varphi ::= g_{a_i} \mid \neg\varphi \mid \varphi_1 \wedge \varphi_2 \mid \varphi_1 \vee \varphi_2 \mid \langle\!\langle A \rangle\!\rangle\, \mathsf{X}\varphi \mid \langle\!\langle A \rangle\!\rangle\, \mathsf{G}\varphi \mid \langle\!\langle A \rangle\!\rangle\, \varphi_1 \mathsf{U}\varphi_2$$

where $a_i \in Agt$, $A \subseteq Agt$. *We may also write* $\langle\!\langle a_1, \ldots, a_i \rangle\!\rangle$ *instead of* $\langle\!\langle \{a_1, \ldots, a_i\} \rangle\!\rangle$.

The atomic formula g_{a_i} expresses the claim that agent a_i has reached the number of resources it needs (given by $d(a_i)$) to achieve its goal. All other boolean connectives and the standard temporal operators *next* X, *always* G, and *until* U have the usual semantics (cf. e.g. [10]).

Definition 8 (Positional Strategy). *Let* $\mathcal{G}$ *be a game, for each agent* $a \in Agt$, *a (positional) strategy for* a *is a function* $\sigma_a : Conf \to Act$ *such that* $\forall c. \sigma_a(c) \in av(c, a)$. *Given* $A = \{a_1, \ldots, a_r\} \subseteq Agt$, *a joint strategy for* A *is a tuple of strategies* $\sigma_A(\sigma_{a_i}, \ldots, \sigma_{a_r})$ *one for each* $a_i \in A$.

The function $out(c, \sigma_A)$ returns the set of all paths in $Conf^{\omega}$ that can occur when agents in A follow the joint strategy σ_A from configuration c on. We denote by $\pi = c_0, c_1, \ldots$ a path in $Conf^{\omega}$ and with $\pi[i]$ the i-th configuration of π.

$$out(c, \sigma_A) = \Big\{ \pi = c_0, c_1, \ldots \mid c_0 = c \wedge$$
$$\forall i \in \mathbf{N}, \exists (act^i_{a_1}, \ldots, act^i_{a_n}), \forall a \in Agt.$$
$$\big(act^i_a \in av(c_i, a) \wedge \big(a \in A \to act^i_a \in \sigma_A\big) \wedge q_i \overset{(act^i_{a_1}, \ldots, act^i_{a_n})}{\to} q_{i+1}\big) \Big\}$$

Formulae of $\mathcal{L}_{\text{GDP}}$ are evaluated over a game configuration graph $\mathfrak{G}$. The satisfaction relation $\mathfrak{G}, c \models \varphi$ is defined inductively on the structure of formulae, for all $c \in \mathfrak{G}$, as follows, where $\text{res}(c, a) = |c^{-1}(a)|$, i.e. the number of resources owned by a in configuration c.

- $\mathfrak{G}, c \models g_{a_i}$ iff $\text{res}(c, a_i) \geq d(a_i)$;
- $\wedge$, $\vee$ and $\neg$ are treated as usual;
- $\mathfrak{G}, c \models \langle\!\langle A \rangle\!\rangle \mathsf{X}\varphi$ iff there is a joint strategy σ_A such that, for every path $\pi \in out(c, \sigma_A)$ we have that $\mathfrak{G}, \pi[1] \models \varphi$;
- $\mathfrak{G}, c \models \langle\!\langle A \rangle\!\rangle \mathsf{G}\varphi$ iff there is a joint strategy σ_A such that, for every path $\pi \in out(c, \sigma_A)$ and for every $i \in \mathbf{N}$ we have that $\mathfrak{G}, \pi[i] \models \varphi$;
- $\mathfrak{G}, c \models \langle\!\langle A \rangle\!\rangle \varphi_1 \mathsf{U}\varphi_2$ iff there is a joint strategy σ_A such that, for every path $\pi \in out(c, \sigma_A)$ we have that $\exists i \geq 0.\mathfrak{G}, \pi[i] \models \varphi_2$ and $\forall 0 \leq j < i.\mathfrak{G}, \pi[j] \models \varphi_1$.

For every game configuration graph $\mathfrak{G}$ and a formula $\varphi \in \mathcal{L}_{\text{GDP}}$ we define the **extension of φ in $\mathfrak{G}$** to be the set of all configurations in $\mathfrak{G}$ that satisfy φ:

$$[\![\varphi]\!]_{\mathfrak{G}} = \{c \in \mathit{Conf} \mid \mathfrak{G}, c \models \varphi\}.$$

Example 5. The figure shows a graphical representation of c as in Example 2 where red resources are held by the agents in brackets. It is easy to see that the following formulae hold in $\mathfrak{G}, c$:

$\langle\!\langle a_1 \rangle\!\rangle \mathsf{G}\langle\!\langle a_1 \rangle\!\rangle \mathsf{F}g_{a_1}$, meaning that there is a strategy for agent a_1 to reach its goal infinitely often (it can always get resources r_1 and r_2), and $\langle\!\langle a_2, a_3 \rangle\!\rangle \mathsf{F}g_{a_2}$ (there is a strategy for a_2, a_3 to eventually reach the goal of a_2). But, there is no strategy for a_3 alone to eventually reach its goal, i.e. $\mathfrak{G}, c \not\models \langle\!\langle a_3 \rangle\!\rangle \mathsf{F}g_{a_3}$ as there exists a strategy for a_2 which prevents a_3 to acquire r_4 or r_5. Such a strategy by a_2 simply amounts to mimic a_3 requests (notice that the semantics of $\mathcal{L}_{\text{GDP}}$ only require such a counterstrategy to exists, even if in practice that would mean a_2 to guess a_3 requests).

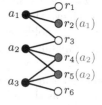

The **global model checking problem** for $\mathcal{L}_{\text{GDP}}$ is a computational problem which amounts to compute $[\![\varphi]\!]_{\mathfrak{G}}$ given $\mathcal{G}$ and $\varphi \in \mathcal{L}_{\text{GDP}}$.

Lemma 1. *The global model checking problem for $\mathcal{L}_{\text{GDP}}$ has worst-case time complexity exponential in the number of resources.*

Proof. Follows immediately from Proposition 1 and the linear time complexity of the global model checking algorithm for ATL.

Since the number of resources in a GDP game scenarios is generally large, this exponential time bound is bad, and it is clear that the standard model checking algorithm for ATL [4], [10, Chap. 9] would be generally intractable if applied to explicitly generated configuration graphs of GDP games. That is why we develop here an abstract symbolic representation that will eventually keep the worst-case complexity of the model checking problem *polynomial* in the number of resources.

4 Symbolic Representation of Configurations on GDP Games

Our solution is based on an abstraction which represents sets of configurations by means of symbolic expressions. Essentially, we group together configurations which cannot be distinguished by $\mathcal{L}_{\text{GDP}}$-formulae and that behave "similarly" with respect to the $\mathfrak{G}$ transition function. In this way we solve the global model checking problem of $\mathcal{L}_{\text{GDP}}$-formulae without explicitly computing the configuration graph $\mathfrak{G}$, but rather by computing the corresponding symbolic expressions.

In other words, configurations contains more information than needed to answer $\mathcal{L}_{\text{GDP}}$-formulas: indeed, configurations describe which resources each agents holds, while $\mathcal{L}_{\text{GDP}}$-formulas cannot distinguish configurations where agents hold the same number of resources. Therefore, intuition suggests that abstracting from specific resources each agent has and keep only the number of those held by each agent may suffice. Unfortunately this not the case, as resources are in general accessible to subset of agents only (as specified by the *Acc*) thus, given a subset of agents, how many resources *among those accessible to them* are still available is a necessary information to reason on the capability to achieve their goals. To see this, let us consider a situation of $\mathcal{G}$ in Example 1 where a_3 holds one resource only: knowing which one is important to determine the capability of a_3 to reach g_{a_3} in the next state, namely to assess whether $\mathfrak{G}c \models \langle\langle a_3 \rangle\rangle \mathsf{X} g_{a_3}$. Indeed, if the resource owned by a_3 is r_4 (or r_5), then the above formula is true (as a_3 can always request and obtain r_6) whilst if it is r_6 is false, as there is no guarantee it can obtain one among r_4 or r_5.

Our symbolic representation is based on the intuition that only resources that are accessed by the same subset of agents can be safely regarded as undistinguishable. Thus, we start by formally describing such sets of resources. With mild notational abuse, we define $Acc^{-1} : Res \rightarrow 2^{Agt}$ such that $Acc^{-1}(r) = \{a \mid Acc(a, r)\}$ and $Acc : 2^{Agt} \rightarrow 2^{Res}$ such that $Acc(A) = \{r \in Res \mid \forall a \in Agt.a \in A \leftrightarrow Acc(a, r)\}$, namely the set of resources shared by, and only by, the agents in A.

The function Acc^{-1} defined above induces an *equivalence relation* $\sim \subseteq Res \times Res$ among resources as follows:

$$r_i \sim r_j \leftrightarrow (Acc^{-1}(r_i) = Acc^{-1}(r_j))$$

We denote by $\mathcal{R} = \{R_1, \ldots, R_u\}$ the quotient set of *Res* by $\sim$, namely the set of equivalence classes in *Res* induced by $\sim$. Notice that each $R_i \in \mathcal{R}$ uniquely identifies the set of agents which have access to them. Again with mild notational abuse we denote such a set by $Acc^{-1}(R_i)$.

Our symbolic representation defines for each agent a_i and each equivalence class of resources R_j, the range on the number of resources in R_j that a_i holds.

Definition 9 (Interval Expressions). *Let $\mathcal{G}$ be a game, based of an intuitive understanding of the notion of 'intervals', interval expressions α are defined as follows:*

$$\alpha ::= \bigwedge_{a \in Agt} \bigwedge_{R \in \mathcal{R}} (a, R)[l_R^a, l_R^a] \mid \alpha_1 \vee \alpha_2$$

Intuitively we call each $(a, R)[l_R^a, l_R^a]$ an **interval**, and it constrains the number of resources held by a and belonging to R to be between l_R^a and L_R^a.

Now we define the *formal semantics* of interval expressions. In every GDP game configuration graph $\mathfrak{G}$ each interval expression α defines a set of configurations $\|\alpha\|_{\mathfrak{G}}$, the **extension** of α, as follows:

- $\|(a, R)[l_R^a, L_R^a]\|_{\mathfrak{G}} = \{c \in Conf : l_R^a \leq |c^{-1}(a) \cap R| \leq L_R^a\}$;
- $\|\bigwedge_{a \in Agt} \bigwedge_{R \in \mathcal{R}} (a, R)[l_R^a, l_R^a]\|_{\mathfrak{G}} = \bigcap_{a \in Agt} \bigcap_{R \in \mathcal{R}} \|(a, R)[l_R^a, L_R^a]\|_{\mathfrak{G}}$;
- $\|\alpha_1 \vee \alpha_2\|_{\mathfrak{G}} = \|\alpha_1\|_{\mathfrak{G}} \cup \|\alpha_2\|_{\mathfrak{G}}$.

Example 6. With reference to $\mathcal{G}$ as in Example 1, we notice that there are four equivalence classes in $\mathcal{R}$: $R_1 = \{r_1, r_2\}$, with $Acc^{-1}(R_1) = \{a_1\}$; $R_2 = \{r_3\}$, with $Acc^{-1}(R_2) = \{a_1, a_2\}$; $R_3 = \{r_4, r_5\}$, with $Acc^{-1}(R_3) = \{a_2, a_3\}$ and $R_4 = \{r_6\}$, with $Acc^{-1}(R_4) = \{a_3\}$. The expression α':

$$(a_1, R_1)[0, 1] \wedge (a_1, R_2)[0, 1] \wedge$$
$$(a_2, R_2)[0, 1] \wedge (a_2, R_3)[1, 2] \wedge$$
$$(a_3, R_3)[0, 0] \wedge (a_3, R_4)[0, 1]$$

is such that the configuration c from Example 2 belongs to $\|\alpha'\|_{\mathfrak{G}}$. For the sake of readability we omit intervals for pairs (a', R') such that $R \cap Acc^{-1}(a') = \emptyset$. Let also c' be as c but with agent a_1 not holding r_2. Then, $c' \in \|\alpha'\|_{\mathfrak{G}}$ as well.

Interval expressions, being defined on intervals, are very modular, and they can represent sets of configurations of different sizes. They can be classified from "coarsest" to "finest" depending on the number of configurations in their extensions. The larger the number, the coarser the abstraction. Given a game $\mathcal{G}$ there is a coarsest formula $\alpha_\top$ representing the whole set $Conf$ of configurations which is such that for all $a \in Agt$ and $R \in \mathcal{R}$, $(a, R)[0, min(d(a), |R|)]$. Analogously, it is always possible to define a finest expression $\alpha_\perp$ the extension of which is empty (it is enough to define an interval $(a, R)[L, L]$ where $L > |Res|$). The finer expressions are more important for our purposes, as they provide the smallest abstraction level we can manipulate. Those expressions, of the form: $\bigwedge_{a \in Agt} \bigwedge_{R \in \mathcal{R}} (a, R)[\ell_R^a, \ell_R^a]$ will be called β-**expressions**. Clearly, every interval expression can be transformed into a semantically equivalent (i.e., with the same extension) *expanded normal form*, that is, a disjunction of β-expressions.

Definition 10 (Expanded Normal Form). *An interval constraint expression* α *is in* ***expanded normal form*** *if each interval is of the form* $(a, R)[\ell_R^a, \ell_R^a]$.

We recursively define function EXPD(α) which translates α in its extended normal form as follows:

$$\text{EXPD}(\bigwedge_{a \in Agt} \bigwedge_{R \in \mathcal{R}} (a, R)[l_R^a, L_R^a]) ::= \bigvee_{l_{R_1}^{a_1} \leq \ell_{R_1}^{a_1} \leq L_{R_1}^{a_1}} \cdots \bigvee_{l_{R_u}^{a_1} \leq \ell_{R_u}^{a_1} \leq L_{R_u}^{a_1}}$$

$$\vdots$$

$$\bigvee_{l_{R_1}^{a_n} \leq \ell_{R_1}^{a_n} \leq L_{R_1}^{a_n}} \cdots \bigvee_{l_{R_u}^{a_n} \leq \ell_{R_u}^{a_n} \leq L_{R_u}^{a_n}}$$

$$(\bigwedge_{1 \leq i \leq n} \bigwedge_{1 \leq j \leq u} (a_i, R_j)[l_{R_j}^{a_i}, l_{R_j}^{a_i}])$$

$$\text{EXPD}(\alpha_1 \vee \alpha_2) ::= \text{EXPD}(\alpha_1) \vee \text{EXPD}(\alpha_2)$$

Lemma 2. *Let* α *be an interval expression. Its expanded normal form* EXPD(α) *has, in the worst-case, size which is double exponential in the number of agents and polynomial in the number of resources.*

Proof. Consider $\alpha_\top$ in a game $\mathcal{G}$. Then EXPD($\alpha_\top$) returns all possible β-expressions in $\mathcal{G}$. If we consider $|Res|$ as the complexity parameter, then the worst case is when $d(a_i) = |Res|$ for each $i \in \{1, \ldots, n\}$, which results in a size of EXPD($\alpha_\top$) which is polynomial in the number of resources. If we consider $|Agt|$ as the complexity parameter, then the worst case is given by Acc such that we have all equivalence classes for resources, i.e., $2^{|Agt|}$. Given that each interval in α is $(a, R)[0, |Res|]$, we have size of α double exponential in the number of agents.

5   Symbolic Verification of $\mathcal{L}_{\text{GDP}}$ Formulae in GDP Games

The algorithm for the symbolic verification of $\mathcal{L}_{\text{GDP}}$-formulae has the same structure as that for ATL formulae [4], [10, Chap. 9]: for the strategic next-time operator it computes the controllable by a given coalition A pre-image of the extension of a given formula, and then exploits the fixpoint characterisations of the temporal operators G and U. Now, instead of manipulating states of the explicit model $\mathfrak{G}$, i.e. the configuration graph, our algorithm works symbolically, with interval expressions. The basic step of the algorithm is to compute, given a coalition A and an interval expression α', the *controllable by A symbolic pre-image* $\alpha = \text{PRE}(\mathcal{G}, A, \alpha')$ which, intuitively, is the set of all interval expressions that only 'contain' configurations from which coalition A has a joint action that forces the outcome to be in the set of configurations defined by the extension of α'.

Definition 11 (Controllability). *Let* $\mathcal{G}$ *be a GDP game,* $A \subseteq Agt$, *and* α, α' *interval expressions. We say that* α ***is in the controllable by*** A ***pre-image of*** α' *iff there exists a strategy* σ_A *such that, for all* $c \in \|\alpha\|_{\mathfrak{G}}$ *and for all* $\pi \in out(c, \sigma_A)$ *we have that* $\pi[1] \in \|\alpha'\|_{\mathfrak{G}}$.

In what follows, we show how to compute the controllable pre-image of a interval expression. Given a GDP game $\mathcal{G}$, we do the following:

1. As pre-processing we compute the set AP of all action profiles in $\mathcal{G}$.
2. Then, we consider the expansion $\text{EXPD}(\alpha') = \beta'_1 \vee \ldots \vee \beta'_s$, where $\beta'_1, \ldots, \beta'_s$ are β-expressions. For each $\beta'_i \in \text{EXPD}(\alpha')$ and each $ap_j \in AP$ we compute set $\{\beta_{i,j,1}, \ldots \beta_{i,j,t}\}$ of symbolic β-expressions such that, for each $c \in \text{EXPD}(\beta_{i,j,k})$, performing ap leads to a configuration $c' \in \text{EXPD}(\beta'_i)$.
3. Lastly, for each $\beta_{i,j,k}$ we check if it belongs in the controllable by A pre-image of α', by verifying if there exists a joint action for agents in A such for every possible joint actions for the other agents, the successor is in α'. This requires to check all possible action profiles. If such a joint action exists, $\beta_{i,j,k}$ will be added in disjunction to a formula $\alpha_{\beta'_i}$ which represents (part of) the controllable pre-image of α' that has been computed, roughly speaking, by considering predecessors of β'_i. The whole controllable by A pre-image α of α' is simply the union of such controllable predecessors, namely $\alpha = \bigcup_{1 \leq i \leq s} \alpha_{\beta'_i}$.

We now present steps (2) and (3) in detail.

Step (2). The difficulty lies in the fact that the transition function $\rho(\mathfrak{G})$ is not *injective*, thus when computing the pre-images of an interval expression β'_i for a specific ap_j, the result consists, in general, of more than one β-interval expression. In order to see why, let us consider the following example.

Example 7 Let α' be as in Example 6 and let β'_i:

$$
\begin{aligned}
&(a_1, R_1)[0,0] \wedge (a_1, R_2)[0,0] \wedge \\
&(a_2, R_2)[0,0] \wedge (a_2, R_3)[2,2] \wedge \\
&(a_3, R_3)[0,0] \wedge (a_3, R_4)[0,0]
\end{aligned}
$$

be one of its expanded β-expressions. Notice that the configuration c' defined at the end of Example 6 belongs to β'_i.

Now, let $\langle \text{rel}^{a_1}_{all}, \text{req}^{a_2}_{r_4}, \text{rel}^{a_3}_{r_6} \rangle$ be an action profile ap_j. Which interval expressions belong to the pre-image of β'_i given ap_j? In other words, from which interval expressions is β'_i reachable when performing ap_j? Issues arise with the *release-all* action $\text{rel}^{a_1}_{all}$, as we do not know which resources a_1 was holding in the previous configuration, hence to which equivalence classes those resources belonged to. Indeed, since $d(a_1) = 2$, a_1 was holding 2 resources that could have been any of the following: $\{\{r_1, r_2\}, \{r_1, r_3\}, \{r_2, r_3\}\}$. This gives rise to two different interval expressions in the pre-image of β' given ap_j:

– in case a_1 was holding $\{r_1, r_2\}$, it is $\beta_{i,j,1}$ defined as follows:

$$
\begin{aligned}
&(a_1, R_1)[2,2] \wedge (a_1, R_2)[0,0] \wedge \\
&(a_2, R_2)[0,0] \wedge (a_2, R_3)[2,2] \wedge \\
&(a_3, R_3)[0,0] \wedge (a_3, R_4)[0,0]
\end{aligned}
$$

– and, if a_1 was holding either $\{r_1, r_3\}$ or $\{r_2, r_3\}$, it is $\beta_{i,j,2}$, defined as follows:

$$(a_1, R_1)[1,1] \wedge (a_1, R_2)[1,1] \wedge$$
$$(a_2, R_2)[0,0] \wedge (a_2, R_3)[2,2] \wedge$$
$$(a_3, R_3)[0,0] \wedge (a_3, R_4)[0,0]$$

During Step (2) for each $ap_j \in AP$, we compute the set $\{\overline{ap}_{j,1}, ..., \overline{ap}_{j,q}\}$ of *expanded* action profiles for ap_j, where each $\overline{ap}_{j,k}$ is a set of actions where:

(i) resources are replaced by their equivalence classes, and

(ii) the *release-all* actions by each agent a are replaced by a set of (single-resource) *release* actions one for each resource to be released by a in an equivalence relation in all possible ways.

It is easy to see that the number of possible *expanded* action profiles for ap_j is, in the worst case, double exponential in the number of agents. Indeed, if agent a performs a *release-all* action, such that $d(a) = |Res|$ and a has access to all possible $2^{|Agt|-1}$ equivalence classes of resources, defined by sets of agents where a occurs, then we get a predecessor for each way of assigning $|Res|$ resources to these $2^{|Agt|-1}$ equivalence classes, thus $|Res|^{2^{|Agt|-1}}$.

We now show how we actually compute the predecessor. Let

$$\beta' = \bigwedge_{a \in Agt} \bigwedge_{R \in \mathcal{R}} (a, R)[\ell_R^a, \ell_R^a]$$

be the successor expression under consideration and let $(ap_j, \overline{ap}_{j,k})$ as before. We build β such that $\beta_{i,j,k} \xrightarrow{(ap_j, \overline{ap}_{j,k})} \beta'_i$ by first considering the release actions in $\overline{ap}_{j,k}$ and then the request actions in ap_j:

– for each release action $\mathsf{rel}_{\bar{R}}^{\bar{a}} \in \overline{ap}_{j,k}$ we increase by one the number $\ell_{\bar{R}}^{\bar{a}}$;

– for each request action $\mathsf{req}_{\bar{r}}^{\bar{a}} \in ap$ such that $\neg \exists a'$ such that $\mathsf{req}_{\bar{r}}^{a'} \in ap$ we decrease by one the number $\ell_{\bar{R}}^a$ where $\bar{R}$ is the equivalence class of $\bar{r}$.

Step (3). Given a $\beta_{i,j,k}$ obtained in the previous step, for each joint action for A we check if that is the required one-step strategy σ_n by simply verifying if all action profiles extending σ_n lead to α'. Let ap_v be one of such action profiles. We compute $\beta'_{i,j,k,v}$ such that $\beta_{i,j,k} \xrightarrow{ap_v} \beta'_{i,j,k,v}$. The issue is that $\beta_{i,j,k} \xrightarrow{ap_v} \beta'_{i,j,k,v}$ may not be a step, for two reasons:

(i) $\beta_{i,j,k}$ may be inconsistent, i.e., $\|\beta_{i,j,k}\|_{\circledS} = \emptyset$, or

(ii) ap_v may not be executable in $\beta_{i,j,k}$, meaning that there is no $c \in \|\beta_{i,j,k}\|_{\circledS}$ such that $c \xrightarrow{ap_v} c'$ is a step.

We first perform check (i) and then (ii) separately, but with the same technique: by a reduction to the maximal matching problem in a bipartite graph (cf. [13]), defined as follows. Given bipartite graph $G = (V = (X, Y), E)$ where $E \subseteq X \times Y$, the maximum matching problem amounts to find a *maximal* set

$M \subseteq E$ such that for all $(x, y), (x', y') \in M$, $x = x'$ iff $y = y'$, i.e. different edges in M share no vertices. The Hopcroft-Karp algorithm [13] solves it in worst-time complexity linear in the size of G. We now show how to build the bipartite problem which solution guarantees that $\beta_{i,j,k}$ is consistent. From $\beta_{i,j,k}$ and Acc we define $bpg(\beta, Acc) = ((X, Y), E)$ as follows:

- for each $a \in Agt$ and $R \in \mathcal{R}$ we have $\{x(a, R, 1), \ldots, x(a, R, \ell_R^a)\} \in X$ iff $\ell_a \neq 0$;
- $Y = Res$;
- for each $a \in Agt$ and $R \in \mathcal{R}$ and $1 \leq t \leq \ell_R^a$, $(x(a, R, t), r) \in E$ iff $(a, r) \in Acc$ and $r \in R$.

In the worst case, the size of G is double exponential in the number of agents, coming from the number of x-nodes. It is easy to prove that solution M of the maximum matching problem is such that each x-node is covered, i.e., $\forall x \in X, \exists y \in Y . (x, y) \in M$ if and only if $\|\beta_{i,j,k}\|_{\circledS} \neq \emptyset$. The 'only if' direction follows from the fact that M is the largest possible such set.

If $\|\beta_{i,j,k}\|_{\circledS} \neq \emptyset$ then we check if ap_u can be executed in $\beta_{i,j,k}$.

Remark. Notice that given $\beta_{i,j,k}$ consistent, and ap_u, then there is a unique possible *expansion* $(ap_u, \overline{ap}_u)$, as $\beta_{i,j,k}$ provides information on the equivalence classes of resources held by agents performing release-all actions. This entails there is only a successor for each action profile, which guarantees that the transition relation between interval expressions is actually a function.

Now, we will build an instance of the maximum matching problem for the bipartite graph $bpg(\beta_{i,j,k}, Acc, (ap_u, \overline{ap}_u))$ starting from $bpg(\beta_{i,j,k}, Acc) = ((X, Y), E)$ as before and modifying it so as to account for $(ap_u, \overline{ap}_u)$:

1. for each $\text{rel}_R^a \in \overline{ap}_u$ we:
   (a) remove one node among $\{x(a, R, 1), \ldots x(a, R, \ell_R^a)\}$ and its corresponding edges. If no such node exists, then $\beta_{i,j,k}$ is not compatible with $(ap_u, \overline{ap}_u)$ and we discard them both;
   (b) add a node $z(\text{rel}_R^a)$ to X and edges $(z(\text{rel}_R^a), r)$ for all $r \in R$.
2. for each request action $\text{req}_r^a \in ap$: add a node $z(\text{req}_r)$ and edge $(z(\text{req}_r), r)$.

The maximum matching problem for the above has a solution covering all the nodes in X iff $(ap_u, \overline{ap}_u)$ is executable in $\beta_{i,j,k}$. The 'only if' direction is again guaranteed by M being the largest. The (again, unique) successor state $\beta'_{i,j,k,u}$ can be easily computed by modifying the intervals in the intuitive way.

We now present our main result, the soundness and completeness of our symbolic technique for global model checking of $\mathcal{L}_{\text{GDP}}$ formulae. The proof is constructive and provides a model checking algorithm.

Theorem 1. *For every GDP game $\mathcal{G}$ and a formula $\varphi \in \mathcal{L}_{\text{GDP}}$ there exists an interval constraints expression $\alpha(\mathcal{G}, \varphi)$ such that:*

$$[\varphi]_{\circledS} = \|\alpha(\mathcal{G}, \varphi)\|_{\circledS}$$

The proof is in the Appendix.

Theorem 2. *For each $\varphi \in \mathcal{L}_{\text{GDP}}$, computing $\|\alpha(\mathcal{G}, \varphi)\|_\circledast$ can be done in time which is worst-case double exponential in the number of agents and polynomial in the number of resources.*

Proof. Let us first estimate the time complexity of the controllable pre-image subroutine. As pointed out in the description of the algorithm, Step (2) generates all possible distinct β-expressions, which are, as stated in Lemma 2, double exponential in the number of agents and polynomial in the number of resources. Given a formula $\varphi \in \mathcal{L}_{\text{GDP}}$, the number of times the controllable pre-image subroutine is called is linear in the number of possible distinct β-expressions for each fixpoint computation and the number of fixpoint computations is linear in the size of the formula.

6   Conclusions and Outlook to Future Work

In this paper we have introduced the Generalised Dining Philosophers games, which are a substantial generalisation of the original dining philosophers problem proposed as a modelling framework for distributed multi-agent dynamic resource allocation problems. We have developed a symbolic algorithm for the verification of properties of GDP games specifiable in a version of ATL, built over atomic propositions stating that an agent's goal is achieved. We have showed that this symbolic algorithm works in time which is polynomial in the number of resources, though worst-case double exponential in the number of agents while the standard ATL model checking algorithm, applied on the explicit configuration graph of the game generally works in time exponential in both the number of agents and the number of resources. Thus, both algorithms are generally incomparable in their efficiency, but the symbolic algorithm is significantly more efficient than the explicit one in cases where the number of resources is much larger than the number of agents.

From theoretical perspective, GDP games are amenable to various natural modifications of the operational semantics, e.g. allowing agents to request multiple resources at a time, or assuming basic individual rationality according to which a rational agent will always attempt to acquire more resources rather than idle, until it reaches its goal. These considerations are left to future work, in which we also intend to apply our framework and techniques to more realistic scenarios arising in 'classical' Computer Science (e.g. operating systems). Also, both narrowing and broadening of our approach are worth exploring. The former looks for classes of formulas, or classes of models (e.g., those with a specific accessibility graph) that allow for more compact symbolic representations, while the latter investigates to which extent our ideas can be successfully applied to more general settings.

From practical perspective, a future implementation of the symbolic technique described here will allow to compare the effectiveness of our approach against explicit or BDD-based model checkers for MAS.

Acknowledgements. The work of Valentin Goranko and Riccardo De Masellis was supported by a research grant 2015-04388 of the Swedish Research Council, which also partly funded a working visit of Nils Timm to Stockholm. Valentin Goranko was also partly supported by a visiting professorship grant by the University of Pretoria.

A   Appendix: Proof of Theorem 1

The proof is by induction on φ.

- $\varphi = g_a$. Then $[\![\varphi]\!]_{\mathfrak{G}} = \{c : \mathsf{res}(c, a) = d(a)\}$. The required formula is $\alpha = \alpha_1 \vee \ldots \vee \alpha_s$ where each α_i is the conjunction of intervals for agents $a' \neq a$ and the conjunction of intervals for a. The conjunction of intervals for $a' \neq a$ is the same for every α_i, viz. $(a', R)[0, d(a')]$, for each $a' \neq a$ and for each $R \in \mathcal{R}$. The conjunction of intervals for a is different for each α_i, has the form $(a, R)[\ell^a_R, \ell^a_R]$ and is a solution of the constraint $\sum_{R \in \mathcal{R}} \ell^a_R = d(a)$.
- $\varphi = \langle\!\langle A \rangle\!\rangle \mathsf{X}\psi$. Then $[\![\varphi]\!]_{\mathfrak{G}}$ is the set of configurations from which there exists a collective strategy σ_A for A such that for all $\pi \in out(c, \sigma_A)$, $\mathfrak{G}, \pi[1] \models \psi$. We show that $\mathrm{PRE}(\mathcal{G}, A, \alpha(\mathcal{G}, \psi))$ satisfies the claim. We prove separately $[\![\varphi]\!]_{\mathfrak{G}} \supseteq \|\mathrm{PRE}(\mathcal{G}, A, \alpha(\mathcal{G}, \psi))\|_{\mathfrak{G}}$ (soundness) and $[\![\varphi]\!]_{\mathfrak{G}} \subseteq \|\mathrm{PRE}(\mathcal{G}, A, \alpha(\mathcal{G}, \psi))\|_{\mathfrak{G}}$ (completeness).

Soundness. We have to prove that for every $c \in \|\mathrm{PRE}(\mathcal{G}, A, \alpha(\mathcal{G}, \psi))\|_{\mathfrak{G}}$ there exists one step strategy to reach a state where ψ holds. By the inductive hypothesis, all (and only) configurations in $\alpha(\mathfrak{G}, \psi)$ satisfy ψ, thus it is enough to show that a one step strategy reaches $\alpha(\mathfrak{G}, \psi)$. Let us then consider a generic $c \in \|\mathrm{PRE}(\mathcal{G}, A, \alpha(\mathcal{G}, \psi))\|_{\mathfrak{G}}$. If $c \in \|\mathrm{PRE}(\mathcal{G}, A, \alpha(\mathcal{G}, \psi))\|_{\mathfrak{G}}$ then, by the pre-image algorithm, there exists β, β' and $(ap, \overline{ap})$ such that:

- $c \in \|\beta\|_{\mathfrak{G}}$;
- β is in the predecessor of β', thus $\beta \overset{(ap, \overline{ap})}{\rightarrow} \beta'$ from soundness of step (2) in the pre-image computation;
- there exists a one-step strategy σ_A from β which leads to $\alpha(\mathcal{G}, \psi)$.

We have to prove that such a strategy satisfies Definition 11.

Let $(ap, \overline{ap})$ be a generic action profile consistent with strategy σ_A such that $\beta \overset{(ap, \overline{ap})}{\rightarrow} \beta''$, with clearly $\beta'' \in \mathrm{EXPD}(\alpha(\mathcal{G}, \psi))$. We show that for all $c \in \|\beta\|_{\mathfrak{G}}$ there exists $ap' \in AP$ and there exists $c' \in Conf$ such that $c \overset{ap'}{\rightarrow} c'$, $c' \in \|\beta''\|_{\mathfrak{G}}$ and the *expansion* of ap' is $\overline{ap}$.

Constructively, we build ap' agent-wise from $\overline{ap}$. For each $a \in Agt$, its action in $\overline{ap}$ can be one of the following:

(1) idle^a;
(2) (single release) rel^a_R;
(3) (multiple release) $\{\mathsf{rel}^a_{R_1}, \ldots, \mathsf{rel}^a_{R_t}\}$; or
(4) req^a_R.

If (1) then the $ap'(a) = \text{idle}^a$. If (2) then $ap'(a) = \text{rel}_r^a$ for a random $r \in R$ such that $c(r) = a$. Notice that this is always possible, given that $\beta \overset{(ap,\overline{ap})}{\to} \beta''$, meaning that a does have the number of resources necessary to perform a release. If (3), then $ap'(a) = \text{rel}_{all}^a$. This is always possible again from the same observations in (2). If (4), then it is not enough to pick a random available $r \in R$ and set $ap'(a) = \text{req}_r^a$, as, depending on the requests actions performed by the other agents, a can own or not r in the next configuration. Thus, first the requested r must be such that $c(r) = null$ (there exists one otherwise $\beta \overset{(ap,\overline{ap})}{\to} \beta''$ would not be a step) but also the request actions of all agents have to be considered together when building ap'. It is, however, enough to look at the interval $(a, R)[\ell_R^{\prime a}, \ell_R^{\prime a}]$ in β'' in order to determine whether the request of a has been successful or not. If $\ell_R^{\prime a} = \ell_R^a$, then it was successful, and there is another agent a' performing a request for the same resource (this is guaranteed by how the intervals of β'' are computing from β by performing $(ap, \overline{ap})$). Otherwise, if $\ell_R^{\prime a} = \ell_R^a + 1$, then the request by a of r has been successful and, when building ap' for the other agents, we have to be sure that no other agent requests r (again, this is always guaranteed by how the intervals of β'' are computing from β by performing $(ap, \overline{ap})$). Finally, we have to show that $c' \in \|\beta''\|_{\mathfrak{G}}$. This is immediate, as the expansion of ap' is $\overline{ap}$ and from the way intervals are updated when computing $\beta \overset{(ap,\overline{ap})}{\to} \beta''$.

Completeness. We have to prove that if $c \in Conf$ and there exists a strategy σ_A which in one step reaches a configuration satisfying ψ, then $c \in \| \text{PRE}(\mathcal{G}, A, \alpha(\mathfrak{G}, \psi)) \|_{\mathfrak{G}}$. If c satisfies the above, then $\exists ap, c'.c \overset{ap}{\to} c'$ with $ap \in \sigma_A(c)$ and $c' \in [\![\psi]\!]_{\mathfrak{G}}$. By inductive hypothesis, $c' \in \|\alpha(\mathcal{G}, \psi)\|_{\mathfrak{G}}$, meaning that there exists $\beta' \in \alpha(\mathcal{G}, \psi)$ and $c' \in \|\beta'\|_{\mathfrak{G}}$. The pre-image algorithm tries all action profiles in AP that could lead to β', thus also ap. We have to show that there exists $\overline{ap}_k$ such that $\beta \overset{(ap,\overline{ap}_k)}{\to} \beta'$ and $c \in \|\beta\|_{\mathfrak{G}}$. Constructively, the required $\overline{ap}_k$ is built by looking at configuration c, which tells us to which classes the resources of the agents performing the release-all actions belong. Since it exists, the algorithm finds it as it explores all possible $\overline{ap}_k$ for any ap. Also, $c \in \|\beta\|_{\mathfrak{G}}$ by inspecting how we modify the intervals when computing the predecessors. It remains to prove that β is controllable. This is straightforward, as the required controllable joint action for A in Definition 11 is easily obtained from ap, which is a one-step strategy by hypothesis.

- $\varphi = \neg\psi$. From $\alpha(\mathcal{G}, \psi) = \beta_1 \vee \ldots \beta_s$ we compute the negation $\overline{\beta_i}$ of each β_i, complementing its intervals. Such operation produces at most two intervals, thus each $|\overline{\beta_i}|$ is at most $2 \cdot |\beta_i|$. Then we produce the intersection of those β_i by simply "projecting" on intervals common to each β_i.
- $\varphi = \langle\!\langle A \rangle\!\rangle \, \mathsf{G}\psi$. Recursively, we start from $\alpha(\mathcal{G}, \psi)$ and compute the conjunction with its pre-image until a fixpoint is reached, just like in the explicit model checking algorithm for ATL (cf. [4] or [10, Chap. 9]). Sound and completeness follows from the fixpoint characterisation of the G operator.

– $\varphi = \langle\!\langle A \rangle\!\rangle \, \psi_1 \mathsf{U} \psi_2$. Similarly to the previous case, but each iteration computes the disjunction of $\alpha(\mathcal{G}, \psi_2)$ with the conjunction of the pre-image of the expression obtained at the previous step with $\alpha(\mathcal{G}, \psi_1)$, again just like in the explicit model checking algorithm for ATL. Soundness and completeness follow from the fixpoint characterisation of the operator U.

References

1. Alechina, N., Logan, B., Hoang Nga, N., Rakib, A.: Resource-bounded alternating-time temporal logic. In: Proceedings of the 9th International Conference on Autonomous Agents and Multiagent Systems: Volume 1, AAMAS 2010, vol. 1, pp. 481–488. International Foundation for Autonomous Agents and Multiagent Systems, Richland (2010)
2. Alechina, N., Logan, B., Hoang Nga, N., Raimondi, F.: Symbolic model checking for one-resource RB±ATL. In: Proceedings of the 24th International Conference on Artificial Intelligence, IJCAI 2015, pp. 1069–1075. AAAI Press (2015)
3. Alechina, N., Logan, B., Hoang Nga, N., Raimondi, F., Mostarda, L.: Symbolic model-checking for resource-bounded ATL. In: Proceedings of the 2015 International Conference on Autonomous Agents and Multiagent Systems, AAMAS 2015, pp. 1809–1810. International Foundation for Autonomous Agents and Multiagent Systems, Richland (2015)
4. Alur, R., Henzinger, T.A., Kupferman, O.: Alternating-time temporal logic. J. ACM 49(5), 672–713 (2002)
5. Bhargava, D., Vyas, S.: Agent based solution for dining philosophers problem. In: 2017 International Conference on Infocom Technologies and Unmanned Systems (Trends and Future Directions), ICTUS, pp. 563–567, December 2017
6. Chandy, K.M., Misra, J.: The drinking philosophers problem. ACM Trans. Program. Lang. Syst. 6(4), 632–646 (1984)
7. Chevaleyre, Y., et al.: Issues in multiagent resource allocation. Informatica 30, 3–31 (2006)
8. Choppella, V., Kasturi, V., Sanjeev, A.: Generalised dining philosophers as feedback control. CoRR abs/1805.02010 (2018)
9. Datta, A.K., Gradinariu, M., Raynal, M.: Stabilizing mobile philosophers. Inf. Process. Lett. 95(1), 299–306 (2005)
10. Demri, S., Goranko, V., Lange, M.: Temporal Logics in Computer Science. Cambridge Tracts in Theoretical Computer Science. Cambridge University Press, Cambridge (2016). http://www.cambridge.org/core_title/gb/434611
11. Dijkstra, E.W.: Hierarchical ordering of sequential processes. Acta Informatica 1(2), 115–138 (1971)
12. Herescu, O.M., Palamidessi, C.: On the generalized dining philosophers problem. In: Proceedings of the Twentieth Annual ACM Symposium on Principles of Distributed Computing, PODC 2001, pp. 81–89. ACM, New York (2001)
13. Hopcroft, J.E., Karp, R.M.: An $n^{5/2}$ algorithm for maximum matchings in bipartite graphs. SIAM J. Comput. 2(4), 225–231 (1973)
14. Papatriantafilou, M.: On distributed resource handling: dining, drinking and mobile philosophers. In: Proceedings of the First International Conference on Principles of Distributed Systems, OPODIS, pp. 293–308 (1997)
15. Sidhu, D.P., Pollack, R.H.: A robust distributed solution to the generalized Dining Philosophers problem. In: 1984 IEEE First International Conference on Data Engineering, pp. 483–489, April 1984

Interpreting Information in Smart Environments with Social Patterns

Rubén Fuentes-Fernández[✉] and Jorge J. Gómez-Sanz

Research Group on Agent-Based, Social and Interdisciplinary Applications
(GRASIA), Universidad Complutense de Madrid, Madrid, Spain
{ruben,jjgomez}@fdi.ucm.es
http://grasia.fdi.ucm.es

Abstract. Smart Environments (SEs) work in close interaction with their users. To perform properly, these systems need to process their information (both from sensing and to act) considering its meaning for people. For instance, when they manage workflows that represent users' activities or consider the tradeoffs between alternative actions. Introducing *social knowledge* about the human context helps SEs to better interpret information, and thus people' needs and actions. However, working with this knowledge faces several difficulties. Its level of abstraction is different from that directly related to system components. Moreover, it belongs to a background that is not frequent among engineers. In order to address these issues, this paper proposes the use of Social Context-Aware Assistants (SCAAs). These Multi-Agent Systems (MASs) manage explicitly social information using specifications conform to a domain-specific Modelling Language (ML). The ML aims at describing human aspects and their changes in a given context related to a SE. *Social properties* describe reusable knowledge using a template with these specifications and textual explanations. Working with the ML facilitates the semi-automated transformation of specifications to integrate social and other system information, derive new one, and check properties. Specific agents are responsible for managing information, and translating data from sensors to the ML, and from this to data for actuators. A case study on an alert system to monitor group activities, extended with social knowledge to interpret people' behaviour, illustrates the approach.

Keywords: Smart environment · People' behaviour ·
Human environment · Social knowledge · Social property ·
Semi-automated verification · Multi-agent system ·
Social Context-Aware Assistant

1 Introduction

Smart Environments (SEs) integrate multiple technologies and devices to support their users' activities in sensitive and unobtrusive ways [9]. Their functionality is based on an information flow. Sensors in the environment capture raw

© Springer Nature Switzerland AG 2019
M. Slavkovik (Ed.): EUMAS 2018, LNAI 11450, pp. 48–61, 2019.
https://doi.org/10.1007/978-3-030-14174-5_4

data, which other components filter and aggregate to create more abstract information. Application components process this information to make decisions on how to act, which actuators translate to specific actions over the environment.

This process usually involves building and updating a representation of the environment, which is known as the *context* [10]. This is populated from sensor data, and additional knowledge makes possible to interpret and transform those data to derive new information. This knowledge works at different levels of abstraction, from the low-level component-oriented data to the high-level user-oriented information [15,25].

Most of current approaches to manage that knowledge are focused on providing mechanisms for its representation and transformation (e.g. [15,22,25]). There are fewer works dealing with specific knowledge to interpret information, particularly regarding the human environment. For instance, the differences between people moving in a walk or an emergency. This limits the capability to abstract information and identify what actions should be triggered based on the understanding of the situation. When such knowledge is considered, it is usually in ad-hoc ways (e.g. coded) and for specific issues (e.g. for non-verbal communication in [5] or social and physical proximity in [15]). These works are difficult to extend with additional knowledge or to cover other aspects. Some efforts focus on more general, reusable and flexible representations (e.g. ontologies and rule-based [7]), with some of them considering human aspects or social contexts (see some examples in [23]). The usual problem here is that many of them lack of a general underlying theoretical background in Social Sciences. This would allow a guided application to study the human context, and inform processes to translate the obtained information to actual SE functioning.

Our work addresses these issues with the development of a framework for Social Context-Aware Assistants (SCAAs). These are Multi-Agent Systems (MASs) [14] that allow introducing the human meaning of observations and actions in the information flow of SEs. The approach comprehends two elements: means to represent and manage that information; agents and resources to work with them and interact with the rest of the SE.

The management of social information is based on *social properties* [13]. These have been proposed as a way to formalise social knowledge for automated processing based on the Activity Theory (AT) framework [18]. The core of their specification is diagrams conform to a specific Modelling Language (ML) for social activities in context. A pattern matching algorithm allows looking for complete or partial instances of query specifications among available information. This mechanism is the basis to check properties or generate new information. This allows for instance deriving social bonds from the observation of some behavioural patterns.

Our work extends that ML to address specific issues related to SEs. It considers low-level information (without social meaning) with the introduction of specific types (e.g. *data* versus *information*). Their management requires certain attributes, for instance related to tags, sources and reliability. The transformation of this information is described through the added *operations* and their related extensions. The result is the SCAA-ML.

The architecture of a SCAA MAS introduces *social agents* that manage the resources to work with that information: *sensing agents* transform low-level data from sensors into social information, while *observer agents* send to the rest of the system relevant information and support queries; *reasoner agents* derive new social information from available one. This organisation follows common practices in information fusion [3] and for agents as resource managers in SEs [2,10]. This links social agents with semantic ones in literature [7,23], though the specific focus of the former is on the use of social properties [13].

This overall approach aims at the explicit and flexible management of social information, with a theoretical background that guides its interpretation. This facilitates the integration of SCAAs with other components from SEs and their information, as well as the adaptation of these assistants to different needs.

A case study on a care system for disabled and elderly people in group activities exemplifies the approach. It extends the work in [24], where smartphones support tailored alarms depending on participants and environment features. The alarms can be scaled up to different smartphones in the group or the emergency services looking for assistance. Here, a SCAA improves the information about the context of activities (e.g. type and place), and thus of people' potential behaviours and constraints. This allows providing more tailored alarms.

The rest of the paper is organised as follows. Section 2 introduces the background regarding *social properties* [13]. Then, Sect. 3 describes the SCAA framework, including its architecture and ML. The case study in Sect. 4 illustrates the approach with the distributed collaborative monitoring system with social knowledge. Section 5 compares this approach and related work. Finally, Sect. 6 discusses some conclusions and future work.

2   Background

The work on *social properties* [13] pursues applying knowledge from Social Sciences in effective ways for software engineering and with tool support. Its theoretical background comes from the AT [18]. This provides the basis for the definition of a specific ML called UML-AT, and the rules that govern the evolution of its systems.

The AT is organised around the key concept of *activity*. It is a transformation act (mental or physical) that is both intentional and social. A *subject* carries out the activity in order to satisfy her/his *objectives*, which brings the intentional perspective. The execution of the activity transforms an *object* using *tools* into an *outcome*, which should satisfy the objectives. The social dimension appears with the *community* where any activity happens. It is the set of subjects that share the environment and meanings both now and historically. The community affects the activity through two mediating artefacts. The *division of labour* specifies the organisation of the community related to the execution of the activity. The *rules* come from the society and influence the activity, for instance with norms and culture. All the previous elements constitute the *activity system* of an activity.

Social contexts are described as networks of activity systems linked by shared elements. These elements can change their role between activity systems, e.g. the outcome of an activity system becomes a tool or a rule in other.

UML-AT formalises this conceptual framework as a Unified Modelling Language (UML) profile [20]. It represents these concepts using stereotypes. It also adds additional elements. The type *artefact* to represent any type in an activity system, and the relationship *change of role* (stereotype) to indicate that a given concept adopts different types. Following the UML, it has relations to represent aggregation (with a diamond shape) and inheritance (a hole triangle), and cardinality adornments.

This ML is used to specify information, and in particular the *social properties*. They represent knowledge on human activities applicable in different settings. Their specification applies a template organised around UML-AT diagrams explained with text.

A pattern matching algorithm uses the diagrams to check properties or derive information. It looks for complete or partial instances of the searched specifications (translated from the UML-AT diagrams) in available information. For instance, the appearance of an instance of a property is used to explain the meaning of the located information. The algorithm has been implemented in different ways, e.g. object-oriented, rule-based, and logics programming.

3   The SCAA Framework

The SCAA framework is aimed at providing SEs with means to integrate in a flexible way social information. This allows a better interpretation of the raw data in order to get a more meaningful context. For instance, considering how different types of activity change people' behaviour or some expected parameters of their actions (e.g. applicable norms or collaborators).

The framework supports that functionality with means to manipulate social information and knowledge (see Sect. 3.1). Specific agents integrate those means and provide the interfaces with the rest of an SE (see Sect. 3.2).

3.1   Social Information

The SCAA-ML specifies the primitives to characterise people, their activities, and the required capabilities to participate in them. The manipulation of this information relies on the AT concepts of activity and operation [18]. Figure 1 shows some of these elements.

The root of the inheritance hierarchy of concepts is the type *element* (not shown in the figure), which only has a unique *id* attribute. Its sub-types are *data* to represent observations from sensors, and *information* for abstracted social information. The latter has as sub-types *object* and *transformation*.

The type *object* represents elements from the external environment or the SE. They can be *located* (a physical object like a person or a sensor), *non-located* (an intangible object such a norm or an interaction), and *mixed* (e.g. a museum with

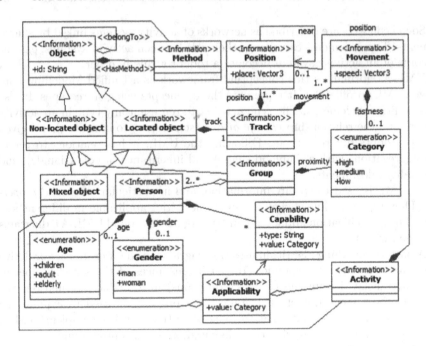

Fig. 1. Partial SCAA-ML specification with primitives to specify information related to type *object*.

physical assets and a website). Located objects have a *track* that characterises their *position* and *movement* (i.e. speed and address) over time.

Transformations (not included in the figure) are information types that represent functional capabilities to transform *data* or *information*. Following the AT [18], they can be *activities* or *operations* (more on them later in this section). *Activities* can be located in a position, so they are also mixed objects. *Methods* are operations attached to objects using a *HasMethod* relationship.

The type *person* is a located object with some common attributes (e.g. *age* and *gender*) and *capabilities*. These are general descriptions of mental and physical skills. They take a value in the *category* enumeration, where *medium* corresponds to the average capability of a middle-aged adult without any particular condition, and *low* indicates some difficulties.

To describe people' needs when participating in activities, the ML incorporates the concept of *applicability*. Given an activity and some participant' capabilities, certain methods can or should be used. For instance, when a group is performing the activity of walking, and some people have a low visual capability, a method of noise alarm is preferred to a visual one.

The ML also supports representing instances of these types. For every type *name*, the stereotype *Iname* represents its instances.

Both for types and instances, attributes can have or not values, and these can be constants or variables. For instance, a *person* type may not indicate a value for the attribute *age* (which corresponds to instances of any age), or may specify a constant (only instances with that value) or a variable (to hold the value for later use). As a convention, constants are written with their literal, and variable names start with an underscore. For instance, a value of "10" specifies people who are children, and a variable "*_AGE*" to specify people with the same age.

Elements can also have attached an arbitrary number of string *tags* to characterise them. For instance, to describe a position as a park or a museum. *Elements* also have a *reliability* attribute to describe the level of confidence in their information as a value of *category*. For instance, in the care system, when somebody walks away from the group in a museum, there is a *medium* confidence that s/he is going to see some piece, but when s/he has a mental condition, there could be a *low* confidence that s/he is wandering and could become lost.

For *data* types, there is an additional attribute *source*. It identifies the generator (e.g. sensor or software) of that information.

As introduced before, *transformations* represent changes in social information. Their semantics is that when instances of all its input types are available, they generate instances of the output types. They can have attached a "snippet" to further specify its transformation, written for instance in code or logics. According to the AT [18], activities have an intentional and social meaning for their participants, while operations only depend on their information context. Thus, operations just indicate inputs and outputs, but activities indicate specific roles of inputs (e.g. subject, object or tool) through the types of relationships.

Inputs and outputs of transformations can have several adornments. Conditional adornments are related to the presence of instances of the input types among the available information. They can indicate the absence of instances (i.e. "NOT"), the presence of at least one instance (i.e. "ANY"), or that all the available instances meet the specified constraints (i.e. "ALL"). There are also cardinality adornments similar to those in the UML [20]. A number indicates that given number of instances, an * any number of instances (including 0), and a range N..M indicates at least N instances and at most M.

In this context, *social properties* are SCAA-ML specifications that describe recurrent situations coming from social knowledge (e.g. from the AT [18]). For instance, how people behave on certain activities or how social bonds change their behaviour in certain places. Figure 2 shows an example of these properties.

Information on potential social bonds among persons can sometimes be inferred from observations. The property derives from the facts that some persons (i.e. instances *Person1* and *Person2* of type *person*) have participated together in several *group activities*, that they must have become at least acquaintances. Those activities are characterised as instances of a type *activity* with an explicit *community* (here *Group1*). The social bond is specified as an *acquaintance* group with a value *low* for *proximity*. The *bond type* tag represents its specific type.

The variables *_ID1* and *_ID2* in the *id* attribute of the instances of *Person* allow indicating that they must be different (i.e. different ids). This condition cannot be expressed with the modelling primitives of the SCAA-ML. It is specified in the snippet of the operation.

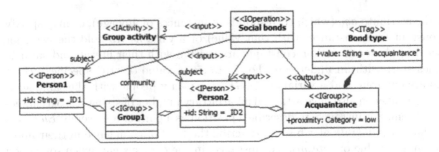

Fig. 2. Operation to derive information on social bonds from observations.

Working at the level of specifications facilitates users interpreting and changing the information. Modifying the analysis to perform only needs to consider other social properties as far as no new external data are required.

3.2   Architecture

The SCAA approach also considers the integration with SEs at the level of components. It adopts an approach based on the MAS paradigm [14], and the observer pattern and a shared representation of the context according to [10].

Sensing agents get external observations and transform them into SCAA-ML facts. They manage *social sensors* with *notify* methods to receive events and *perceive* methods to implement queries to other SE components. These social sensors actually work as interfaces to SE components that provide data, as common sensors.

Reasoner agents carry out the actual manipulation of SCAA-ML information. They manage a *social engine*. This offers *assert* methods that social sensors invoke to add facts into its base. The engine uses these facts to trigger transformations (i.e. activities and operations) and check properties and derive information. The engine also offers *report* methods to notify *observer agents* from changes in the base, and *consult* methods to query it.

Observer agents manage *social observer* resources and transform SCAA-ML facts into suitable formats for the rest of the SE. These resources offer *notify* methods to be reported on changes in the engine base, and *consult* methods to support information requests to the *reasoner*. These resources also participate in observer patterns for other SE components. They are the elements that can be observed.

The inner representation and processing of SCAA-ML information in these components currently adopts a rule-based approach inspired by [26]. This is well-aligned with the proposed use of transformations in SCAAs (see Sect. 3.1).

4   Case Study: Distributed Collaborative Care System

The case study considers a care system for participants in group activities. It takes into account people' specific needs and how to address them, according to their features, those of other members of the group, and the kind of activity.

The basis is the collaborative system in [24]. It is aimed at providing tailored alarms during group activities with disabled and elderly people. The system knows their profiles (e.g. cognitive issues and physical impairments). In a group activity, it runs on the participants' smartphones. Their sensors allow perceiving the environment and situation of participants. When there is a situation to warn, the system considers the people involved and the environment to select suitable actuators. For instance, it uses a visual and vibratory notification to warn a hearing-impaired person or in a noisy environment. If it gets no answer, it can scale up notifications. It starts with the smartphones of the involved users, and continues in order with those of other participants, supervisors, and emergency services.

The use of a SCAA with this SE is intended to improve the adaptation of alarms. The SCAA can help to select what alarms to notify or the way to do it using its social knowledge about the context (e.g. type and place of activity).

To support this functionality, the original system data flows are changed. The architecture of this SCAA is very close to the generic one (see Sect. 3.2), including social *sensing*, *reasoner*, and *observer* agents.

Sensing agents receive information from the smartphone sensors and from the original components that raise the alarms (here referred as *generators*). They assert this information in the social engine as SCAA-ML facts. In the case of alarms, they define the information to be communicated using the methods of an actuator or device. According to the SCAA model (see Fig. 1), this corresponds to a *method* of an *object*. The information about alarms is diverted from the generators in the modified system, so it only arrives to the SCAA components.

The reasoner agent runs the engine to trigger the transformations using the asserted and derived information. The new derived information can modify requested alarms (see transformations later in this section).

Finally, observer agents send the modified alarms to the original components that notify them to users (here the *notifiers*). This can imply a redirection of the alarm to actuators or methods different to those initially selected by the generators, or with another parametrisation.

From the point of view of the original components, the only modification is sources and targets in the data flow. SCAA components receive the information from sensors and generators, and are the source of alarms for notifiers.

The characterisation of the context for the SCAA is done using the observed and derived information. To produce the latter, the engine base includes social properties with transformations. In this case, it uses the property about social bonds seen previously (see Fig. 2) and two other about tailored alarms considering the place (see Fig. 3) and means (see Fig. 4) to notify them.

In this domain, social knowledge indicates that, beyond the user capabilities, not all the ways of notifying alarms are suitable in every place and situation. For

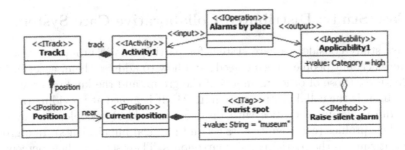

Fig. 3. Operation on suitable ways to notify an alarm according to the place.

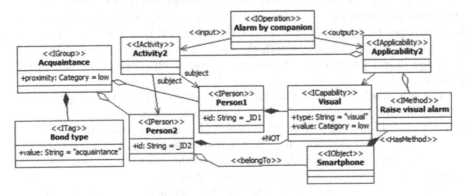

Fig. 4. Operation on suitable ways to notify an alarm with accompanying person.

instance, there are places where noisy alarms should be avoided (e.g. a museum or cinema), and visual notifications should not be used when driving.

Figure 3 shows a property that considers that silent alarms must be used in museums. The element that generates the new information is the instance of operation called *alarms by place*. It shows an activity occurring in a given position *Position1*. According to the SCAA-ML, activities are *mixed objects* and thus have a position. That position is close (maybe inside) another one called *Current position*, which is tagged as a museum. The instance of applicability *Applicability1* determines a kind of method applicable in this context, in this case *Raise silent alarm*.

The care system could also take into account the social bonds existing among people in the group. The property in the previous section (see Fig. 2) calculated acquaintance bonds as those appearing among people that go together to make activities. Usually, these persons are prone to help each other.

The second property (see Fig. 4) exploits social bonds to choose the best targets when scaling up alarms to neighbours. In this case, it considers two persons (i.e. *Person1* and *Person2*) who are acquaintances (shown by the instance of group *Acquaintance*). The instance of capability specifies that one of these persons is visually impaired (i.e. *type visual* and *value low*), and the other not (i.e. adornment "NOT").

In this case, the system relies on the presence of the acquaintance without visual impairment (i.e. *Person2*) to use visual notifications. Nevertheless, it needs to redirect the notification to that person. Thus, the applicability instance is linked to a method that belongs to the object *Smartphone* of this person (i.e. the relationship *belongTo*).

The previous information allows having a more precise description of the SE context than in the original work. The SE does not only consider the information provided explicitly to it, but can also learn from the observations according to the social knowledge. For instance, it can use information on people with the closest bonds, or consider the type of activity. This could even be suitable to anticipate alarms, for instance on mobility issues when the path of the group includes stairs.

The information exchanged between the original system and the SCAA continues to be the original one. Nevertheless, the architecture facilitates the integration of other sources. For instance, smartphones have problems to manage location in closed spaces. The issue could be solved adding a new social sensor able to connect to some indoor location system (e.g. beacon sensors). Also, the system could be scaled to include different properties just adding them to the base of the engine.

5 Related Work

The work on the SCAA approach is related to several fields of research. First, works about the information in the social context and the way to elicit it. Second, how SEs represent and manage it. Third, its use on these SEs.

SEs consider of paramount importance their interaction with users. Given the difficulties to provide a seamlessly and unobtrusive general interaction, they usually narrow the problem to some specific situations, e.g. home, driving or e-learning. In such domains, systems need to deal only with limited information (e.g. [9,25]). Designers mainly rely on experts and users to get knowledge about it, so there is not an underlying theoretical framework. For instance, the system in the case study [24]. In some cases, designers resort to specific theories because of the complexity of the domain. For instance, in e-learning [16] or recommender [1] systems. However, the narrow scope of these theories limits the possibility of extending these systems to work with other aspects.

In the wider scope of social information [19], the problems appear with the lack of widely applicable general theories. For instance, systems that work with individual profiles and group identification make it in ad-hoc ways, maybe using independent theories for the different aspects. There have been efforts to provide unifying frameworks (e.g. [7,21]), but these have been mainly from the point of view of system engineering and not specifically of their knowledge.

Regarding this aspect, the SCAA approach bases is conceptual framework and ML in the AT [18] from Social Sciences. It has been applied in different disciplines, and integrates the human activity with its environment. Thus, it offers a suitable theoretical framework to support this kind of work. There have been discussions on its use for SEs and their social context, but mainly from a theoretical social-oriented perspective [17].

The representation of information in SEs depends on the kind of considered data and their manipulation. Multiple general representation techniques have been applied, e.g. code, tuples, entity-relation, logics, rules, ontologies, MLs or classifiers. [4,7]. There is also research on more specific languages. For instance, those focused on system-related aspects (e.g. sensors, actuators and methods in smart transport [11,12]), time and spatial issues (e.g. [6]), or workflows (e.g. [2,10]). The social context has been the object of some specific modelling works [7,23]. These have mainly been focused on the conceptual framework, but they largely missed guidelines on to use it to study the human context. This is something that the SCAA approach gets through its basis in the AT [18] and previous works in the topic [13].

Regarding the use of social information, efforts are again focused on specific issues (e.g. non-verbal communication [5], interaction detection based on space position [8], or social and physical proximity [15]). Thus, the mechanisms to exploit the information are too specific to be extended to other aspects. On the contrary, the SCAA approach pursues providing more general mechanisms to integrate different uses of social knowledge in a single SE. This is achieved through the use of social properties, which change allows modifying the type of analysis considered.

6   Conclusions

The SCAA approach aims at providing support in SEs to integrate social information and specific knowledge to manage it. With that, SEs can have a better understanding of the human context where they operate, and thus delivering to their users more tailored services.

The SCAA framework comprehends two elements. A MAS provides the functionality and allows the integration with the rest of the SE. The SCAA-ML is used to specify social information, knowledge, and their transformations. A pattern matching algorithm uses specifications with this language both to derive and check information from observations. It is based on the identification of types and values of attributes.

The ML currently supports the specification of information about people and their needs (i.e. capabilities), groups, and context (i.e. objects). For them, information on functionality, position and arbitrary tags can be added. Transformations are defined as social activities or operations that manipulate information.

The architecture includes sensing and observer agents that make the interface with the SE. Reasoner agents manage the engines where social information is manipulated. These engines used an underlying rule-based framework to implement the processing.

The overall approach pursues facilitating user and expert involvement through formats for the specification and processing of information suitable for both people and systems. Besides, it makes systems more flexible, as relevant changes in information analysis just need to modify the considered properties.

The case study introduced a SCAA in an existing care system [24]. It used the social information to improve the alarm notifications. For instance, considering the place where the activity happens, or the social bonds and capabilities of people in the group. These improvements were introduced as social properties. Thus, the system functionality could be altered by merely changing them.

The SCAA research still has several open issues. First, the ML needs to support a more complete description of the social context. The considered concepts of the underlying AT framework are those at the highest level of abstraction. To facilitate semi-automated processing, there is need to provide further details, for instance regarding types and their characterisation. Second, the temporal dimension of information is not considered. Currently, only precedence of transformations can be expressed with produce-consume relations. The validity of information should also be included. For instance, a social bond could render less strong after some time. Finally, more experiments on the integration with SEs must be performed in order to get a better picture of the requirements regarding information and system architecture.

Acknowledgment. This work has been done in the context of the mobility plan for the mobility of researchers "Subprograma de Movilidad del Programa Estatal de Promoción del Talento y su Empleabilidad, en el marco del Plan Estatal de Investigación Científica y Técnica y de Innovación" (grant PRX17/00613) supported by the Spanish Ministry for Education, Culture, and Sports, the projects "Collaborative Ambient Assisted Living Design (ColoSAAL)" (grant TIN2014-57028-R) supported by the Spanish Ministry for Economy and Competitiveness, MOSI-AGIL-CM (grant S2013/ICE-3019) supported by the Autonomous Region of Madrid and co-funded by EU Structural Funds FSE and FEDER, and the "Programa de Creación y Consolidación de Grupos de Investigación" (UCM-BSCH GR35/10-A).

References

1. Adomavicius, G., Tuzhilin, A.: Context-aware recommender systems. In: Ricci, F., Rokach, L., Shapira, B., Kantor, P.B. (eds.) Recommender Systems Handbook, pp. 217–253. Springer, Boston, MA (2011). https://doi.org/10.1007/978-0-387-85820-3_7

2. Fernández-de Alba, J.M., Campillo, P., Fuentes-Fernández, R., Pavón, J.: Opportunistic control mechanisms for ambience intelligence worlds. Expert Syst. Appl. **41**(4), 1875–1884 (2014). https://doi.org/10.1016/j.eswa.2013.08.084

3. Alfonso-Cendón, J., Fernández-de Alba, J.M., Fuentes-Fernández, R., Pavón, J.: Implementation of context-aware workflows with multi-agent systems. Neurocomputing **176**, 91–97 (2016). https://doi.org/10.1016/j.neucom.2014.10.098

4. Baldauf, M., Dustdar, S., Rosenberg, F.: A survey on context-aware systems. Int. J. Ad Hoc Ubiquit. Comput. **2**(4), 263–277 (2007). https://doi.org/10.1504/IJAHUC.2007.014070

5. Baur, T., et al.: Context-aware automated analysis and annotation of social human-agent interactions. ACM Trans. Interact. Intell. Syst. (TiiS) **5**(2), 11:1–11:33 (2015). https://doi.org/10.1145/2764921

6. Bettini, C., et al.: A survey of context modelling and reasoning techniques. Pervasive Mob. Comput. **6**(2), 161–180 (2010). https://doi.org/10.1016/j.pmcj.2009.06.002

7. Bikakis, A., Patkos, T., Antoniou, G., Plexousakis, D.: A survey of semantics-based approaches for context reasoning in ambient intelligence. In: Mühlhäuser, M., Ferscha, A., Aitenbichler, E. (eds.) AmI 2007. CCIS, vol. 11, pp. 14–23. Springer, Heidelberg (2008). https://doi.org/10.1007/978-3-540-85379-4_3

8. Cook, D.J., Crandall, A., Singla, G., Thomas, B.: Detection of social interaction in smart spaces. Cybern. Syst. Int. J. **41**(2), 90–104 (2010). https://doi.org/10.1080/01969720903584183

9. Cook, D.J., Das, S.K.: How smart are our environments? An updated look at the state of the art. Pervasive Mob. Comput. **3**(2), 53–73 (2007). https://doi.org/10.1016/j.pmcj.2006.12.001

10. Fernández-De-Alba, J.M., Fuentes-Fernández, R., Pavón, J.: Architecture for management and fusion of context information. Inf. Fusion **21**, 100–113 (2015). https://doi.org/10.1016/j.inffus.2013.10.007

11. Fernández-Isabel, A., Fuentes-Fernández, R.: Extending a generic traffic model to specific agent platform requirements. Comput. Sci. Inf. Syst. **14**(1), 219–237 (2017). https://doi.org/10.2298/CSIS161010001F

12. Fernández-Isabel, A., Fuentes-Fernández, R.: An integrative modelling language for agent-based simulation of traffic. IEICE Trans. Inf. Syst. **99**(2), 406–414 (2016)

13. Fuentes-Fernández, R., Gómez-Sanz, J.J., Pavón, J.: Modelling culture through social activities. In: Dignum, V., Dignum, F. (eds.) Perspectives on Culture and Agent-based Simulations. SPS, vol. 3, pp. 49–68. Springer, Cham (2014). https://doi.org/10.1007/978-3-319-01952-9_4

14. Gomez-Sanz, J.J., Fuentes-Fernández, R.: Understanding agent-oriented software engineering methodologies. Knowl. Eng. Rev. **30**(4), 375–393 (2015). https://doi.org/10.1017/S0269888915000053

15. Guo, B., Zhang, D., Wang, Z., Yu, Z., Zhou, X.: Opportunistic IoT: exploring the harmonious interaction between human and the internet of things. J. Netw. Comput. Appl. **36**(6), 1531–1539 (2013). https://doi.org/10.1016/j.jnca.2012.12.028

16. Hwang, G.J., Tsai, C.C., Yang, S.J.: Criteria, strategies and research issues of context-aware ubiquitous learning. J. Educ. Technol. Soc. **11**(2), 81–91 (2008)

17. Kofod-Petersen, A., Cassens, J.: Using activity theory to model context awareness. In: Roth-Berghofer, T.R., Schulz, S., Leake, D.B. (eds.) MRC 2005. LNCS (LNAI), vol. 3946, pp. 1–17. Springer, Heidelberg (2006). https://doi.org/10.1007/11740674_1

18. Leontiev, A.: Activity, Consciousness, and Personality. Prentice Hall, Upper Saddle River (1978)

19. Liang, G., Cao, J.: Social context-aware middleware: a survey. Pervasive Mob. Comput. **17**, 207–219 (2015). https://doi.org/10.1016/j.pmcj.2014.12.003

20. OMG: OMG Unified Modeling Language (OMG UML), Version 2.5.1. http://www.omg.org/. December 2017

21. Padovitz, A., Loke, S.W., Zaslavsky, A.: Towards a theory of context spaces. In: Proceedings of the Second IEEE Annual Conference on Pervasive Computing and Communications Workshops, 2004, pp. 38–42. IEEE (2004). https://doi.org/10.1109/PERCOMW.2004.1276902

22. Perera, C., Zaslavsky, A., Christen, P., Georgakopoulos, D.: Context aware computing for the internet of things: a survey. IEEE Commun. Surv. Tutor. **16**(1), 414–454 (2014). https://doi.org/10.1109/SURV.2013.042313.00197

23. Rodríguez, N.D., Cuéllar, M.P., Lilius, J., Calvo-Flores, M.D.: A survey on ontologies for human behavior recognition. ACM Comput. Surv. (CSUR) **46**(4), 43:1–43:33 (2014). https://doi.org/10.1145/2523819

24. Sendra, S., Granell, E., Lloret, J., Rodrigues, J.J.: Smartcollaborative mobile system for taking care of disabled and elderlypeople. Mob. Netw. Appl. **19**(3), 287–302 (2014). https://doi.org/10.1007/s11036-013-0445-z

25. Snidaro, L., García, J., Llinas, J.: Context-based information fusion: a survey and discussion. Inf. Fusion **25**, 16–31 (2015). https://doi.org/10.1016/j.inffus.2015.01.002

26. Xavier, D., Crespo, B., Fuentes-Fernández, R.: A rule-based expert system for inferring functional annotation. Appl. Soft Comput. **35**, 373–385 (2015)

Learning Hedonic Games via Probabilistic Topic Modeling

Athina Georgara[✉], Thalia Ntiniakou, and Georgios Chalkiadakis

School of Electrical and Computer Engineering, Technical University of Crete, Chania, Greece
{ageorgara,tntiniakou,gehalk}@intelligence.tuc.gr

Abstract. A usual assumption in the hedonic games literature is that of complete information; however, in the real world this is almost never the case. As such, in this work we assume that the players' preference relations are hidden: players interact within an unknown hedonic game, of which they can observe a small number of game instances. We adopt probabilistic topic modeling as a learning tool to extract valuable information from the sampled game instances. Specifically, we employ the online Latent Dirichlet Allocation (LDA) algorithm in order to learn the latent preference relations in Hedonic Games with Dichotomous preferences. Our simulation results confirm the effectiveness of our approach.

Keywords: Adaptation and learning · Cooperative game theory

1 Introduction

Hedonic games constitute a class of cooperative games that intuitively attempts to capture the interpersonal relations amongst the players and the social bonds of the formed coalitions. That is, such games model settings where each player defines an ordinal preference relation over all the possible groups she can participate in, depending *exclusively* on the identity of her co-partners. Though most previous works on hedonic games considers complete information over the game, in a more realistic framework that would not be a plausible assumption. In real-life settings that can be modelled as hedonic games, we face the problem of uncertainty, i.e. the agents have little or no information about the overall game. For instance, the players find themselves in a new, unknown environment, and need *to discover their own preferences* through the interaction with others. For this reason, it is essential for an agent to be able to *learn* the underlying hidden game. This was in fact the motivation for the authors in [18] to explore the *Probably Approximately Correct (PAC) learnability* of several classes of hedonic games: it studied how good probabilistic "hedonic" utility function approximations can be derived from a (polynomial) number of samples, and proposed algorithms to do so for specific problem instances in the process.

In this work, we use *probabilistic topic modeling* in order to extract information about each agent's preferences, and thus learn the underlying game. Probabilistic topic models (PTMs) is a statistical approach used in analyzing words of

© Springer Nature Switzerland AG 2019
M. Slavkovik (Ed.): EUMAS 2018, LNAI 11450, pp. 62–76, 2019.
https://doi.org/10.1007/978-3-030-14174-5_5

documents that was originally used in data mining to discover a distribution over topics related to a given text document. Here, instead, we are inspired by recent work of [15], and employ a widely used PTM algorithm, *online Latent Dirichlet Allocation (LDA)*, to operate on instances of formed coalitions, in order to discover the ordinal preferences relations of each agent.

As such, our contributions in this paper are as follows. We propose a novel way to learn the player utility functions, capturing their ordinal preferences. In order for our method to work, we need to prescribe a way to represent coalitions and preferences orderings into text documents that will be the input of the algorithm. For this reason, we present in Sect. 3 a novel procedure to interpret a pair of coalition-preference order into a 'bag-of-words', which can then be channelled into the PTM algorithm. We conducted a preliminary yet systematic evaluation of our approach. Our results show that given a small number–compared to the coalitional space of a game–of observations, we can form beliefs that to a great extent reflect the agents' real preference relations. Note that a concrete benefit derived from our work here, is that once the validity of a topic has been established, an agent can use this information in the future to propose coalitions during some coalition formation protocol. Moreover, our approach can be used by, e.g., recommender systems, to promote bundles of goods.

2   Background and Related Work

In this section we discuss the fundamental notions of our work, hedonic games and probabilistic topic modeling, along with previous related works.

2.1   Hedonic Games

Hedonic Games were initially introduced by [9] to describe economic situations where individuals act in collaboration, and have personal preferences for belonging in a specific coalition. In [4], the authors studied concepts of stability on hedonic coalitions, while [2] provides a thorough and extensive study of hedonic games. To begin, a hedonic game G is given by a tuple $\langle N; \succsim \rangle$. $N = \{1, 2, \cdots, n\}$ is a finite set of agents, and $\succsim = (\succsim_1, \cdots, \succsim_n)$ is a preference relation, where each $\succsim_i \subseteq N_i \times N_i$ is a complete, reflexive and transitive preference relation over the coalitions i can possibly belong to. Given a hedonic game $G = \langle N, \succsim \rangle$, agent $i \in N$ *strongly* prefers coalition $S \in N_i$ over coalition $T \in N_i$ if and only if $S \succ_i T$, and agent i is indifferent between S and T *if and only if $S \sim_i T$*; thus, agent i *weakly* prefers (or just prefers) coalition S over T *if and only if $S \succsim_i T$*.

The outcome of a hedonic game is a coalition structure, i.e. a partition of N. Unlike most class of cooperative games, in hedonic games there is no payoff (reward) assigned to the participants. However, since hedonic games is considered to be a subclass of *non-transferable utility (NTU) games*, one can naturally assume that in such games each agent earns some individual 'moral satisfaction' by simply being part of a certain coalition; that is, each agent is rewarded with personal satisfaction by collaborating with specific other agents. In this light, we can intuitively claim, as Sliwinski and Zick do in their work in [18], that each agent forms a personal utility function v_i which mirrors her personal relation $\succsim_i$; in other words, we translate the abstract notion of preferences over coalitions into a mathematical representation. Formally, given a game $G = \langle N, \succsim \rangle$, for each agent $i \in N$ there is a utility function $v_i : N_i \to \mathbb{R}$ such that for all coalitions $S, T \in N_i$ holds $v_i(S) \geq v_i(T)$ if and only if $S \succsim_i T$.

A particularly appealing class of hedonic games is when agent preferences are *dichotomous*. The notion of dichotomous preferences was initially studied within economic settings, as in [5] to describe situations where outcomes can deterministically be distinguished into *good* and *bad*. In the context of game theory, the concept of dichotomous preferences within hedonic games was introduced in [1] through Boolean Hedonic Games, and several aspects regarding their complexity were thoroughly studied in [17]. In hedonic games with dichotomous preferences, or in boolean hedonic games (BHGs), each agent partitions the coalitional space into two disjoint sets. Formally, let $N_i = \{S \subseteq N : i \in S\}$ be the set of all coalitions that contains agent i; thus, N_i can be partitioned into N_i^+, and N_i^- such that $N_i^+ \cup N_i^- = N_i$ and $N_i^- \cap N_i^+ = \emptyset$. Intuitively, N_i^+ is the set of *desirable* coalitions corresponding to good outcomes and N_i^- is the set of *non-desirable* coalitions. That is, agent i strictly prefers all coalition in N_i^+ to those in N_i^-, and she is indifferent otherwise; i.e. $S \succ_i T$ if and only if $S \in N_i^+$ and $T \in N_i^-$, and $S \sim_i T$ if and only if $S, T \in N_i^+$ or $S, T \in N_i^-$. We refer to coalitions in N_i^+ as *satisfactory*, and to coalitions in N_i^- as *dissatisfactory*.

2.2 Probabilistic Topic Modeling

Probabilistic topic models (PTMs), are statistical methods, introduced in the linguistic scenario of uncovering underlying (latent) topics in a collection of documents. Topic modeling algorithms have been also adapted to other scenarios as well, for example in genetic data, images and social networks. A widely known and successful PTM is that of Latent Dirichlet Allocation (LDA) [3].

Latent Dirichlet Allocation. We first describe the basic terms behind latent Dirichlet allocation (LDA) following [3] and [15]:

▷ A *word* is the basic unit of discrete data. A vocabulary consists of words and is indexed by $\{1, 2, \ldots, V\}$. The vocabulary is fixed and is fed as input to the LDA model.
▷ A *document* is a series of L words, $(w_1, w_2, \ldots, w_L)$.
▷ A *corpus* is a collection of D documents.
▷ A *topic* is a distribution over a vocabulary.

LDA is a Bayesian probabilistic topic model, in which each document can be described by a mixture of topics. A generative process for each document in the collection is assumed, where a random distribution over topics is chosen, and for each word in a document a topic is chosen from the distribution; finally, a word is drawn from the chosen topic. The same set of topics is shared by all documents in corpus, but each document exhibits topics in different proportions.

Latent variables, describing the hidden structure LDA indents to uncover, are assumed to be included to the generative process. The topics are $\beta_{1:K}$, where K is the dimensionality of the topic variable, which is known and fixed. Each topic β_k, is a distribution over the vocabulary of the corpus, where $k \in \{1, \ldots, K\}$; and β_{kw} is the probability of word w in topic k. For the dth document, θ_d is the distribution over topics; and θ_{dk} is the topic proportion of topic k in d. The topic assignments for the dth document are indicated by z_d, and the topic assignment for the lth word of the dth document is denoted by z_{dl}. Consequently, w is the only observed variable of the model and w_{dl} represents the lth word seen in the dth document, while β, θ and z are latent variables. The posterior of the topic structure given the documents is:

$$p(\beta_{1:K}, \theta_{1:D}, z_{1:D} \mid w_{1:D}) = \frac{\beta_{1:K}\theta_{1:D}z_{1:D}w_{1:D}}{p(w_{1:D})}$$

where D is the number of documents, and the computation of the denominator, i.e. the probability of seeing the given document under any topic structure, is intractable [3]. Moreover, LDA includes priors, so that β_k is drawn from a Dirichlet distribution with parameter η and θ_d is drawn respectively from a Dirchlet also, with parameter α.

As mentioned, the posterior cannot be computed. To approximate it, the two most prominent approaches are (a) variational inference introduced in [13] and (b) Markov Chain Monte Carlo sampling methods proposed in [12]. In variational inference, the true posterior is approximated by a simpler distribution q, which depends on parameters $\phi_{1:D}, \gamma_{1:D}$ and $\lambda_{1:K}$ defined as:

$$\phi_{dwk} \propto \exp\{E_q[\log \theta_{dk}] + E[\log \beta_{kw}]\},$$

$$\gamma_{dk} = \alpha + \sum_w n_{dw}\phi_{dwk} \qquad \lambda_{kw} = \eta + \sum_d n_{dw}\phi_{dwk}$$

The probability that topic assignment of word w in d is k, under distribution q, is denoted by ϕ_{dwk}. Variable n_{dw} represents how many times the word w has been seen in document d. The variational parameters $\gamma_{1:D}$ and $\lambda_{1:K}$ are associated with variable n_{dw}. The variational inference algorithm's intuition is to minimize the *Kullback-Leibler(KL) divergence* between the variation distribution and the true (intractable) posterior. This is accomplished by iterating between assigning values to document-level variables, and updating topic-level variables.

Online Latent Dirichlet Allocation. In the online version of LDA topic model [11], the documents are received on streams rather than a single batch of the original LDA algorithm. In this approach, the exact number of documents is not required to be known, though an estimation is at least required. As a result, online LDA can adapt to very large corpora. The value of the variational parameter $\lambda_{1:K}$ is updated every time a new batch arrives, while the rate at which the documents of batch t actually affects the value of $\lambda_{1:k}$ is controlled by $\rho_t = (\tau_0 + t)^{-k}$. The variational inference for the online version of LDA is shown in Algorithm 1, where it eventuates that α and η are assigned to a value once, and remain fixed. Ultimately, the estimated probability of the term w in topic k is $\beta_{kw} = \frac{\lambda_{kw}}{\sum \lambda_k}$.

---

1 Initialize λ randomly
2 **for** $t = 1$ **to** ∞ **do**
3 | $\rho_t = (\tau_0 + t)^{-k}$
4 | *Expectation* step:
5 | Initialize γ_{tk} randomly
6 | **repeat**
7 | | set $\phi_{twk} \propto \exp\{E_q[log\theta_{tk}] + E_q[log\beta_{kw}]\}$
8 | | set $\gamma_{tk} = \alpha + \sum_w n_{tw}\phi_{twk}$
9 | **until** $(\frac{1}{k}\sum| \text{ change in } \gamma_{tk}| \le \epsilon)$;
10 | *Maximization* step:
11 | compute $\tilde{\lambda}_{kw} = \eta + Dn_{nt}\phi_{twk}$
12 | set $\lambda = (1 - \rho_t)\lambda + \rho_t\tilde{\lambda}$
13 **end**

Algorithm 1: Online Variational Inference for LDA [11]

2.3 Uncertainty

In most real-life settings, a common phenomenon is the absence of partial or complete information. That is, in situations modeled by a game, the participants (players) usually lack knowledge regarding other players (especially if the number of agents is very large), or are uncertain about even their own contribution in the overall game. For this reason, it is essential for the agents to be able to learn the unknown aspects of the game. That is, the agents should be capable of correctly estimating the outcome (e.g. the potential gain or loss) of a given synergy.

A recent work that studies the uncertainty problem is that of [15]. In this study, the authors exploit online LDA model, a probabilistic topic model, to discover coalitional values during an iterative overlapping coalition formation process. The coalitional values are based on an underlying collaboration structure, emerging due to existing but unknown relations among the agents. All agents extract, from a number of sample text documents, information regarding the formed coalitions' utilities. Specifically, they proposed the OVERPRO method which allows each agent to refine significant agents, and therefore characterize coalitions as profitable and unprofitable. In the model of [15], each agent trains her own online LDA probabilistic topic model, during the iterative process, and discover coalitions significant to her. This is the only work we are aware of that explicitly tries to learn the underlying inter-agent synergies or the underlying cooperation structure, and actually does so using probabilistic topic modeling.

Of course, there are many other papers that attempt to deal with various aspects of uncertainty in cooperative game settings. For instance, Kraus et al. in [14] study a series of strategies for revenue distribution in environments with incomplete information. In [7], the authors propose a Bayesian reinforcement learning model, in order to reduce uncertainty regarding coalitional values.

Sliwinski and Zick in [18] tackle the uncertainty problem in hedonic games, by adopting *probably approximately correct* algorithms in order to approximate agents' preference relations and discover core stable partitions. The authors work on discovering the underlying hedonic game, by estimating each agent's personal utility; and to that extend, find stable coalitions, i.e. find coalitions where agents are less (stable) or more (unstable) probable to diverse. [18] exploits the characteristics a hedonic game should have in order to be PAC learnable and/or PAC stabilizable. In particular, the authors show that certain classes of hedonic games, such as Additevelt Separable Hedonic Games and $W-$Hedonic Games, are PAC learnable; and therefore focus on PAC stabilizability of hedonic games.

In our work, we adopt the probabilistic topic modeling framework of [15] and apply it to hedonic games. In particular, we lay our interest on a special class of hedonic games, that of *hedonic games with dichotomous preferences* (also known as boolean hedonic games, discussed in Sect. 2.1). Of course, for our approach to work, we need first to devise a way for coalitional instances to be transformed into input "documents" the LDA can operate on. In Sect. 3 we describe in detail how this is achieved.

3   Game Interpretation as Documents

Let $G = \langle N, \phi_1, \cdots, \phi_n \rangle$ be a hedonic game with dichotomous preferences, where N is a set of players and $n = |N|$. ϕ_i represents a logic formula correlated to agent $i \in N$, which allows agent i to 'approve' or 'disapprove' a given coalition. The formula ϕ_i consists a concise representation of the preference relation $\succsim_i$ (the preference relation $\succsim_i$ may be of size exponential in n, as opposed to formula ϕ_i which may be significantly shorter). In hedonic games with dichotomous preferences, each agent classifies the coalitions related to her into satisfactory coalitions and dissatisfactory ones. Intuitively, formula ϕ_i expresses agent i's goal, and agent i is satisfied if her goal is achieved, or dissatisfied otherwise.

We define an instance κ of game G as a tuple $\langle \pi^\kappa, satisfied_1, \cdots, satisfied_n \rangle$, where π^κ is a partition of N, and $satisfied_i$ is a auxiliary boolean variable that indicates whether agents i's goal is achieved or not. We let each instance κ produce n documents, one document per agent. Since preferences are assumed to be exclusively personal, it is natural to consider n formulae to be independent. Under this assumption, it is natural to train and maintain n different LDA models each of which learns a single formula. Therefore, we assign a single model to each agent, which corresponds to the formula related to that agent; i.e. agent i is responsible for the ith model, which is used to discover ϕ_i.

We sample m instances of the game G. Every instance produces n documents, where each document refers to exactly one's agent formula. Thus, in total we have $m \cdot n$ documents to train n different models. That is, each agent i uses in her own probabilistic topic model exactly m documents, which correspond to her own formula ϕ_i. Thus, the corpus of each model is of size $(1/n)\%$ of the total number of produced documents.

This approach is similar to the one used in [15]. However, in [15] agents belonging in the same coalition process identical documents describing this coalition; while in our case, agents belonging to the same coalition process *different* documents describing the same coalition. This divergences from [15] makes our representation distinct, and is due to the nature of the games: the work of [15] concentrates in transferable utility games, where a coalition is related to a single utility; by contrast, we work with hedonic games with dichotomous preferences, where each agent participating in a coalition S characterizes S as *satisfactory* or *disatisfactory*, regardless of the corresponding characterization by her partners.

A document $\mathbf{d}_{i,\kappa}$ is related to agent i in instance κ, and contains the following: an indicative word for each agent in coalition $\pi^\kappa(i)$, the indicative word 'gain' if the coalition $\pi^\kappa(i)$ is satisfactory (i.e. if $\pi^\kappa(i) \in N_i^+$), or the indicative word 'loss' if $\pi^\kappa(i)$ is disatisfactory ($\pi^\kappa(i) \in N_i^-$). For example, with $N = \{1, 2, 3, 4, 5\}$, and an instance $S^\kappa = \langle\{1, 2\}, \{3, 5\}, \{4\}, True, False, True, False, True\rangle$, the produced documents are of the form:

$$\mathbf{d}_{1,\kappa} = \big(ag_1, ag_2, gain\big) \quad \mathbf{d}_{2,\kappa} = \big(ag_1, ag_2, loss\big)$$
$$\mathbf{d}_{3,\kappa} = \big(ag_3, ag_5, gain\big) \quad \mathbf{d}_{4,\kappa} = \big(ag_4, loss\big)$$
$$\mathbf{d}_{5,\kappa} = \big(ag_3, ag_5, gain\big)$$

As we can see, agents belonging in the same coalition may have identical documents, $d_{3,\kappa} \equiv d_{5,\kappa}$; while other agents may have different documents, $d_{1,\kappa} \not\equiv d_{2,\kappa}$.

The approach can easily represent instances of quite complex preferences, where, e.g., agent 1 realizes she is happy to work with agent 2 or 3 alone, however, together they turn out to be insupportable, and thus she dislikes being with both of them. Of course, using a carefully crafted for a domain of interest translation method, and a possibly different unsupervised learning method, one can potentially extract even more subtle preferences (relating, e.g., to the degree by which an agent prefers a coalition or even a specific partner to others).

4   Experimental Evaluation

In order to evaluate the performance of probabilistic topic modeling in discovering good and bad outcomes in hedonic games, we work in environments of hedonic games with dichotomous preferences with 50 agents ($n = 50$). The coding for the simulation framework was in Python 3.5; all the experiments ran on a PC with an i5@3.2 GHz processor and 4 GB of RAM.

4.1   Dataset and Setting Escalation

For the simulations we created several game instances of hedonic games with dichotomous preferences, according to the following procedure:

1. for each game G define the preference relations through ϕ formulae
2. generate randomly partitions of N, π
3. for each agent i decide whether $\pi(i)$ is satisfactory regarding formula ϕ_i
4. interpret and log the instance information $\langle\pi(i), satisfied_i\rangle$ into documents.

Formulae Construction. A formula ϕ_i expresses agent i's goal in a concise and short representation. That is, each ϕ_i consists of two subsets of agents: (a) the "must be included" (denoted below as included) agents and (b) the "must be excluded" (denoted below as excluded) agents. In its simplest form, agent i is satisfied within a coalition where all agents in the included set are members of this coalition, and no agent in the excluded set is participating. However, an agent i may have several pairs of $\langle included, excluded \rangle$ subsets of agents, and be satisfied with a coalition if this coalition is consistent with at least one such pair. Therefore, in the general form we have that: $\phi_i = \bigvee_{l=1:L} \phi_{i,l} = \phi_{i,1} \vee \phi_{i,2} \vee \cdots \vee \phi_{i,L}$ where $\phi_{i,l} = included_{i,l} \wedge excluded_{i,l}$. The complexity of a formula ϕ_i depends on (a) the number of $\phi_{i,l}$ and (b) the number of agents each pair $\langle included, excluded \rangle$ contains. We ran our experiments on settings with escalated complexity of the structure of the formulae ϕ.

▷ **Low Complexity:** ϕ_is with low complexity consist of a single $\langle included, excluded \rangle$ pair, and the total number of agents appearing in this pair of subsets is fixed to $n/10$. That is, each agent has a must include or exclude demand on 10% of the agents.

▷ **High Complexity:**[1] we escalate the complexity by increasing the number of $\langle included, excluded \rangle$ pairs to 3. The number of agents appearing in each pair uniformly ranges in $[n/10, n/5]$.[2]

Instance Generation and Information Logging. For a given hedonic game G, i.e. for a given set of formulae ϕ, we generate a number of game instances. In each instance we randomly partition agents into coalitions. Each coalition in the formed partition is characterized as satisfactory or dissatisfactory according to formula ϕ_i, for every agent i within the coalition. After characterizing the coalition of a certain agent, we log the contained information into a text document using the interpretation described in Sect. 3.[3] The total number of documents produced per game varies depending on the game's complexity, and ranges between $[500K, 2.5M]$. However, the size of the corpus each online LDA model is fed with, corresponds only to the 2% of the total number of produced documents. That is, we use $10,000$ and $50,000$ game instances for the low and the high complexity environments respectively.

Note that the game instances do not have stochasticity. That is, for each generated sample (document) of a game instance, it is deterministically guaranteed that if the sample contains the word 'gain' ('loss') then the respective coalition

---

[1] We have preliminary results showing our approach can be quite effective in even more complex settings.

[2] The number of agents participating in each ϕ_i, along with the total number of formulae per agent within each level of complexity environment were chosen so that the required dataset could be generated within a reasonable time frame; these numbers do not impose any burden on the LDA algorithm itself.

[3] For practical reasons, the logged information is repeated more than once within a document. That is, we boost the term frequency of the agents' indicative words, and the characterization's 'gain'/'loss', to avoid misleading words with low frequencies.

is satisfactory (dissatisfactory). In a real life situation of a boolean hedonic game formation, we would sample instances of the game by letting participants form groups, and then receive by each one a feedback on whether or not they were satisfied in their coalition.

LDA Model. For the implementation, we used the *scikit-learn* Python 3.5 library [16]. As mentioned, the online version of LDA was used for a range of topics, iterations and batch sizes related to the formulae complexity. In Table 1 we present the different parameters used in the simulations.

Table 1. Simulation parameters

Complexity	$|corpus|$	$|\phi_i|$	Number of topics	Batch size	Iterations
Low	10000	1	5–12	5–25 (step 5)	50
High	50000	3	5–19 (step 2)	5–25 (step 5)	50

The number of topics K, is a parameter that the (online) LDA model needs to be provided with. In situations such as the ones we are studying, the exact number of topics cannot be known a priori; however, it can systematically be chosen depending on the problem at hand. Moreover, there exist other LDA variations, such as HDP [19], that are non-parametric on the number of topics.

4.2 Significant Agents and Valid Topics

In order to evaluate our experimental results, we first need to determine the coalition that is primarily described by each topic. For this reason, we define *significant agents* within each topic. Given a topic k, the agents that appear with probability greater than a small number ϵ are considered to be significant: formally, agent i is significant with respect to topic k if $Pr(i|topic = k) \geq \epsilon$. Therefore, the coalition related to topic k is $S = \{j \in N : Pr(j|topic = k) \geq \epsilon\}$.

After establishing the coalition described in each topic, we move to assessing the *validity* of the topic. Intuitively, the validity of a topic signifies whether the topic reflects a sub-formula describing the agent's hedonic preferences. Thus, given the significant agents within a topic, we characterize it as valid or invalid for agent i. A topic k is valid when:

$$\triangleright\ Pr(gain|topic = k) \geq Pr(loss|topic = k) \text{ and}$$
$$S = \{j : Pr(j|topic = k) \geq \epsilon\} \in N_i^+, \textbf{ or}$$
$$\triangleright\ Pr(loss|topic = k) \geq Pr(gain|topic = k) \text{ and}$$
$$S = \{j : Pr(j|topic = k) \geq \epsilon\} \in N_i^-$$

otherwise the topic is invalid. In words, the meaning of the latter characterization is essentially the actual cross-validation of the topics result with the corresponding formula ϕ_i.

Given these definitions, we can then adopt as an evaluation metric the percentages of valid and invalid topics found by the algorithm. Intuitively, we would like the algorithm to discover *valid* topics; that is, they reflect preferences subformulae that correspond to satisfactory/dissatisfactory collaboration patterns.

4.3   Results

Before presenting our results, in Fig. 1 we show two examples of topics (distributions over the "words"), in the low complexity environment. That is, each bar depicts the probability of a specific word belonging to the given topic. To clarify, the x-axis of the graphs depicts the vocabulary of our corpus, i.e. the bars 0–49 correspond to the agent names; and the last two bars 50, 51 represent the words 'gain' and 'loss', respectively. In the left graph, the distribution over the vocabulary is exhibited for a valid topic, which infers a satisfactory coalition–due to the relatively high probability of the word 'gain' (the 50th word in the axis shown with green bar). Similarly, in the plot we see the graph for a dissatisfactory coalition scenario (the red bar corresponds to the word 'loss').

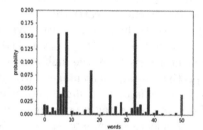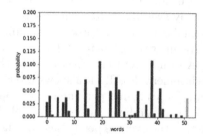

Fig. 1. Environment complexity: *low*. Example of a topic describing 'satisfactory' coalition (left), and a topic corresponding to a 'dissatisfactory' coalition (right). (Color figure online)

We now present our actual results. First, we conducted a series of game simulations for the low complexity environment. Specifically, we constructed 4 different games (denoted $G_{l,1}, G_{l,2}, G_{l,3}$ and $G_{l,4}$ in the Appendix A) that are subject to the characteristics of the low complexity environment. For each one of these games we ran the learning process 5 times, using batches of different size (of the same documents) channeled to the LDA model.[4] The resulting topics were evaluated using the metrics described above, and we computed the average percentage of *valid* and *invalid* topics per game per number of topics.

Table 2 shows our average results in the low complexity setting. As we can see, generally, the model learns correctly at least 88% of the sub-formulae $\hat{\phi}_i$,

---

[4] As we have already mentioned, there is no stochasticity during the dataset creation. However, by employing this repetition of the learning procedure per game, we ensure the robustness of our results.

Table 2. Environment Complexity: *low.* Percentage of *valid* and *invalid* topics over 4 different hedonic games with dichotomous preferences.

Number of topics	(%) valid	(%) invalid
5	89.83%	10.17%
6	91.81%	8.16%
7	96.79%	3.21%
8	92.08%	7.92%
9	93.98%	6.02%
10	92.67%	7.33%
11	95.53%	4.47%
12	88.89%	11.11%

for different numbers of topics.[5] At the same time, the percentage of incorrectly learned sub-formulae does not exceed 11.5%. In Table 6 (in Appendix A), the reader can see the detailed results per game in the low complexity environment. There, we can observe that in some games percentage of the correctly learned topics may even reach 100%, and it is for all consistently greater than 80%.

A similar evaluation process was followed for the high complexity environment. Again, we constructed 4 different games (denoted $G_{h,1}, G_{h,2}, G_{h,3}$ and $G_{h,4}$ in the Appendix A) that are subject to the characteristics of the high complexity environment. So, Table 3 shows the respective average percentage of average percentages per game per number of topics for *valid* and *invalid* topics, while Table 7 in Appendix A shows the detailed average percentages per game. Here, we see that the average percentage of valid topics learned is greater than 78% for various number of topics. The average percentage of incorrectly learned topics does not exceed 22%. As the environment complexity increases along with the number of topics we intend to discover, the accuracy of learned collaboration patterns drops. Looking at the detailed results in Table 7, the correctly learned topics may even reach 100%, and it is for all always greater than 69%.

Topic Significance Constraint. In a more realistic scenario, knowing that the probability of 'gain' is greater than the one of 'loss', and vice versa, within a topic, may not be enough. Intuitively, we would like to be *confident* whether the coalition described within a topic is satisfactory or dissatisfactory. For this reason, we introduce a *topic significance constraint*, according to which a topic is labeled as significant if the absolute difference of the probability of term 'gain' and the probability of term 'loss', exceeds some small number δ, i.e. $|Pr(gain|topic = k) - Pr(loss|topic = k)| \geq \delta$. Thus, each topic is assessed as 'significant' with confidence level δ; and if a topic is 'significant' it can therefore assessed as 'valid' or 'invalid'. Now the definition of a valid topic k becomes:

---

[5] A larger number of topics allows for more preferences sub-formulae to be learned, but it naturally increases complexity.

Table 3. Environment complexity: *high*. Percentage of *valid*, *invalid* and *insignificant* topics over 4 different hedonic games with dichotomous preferences.

Number of topics	(%) valid	(%) invalid
5	84%	16%
7	87%	13%
9	90%	10%
11	90%	10%
13	86.92%	13.08%
15	84.64%	15.36%
17	80.59%	19.41%
19	78.16%	21.84%

$$\triangleright\ Pr(gain|topic = k) \geq Pr(loss|topic = k) + \delta \text{ and}$$
$$S = \{j : Pr(j|topic = k) \geq \epsilon\} \in N_i^+, \text{ or}$$
$$\triangleright\ Pr(loss|topic = k) \geq Pr(gain|topic = k) + \delta \text{ and}$$
$$S = \{j : Pr(j|topic = k) \geq \epsilon\} \in N_i^-$$

By taking into account the topic significance constraint, we re-evaluated the topics arisen from the LDA model for the low and the high complexity environment. The results depicted in Table 4 show an expected drop in the average percentage of valid topics, since we discard a portion of the topics by assessing them as 'insignificant'. It is worth noting that the difference between the average percentage in the unconstrained and constrained cases reaches up to 15.27 percentage points for the assessment of valid topics.

Table 4. Environment complexity: *low—high*. Average percentage of *valid* topics over 4 different games **with** and **without** the significance constraint.

Average (%) valid topics							
Topics	Low complexity			Topics	High complexity		
	(%) unconstr	(%) constr	diff		(%) unconst	(%) constr	diff
5	89.83%	88%	−1.83	5	84%	82%	−2
6	91.81%	91.11%	−0.7	7	87%	82.86%	−4.14
7	96.79%	94.88%	−1.91	9	90%	85.56%	−4.34
8	92.08%	86.77%	−5.31	11	90%	82.27%	−7.73
9	93.98%	86.76%	−7.22	13	86.92%	73.08%	−13.84
10	92.67%	83.83%	−8.84	15	84.67%	71%	−13.67
11	95.53%	83.86%	−11.67	17	80.59%	67.35%	−13.24
12	88.89%	75.35%	−13.54	19	78.16%	62.89%	−15.27

"Anytime" Behaviour. Last but not least, we conducted a simulation experiment to examine how our model behaves during an ongoing learning process.

That is, assume that agents at a certain time t_1 have access to a part of the corpus. The agents train their models with the sub-corpus that is available to them at the time. After this first phase training, agents have some beliefs over satisfactory and dissatisfactory coalitions, that they could use in a decision-making process. At time t_2 a second part of the corpus is revealed to the agents, thus the agents update their already partially trained models with the new documents; and so on and so forth. In each of the later phases, the values of parameters of the model regarding the number of batches and iterations are maintained. Equivalently, these later phases correspond to LDA processes with prior distributions over topics and documents. Intuitively, using priors leads to faster convergence.

In Table 5 we show the results of this procedure, for games in the high complexity environment and for varying numbers of topics. We let the agents train in 3 phases and recorded the time needed and the average percentage of valid topics for each phase. 2500 documents were fed to the model in each phase. As we can see, **Phase 0** requires more time, while the average percentages of valid topics are not particularly encouraging; **Phase 1** requires approximately 75–80% of time required by **Phase 0**, and the percentages rise by up to 22 percentage points; similarly, **Phase 2** requires approximately 40% of the time required by **Phase 0**, and the average percentage of valid topics consistently gets close to or even reaches 100%.

Table 5. Environment complexity: *high*. Anytime behaviour of the training model.

Topics	Phase 0		Phase 1		Phase 2	
	Time (sec)	Valid (%)	Time (sec)	Valid (%)	Time (sec)	Valid (%)
5	43.97 s	80%	35.66 s	100%	24.0 s	100%
7	43.57 s	77.14%	33.04 s	97.14%	25.62 s	100%
9	45.63 s	71.11%	35.45 s	93.33%	27.64 s	97.78%
11	43.91 s	67.27%	35.48 s	90.18%	28.31 s	90.91%

5   Conclusion and Future Work

In this work, we presented a probabilistic topic modeling approach for learning boolean hedonic games. To this end, we presented a novel method for translating game samples into "documents" fed to our algorithm as input. We conducted a systematic evaluation of our approach: first, we studied the performance of online LDA discovering collaboration patterns within environments of different preference relation complexity; and then, we examined the "anytime" performance of our method, by progressively revealing to the model parts of the total corpus. Our results verified the effectiveness of our approach. This work could inspire recommendation methods to be used, e.g., by online advertisers to suggest bundles of goods, or even by travel agencies to promote vacation packages.

As future work, we intend to apply our approach to different classes of hedonic games or non-transferable utility games, in general; and also test it in partition function game environments [8]. We actually already have preliminary results for additively separable hedonic games. Moreover, we intend to compare LDA with different machine learning methods (e.g. neural networks) applied on this domain. It is worth noting that this approach can potentially be used for learning (dichotomous or other) preferences in more generic domains, and not just hedonic games [6,10].

Acknowledgements. *We thank Michalis Mamakos for code sharing.*

A   Appendix: Detailed Results

Table 6. Environment complexity: *low*. Detailed results of average percentage per different game for valid, invalid and insignificant topics for varying number of topics.

Topics	Valid(%)				Invalid(%)			
	$G_{l,1}$	$G_{l,2}$	$G_{l,3}$	$G_{l,4}$	$G_{l,1}$	$G_{l,2}$	$G_{l,3}$	$G_{l,4}$
5	83.33%	84.00%	92.00%	100.00%	16.67%	16.00%	8.00%	0.00%
6	80.56%	96.67%	93.33%	96.67%	19.44%	3.33%	6.67%	3.33%
7	92.86%	100.00%	97.14%	97.14%	7.14%	0.00%	2.86%	2.86%
8	83.33%	97.50%	95.00%	92.50%	16.67%	2.50%	5.00%	7.50%
9	87.04%	100.00%	93.33%	95.56%	12.96%	0.00%	6.67%	4.44%
10	86.67%	100.00%	94.00%	90.00%	13.33%	0.00%	6.00%	10.00%
11	89.39%	100.00%	98.18%	94.55%	10.61%	0.00%	1.82%	5.45%
12	80.56%	90.00%	96.67%	88.33%	19.44%	10.00%	3.33%	11.67%

Table 7. Environment complexity: *high*. Detailed results of average percentage per different game for valid, invalid and insignificant topics for varying number of topics.

Topics	Valid(%)				Invalid(%)			
	$G_{h,1}$	$G_{h,2}$	$G_{h,3}$	$G_{h,4}$	$G_{h,1}$	$G_{h,2}$	$G_{h,3}$	$G_{h,4}$
5	96.00%	84.00%	84.00%	72.00%	4.00%	16.00%	16.00%	28.00%
7	94.29%	85.71%	94.29%	74.29%	5.71%	14.29%	5.71%	25.71%
9	97.78%	82.22%	95.56%	84.44%	2.22%	17.78%	4.44%	15.56%
11	94.55%	78.18%	100.00%	87.27%	5.45%	21.82%	0.00%	12.73%
13	89.23%	75.38%	100.00%	83.08%	10.77%	24.62%	0.00%	16.92%
15	82.67%	84.00%	97.33%	74.67%	17.33%	16.00%	2.67%	25.33%
17	76.47%	82.35%	87.06%	76.47%	23.53%	17.65%	12.94%	23.53%
19	69.47%	69.47%	92.63%	81.05%	30.53%	30.53%	7.37%	18.95%

References

1. Aziz, H., Harrenstein, P., Lang, J., Wooldridge, M.: Boolean hedonic games. In: 15th International Conference on Principles of Knowledge Representation and Reasoning (KR), pp. 166–175 (2016)
2. Aziz, H., Savani, R., Moulin, H.: Hedonic Games. In: Handbook of Computational Social Choice, pp. 356–376. Cambridge University Press (2016)
3. Blei, D.M., Ng, A.Y., Jordan, M.I.: Latent dirichlet allocation. J. Mach. Learn. Res. **3**, 993–1022 (2003)
4. Bogomolnaia, A., Jackson, M.O.: The stability of hedonic coalition structures. Games Econ. Behav. **38**(2), 201–230 (2002)
5. Bogomolnaia, A., Moulin, H., Stong, R.: Collective choice under dichotomous preferences. J. Econ. Theory **122**(2), 165–184 (2005)
6. Brandt, F., Conitzer, V., Endriss, U., Lang, J., Procaccia, A.D.: Handbook of Computational Social Choice. Cambridge University Press, Cambridge (2016)
7. Chalkiadakis, G., Boutilier, C.: Bayesian reinforcement learning for coalition formation under uncertainty. In: Proceedings of the 3rd AAMAS, vol. 3. pp. 1090–1097. IEEE Computer Society, Washington (2004)
8. Chalkiadakis, G., Elkind, E., Wooldridge, M.: Computational Aspects of Cooperative Game Theory (Synthesis Lectures on Artificial Inetlligence and Machine Learning), 1st edn. Morgan & Claypool Publishers, San Rafael (2011)
9. Dréze, J.H., Greenberg, J.: Hedonic coalitions: optimality and stability. Econometrica **48**(4), 987–1003 (1980)
10. Frnkranz, J., Hllermeier, E.: Preference Learning, 1st edn. Springer, Heidelberg (2010). https://doi.org/10.1007/978-0-387-30164-8
11. Hoffman, M.D., Blei, D.M., Bach, F.: Online learning for latent dirichlet allocation. In: Proceedings of NIPS, USA, pp. 856–864 (2010)
12. Jordan, M.I. (ed.): Learning in Graphical Models. MIT Press, Cambridge (1999)
13. Jordan, M.I., Ghahramani, Z., Jaakkola, T.S., Saul, L.K.: An introduction to variational methods for graphical models. Mach. Learn. **37**(2), 183–233 (1999)
14. Kraus, S., Shehory, O., Taase, G.: The advantages of compromising in coalition formation with incomplete information. In: Proceedings of the 3rd AAMAS, pp. 588–595 (2004)
15. Mamakos, M., Chalkiadakis, G.: Overlapping coalition formation via probabilistic topic modeling. In: Proceedings of the 17th AAMAS, pp. 2010–2012 (2018). http://arxiv.org/abs/1804.05235
16. Pedregosa, F., et al.: Scikit-learn: machine learning in Python. J. Mach. Learn. Res. **12**, 2825–2830 (2011)
17. Peters, D.: Complexity of hedonic games with dichotomous preferences. In: AAAI (2016)
18. Sliwinski, J., Zick, Y.: Learning hedonic games. In: Proceedings of the 26th IJCAI-17, pp. 2730–2736 (2017)
19. Teh, Y.W., Jordan, M.I., Beal, M.J., Blei, D.M.: Hierarchical Dirichlet processes. J. Am. Stat. Assoc. **101**, 1566–1581 (2006)

Learning Best Response Strategies
for Agents in Ad Exchanges

Stavros Gerakaris$^{(\boxtimes)}$ and Subramanian Ramamoorthy

School of Informatics, University of Edinburgh, Edinburgh, UK
stevegerak@gmail.com, s.ramamoorthy@ed.ac.uk

Abstract. Ad exchanges are widely used in platforms for online display
advertising. Autonomous agents operating in these exchanges must learn
policies for interacting profitably with a diverse, continually changing,
but unknown market. We consider this problem from the perspective of
a publisher, strategically interacting with an advertiser through a posted
price mechanism. The learning problem for this agent is made difficult
by the fact that information is *censored*, i.e., the publisher knows if an
impression is sold but no other quantitative information. We address this
problem using the Harsanyi-Bellman Ad Hoc Coordination (HBA) algorithm [1,3], which conceptualises this interaction in terms of a Stochastic Bayesian Game and arrives at optimal actions by best responding
with respect to probabilistic beliefs maintained over a candidate set of
opponent behaviour profiles. We adapt and apply HBA to the censored
information setting of ad exchanges. Also, addressing the case of stochastic opponents, we devise a strategy based on a Kaplan-Meier estimator
for opponent modelling. We evaluate the proposed method using simulations wherein we show that HBA-KM achieves substantially better
competitive ratio and lower variance of return than baselines, including
a Q-learning agent and a UCB-based online learning agent, and comparable to the offline optimal algorithm.

Keywords: Ad exchanges · Stochastic game · Censored observations ·
Harsanyi-Bellman Ad Hoc Coordination · Kaplan-Meier estimator

1 Introduction

Real-time ad exchanges (AdX) have become a common marketplace for online
display advertising. These automated transactions take place numerous times a
day, when a user visits a web page whose advertising inventory is managed by an
AdX. The webpage, which is essentially the *publisher*, communicates a *reserve
price* to the ad exchange for the *impression*, which consists of a description of the
webpage, of the user and some other relevant content. The ad exchange offers

---

This work was partially supported from the Greek State Scholarship Foundation by
the project "IKY Scholarship" from resources of ESF and ESPA.

M. Slavkovik (Ed.): EUMAS 2018, LNAI 11450, pp. 77–93, 2019.
https://doi.org/10.1007/978-3-030-14174-5_6

the impression to the bidding agents, or *advertisers*, who compete for it in a second price auction with reserve price, managed by the AdX.

These automated transactions play an important role in the economics of the web, which has meant that advertisers routinely use automated methods to target these impressions to user profiles and characteristics that they are shown. The corresponding situation for publishers appears to be different. As argued in a report from Google [10], who run one such large exchange, publishers are lagging behind in being able to automate the setting of auction parameters such as reserve price. Furthermore, a nontrivial fraction of ad exchange auctions involve a *single* advertiser [5]. When only one advertiser is involved, the ad exchange auction mechanism becomes a posted price auction between the publisher and the advertiser.

In this work, we model this interaction and propose the application of novel learning algorithms to address the problem of adapting behaviour within the interaction. We examine the continuous interaction, over a number of rounds, between the advertiser and the publisher through the posted price auction mechanism. There are two key attributes associated with this posted price mechanism; (a) since there are only two agents involved, the observations the publisher makes from the advertiser's bids are *doubly censored* (i.e., the publisher only learns if a bid is successful and does not gain further quantitative knowledge of the advertiser's utilities) and (b) the publisher is facing an *adaptive player* with a number of possible strategies at his disposal.

Conceptually, the problem faced by the publisher is that of interacting in an *ad hoc* manner, with limited prior knowledge of the opponent. Learning in such situations is made difficult by the fact that the open ended nature of the hypothesis space results in unacceptable complexity of learning. We propose that this problem may be addressed by drawing on recent developments in machine learning, which allow tractable learning despite the incompleteness of models. In particular, we use the Harsanyi-Bellman Ad Hoc Coordination (HBA) algorithm [1,3], which conceptualises the interaction in terms of a space of 'types' (or opponent policies), over which the procedure maintains beliefs and uses the beliefs to guide optimal action selection. The attraction of this algorithm is that it can be shown to be optimal even when the hypothesised type space is not exactly correct but only approximates the possible behaviours of the opponent. In this paper, we adapt HBA to the situation where observations are censored, and demonstrate its usefulness in the AdX domain. In addition, addressing the case when opponents are playing essentially randomly (a situation where HBA's belief update process would be inadequate), we propose the use of a Kaplan-Meier estimator to approximate the opponent's stochastic behaviour to choose actions based on that estimate.

We model the interactions between the two agents as a Stochastic Bayesian Game Γ of censored observations. The publisher's goal is to maximise his expected revenue over the T rounds of the game. In order to do so he needs to infer the bidding strategy of the advertiser. We define a space of behaviours for the advertiser, including various distributions and adaptive procedures, such as Q-learning and learn-then-bid strategies [12]. So, a publisher using HBA maintains a belief about the advertiser's behaviour, defined as a probability distribution over this space of types, Θ_A, and plays best response to it. It is worth noting that the *offline optimal* algorithm for the publisher, that serves as an upper bound on the expected revenue of our method, is the strategy that has prior knowledge of the buyer's strategy (something that is unrealistic in practice, but illustrative for algorithm analysis) and plays optimally against it from the first turn of the game.

There is a substantial body of work in the AdX literature, which focuses on finding a publisher's reserve price policy to optimise his revenue in second price auctions with reserve price [8,15]. However, these works, which mainly focus on the theory, often restrict attention to situations such as where an advertiser is only an unknown random distribution (hence, not adaptive), and where the publisher has access to uncensored samples. From a practitioner's perspective, it is interesting to ask if we can go beyond some of these assumptions and devise learning algorithms for the publisher that only uses the censored observations available online (hence making it robust with respect to model mismatch), and allow for more generality in the behaviour of the advertiser, specifically allowing that agent to adapt (which is very realistic in the scenarios we mentioned earlier). In our experiments, we show that the proposed procedure is able to adapt better than baselines such as Q-learning or a UCB-based online learning procedure, and that it approaches the offline optimal benchmark in many situations. In order to understand the behaviour of this algorithm under model mismatch, we also present experiments with an adaptive adversary which is a neural network, looking both at the transient behaviour of HBA when the adversary is actively learning and is non-stationary, and also the case where the adversary is a mixture that is different from any individual element in the type space over which HBA maintains beliefs.

2   Related Work

Much has been written about ad display and sponsored search auctions, for each participating agent of the auction, either as publisher or advertiser. A key paper in the area of ad exchanges is that of Muthukrishnan [16], who laid out several research directions in this domain, informed by exposure to the practice in this domain.

More specific related work, viewing the problem from the perspective of the publisher, are the following. Mohri and Medina [15] discuss selecting the reserve price to optimise revenue in a second price auction with reserve price. They consider a supervised learning setting and assume access to a training sample of uncensored historical data. A similar formulation of revenue is seen in the work of Cesa-Bianchi et al. [8], where the authors assume no historical data, but they get direct observations based on the assumption that every bidder in this market draws his valuation from the same fixed and unknown distribution D. Then they proceed by showing a regret minimisation algorithm, achieving sublinear regret.

In other recent work, Amin et al. [5] define the notion of the *strategic regret* and present no-regret algorithms with respect to that notion. Finally, Huang et al. [13] study the problem of setting a price for a potential buyer with a valuation drawn from an unknown distribution D and prove tight sample complexity bounds for regular distributions, bounded-support distributions, and a wide class of irregular distributions. This work is preceded by Cole and Roughgarden [9], who also analyse sample complexity of revenue maximisation, this time as a function of the number of bidders in an auction.

From the advertiser's perspective, Amin et al. [4] study budget optimisation for sponsored search auctions. The authors cast the problem of budget optimisation as a Markov Decision Process (MDP) with censored observations and propose a learning algorithm based on Kaplan-Meier estimators. The authors also perform a large scale empirical demonstration on auction data from Microsoft, in order to demonstrate that their algorithm is extremely effective in practice. Another take on the advertiser's optimisation problem is the one by Ghosh et al. [12] who study the design of a bidding agent in a marketplace for displaying ads. They provide algorithms and performance guarantees for both settings, while also experimentally evaluating their performance on a fitted log-normal distribution from data observed from the Right Media Exchange.

Another literature that is closely relevant pertains to learning to interact in multiagent domains, with limited or no prior coordination. Related work includes [1,7,19]. A key concept arising from this literature is that of modelling the opponent in terms of a hypothesis space of policies, that in a certain sense approximate the space from which that agent herself chooses the true policy.

3   Model for the Publisher in an Ad Exchange

We start by presenting our model of how we conceptualise this interaction with the (model of an) advertiser, agent A, as a Stochastic Bayesian Game. The advertiser is characterised by a discrete state space $\mathcal{S}$, defined by his own budget B_A and the auction round t, $s^t = \{B_A^t, t\}$. He has a set of actions $\mathcal{A}_A = \{0, \ldots, v_{max}\}$ which are the possible prices he can bid for an impression (his valuation vector) and his strategy is chosen from a well-defined type space Θ_A. A payoff function $u_A : \mathcal{S} \times \mathcal{A} \times \Theta \mapsto \mathbb{R}$ maps his state, type and actions to specific payoff and a strategy $\pi_A : \mathbb{H} \times \mathcal{A}_A \times \Theta_A \mapsto [0,1]$ defines a probability vector over his possible actions. The history vector $\mathbb{H}$ contains all histories $\mathbb{H}^t = \langle s_0, a_0, s_1, a_1, \ldots, s_{t-1}, a_{t-1}, s^t \rangle$.

We realise the interaction between the advertiser and the publisher as a Stochastic Bayesian Game Γ of censored information between two players. The game Γ consists of:

- An advertiser A and a publisher P
- State space $\mathcal{S}$, action space $\mathcal{A} = \{\mathcal{A}_A, \mathcal{A}_P\}$, type space $\Theta = \{\Theta_A, \Theta_P\}$
- Transition function $T : \mathcal{S} \times \mathcal{A} \times \mathcal{S} \mapsto [0,1]$
- Γ starts at time $t = 0$ and state s^0. At each time step $t \in 0, 1, 2, \ldots$:

1. An impression i^t arrives
2. The advertiser chooses his action (bid) $b^t = a_A \in \mathcal{A}_A$ with probability $\pi_A(H^t, a_A^t, \theta_A^t)$
3. The publisher chooses his action (reserve price) $r^t = a_P \in \mathcal{A}_P$ with probability $\pi_P(H^t, a_P^t, \theta_P^t)$
4. The game transitions into state $s^{t+1} \in S$ with probability $T(s^t, a^t, s^{t+1})$
5. If $b^t \geq r^t$ the impression is sold at price r^t; otherwise the impression doesn't get sold
6. The immediate payoff of the advertiser A is $u_A^t = b^t - r^t$
7. The immediate payoff of the publisher P is $u_P^t = r^t$
   - The process is repeated until a terminal state s^t is reached.

In this problem setting, the publisher does not have knowledge of the advertiser's individual strategy $\theta_A \in \Theta_A$, only of his type space Θ_A which is the set of all of his possible strategies. He needs to infer that strategy θ_A from the censored observations he makes at each auction round in order to play his best response strategy against it. As mentioned earlier, we utilise the Harsanyi-Bellman Ad Hoc Coordination Algorithm [1,3]. HBA, and adapt it to the setting of AdX. The main steps of this procedure are as described in Algorithm 1. A key adaptation from the formulation in [1,3] is the incorporation of an estimate of the opponent's actions and to allow for the KM estimator (to be explained in more detail in the next section) for the case of a randomised adversary.

We make the following assumptions for our setting:

Assumption 1. *We control the publisher P and choose his strategy π_P. P has a single type θ_P known to us.*

Assumption 2. *Given a stochastic game Γ we assume all the elements of Γ, except of the type θ_A of the opponent, is common knowledge.*

Assumption 3. *We only have partial observability of states and actions.*

In the HBA algorithm, which maintains a posterior probability of an agent being a specific type based on observing the history of actions, the action is selected by determining a best response, within the game Γ, which implicitly uses in the value calculations Q-values based on the Bellman optimality principle.

The posterior probability of an agent being a specific type is calculated with the use of sum posterior, defined in [2] as:

$$L(H^t) = \sum_{\tau=0}^{t-1} \pi_j(H^\tau, a_j^t, \theta_j)$$

By the term *censored observations* we refer to the type of the information perceived by the publisher. As is the case in posted price auctions, the publisher only gets to observe the outcome of any round t of a sequential auction, which is if he sold the impression or not, but he doesn't get to observe the bid that actually won; he only knows that this bid is greater or equal than the reserve price he specified ($b^t \geq r^t$). Otherwise, he knows that this bid was strictly less than his reserve price ($b^t < r^t$).

Algorithm 1. HBA Censored

Input: SBG Γ, player P, user defined type spaces Θ_A^*, history H^t, discount factor $0 \le \gamma \le 1$

Output: Action probabilities $\pi_P(H^t, r)$

1: **for all** $\theta_A^* \in \Theta_A^*$ **do**

2:     Compute $Pr(\theta_A^*|H^t) = \frac{L(H^t|\theta_A^*)P(\theta_A^*)}{\sum_{\hat{\theta}_A^* \in \Theta_A^*} L(H^t|\hat{\theta}_A^*)P(\hat{\theta}_A^*)}$

3: **end for**

4: **for all** $r \in \mathbf{v}$ **do**

5:     Compute expected payoff $E_{s^t}^r(H^t)$ as follows:

$$E_s^r(\hat{H}) \leftarrow \sum_{\theta_A^* \in \Theta_A^*} Pr(\theta_A^*|H^t) \sum_{b \in \mathbf{v}} Q_s^r(\hat{H}) \pi_A(\hat{H}, b, \theta_A^*)$$

$$Q_s^r(\hat{H}) \leftarrow \sum_{s' \in S} T(s, r, s') \left[u(s, r) + \gamma \max_{r'} E_{s'}^{r'}(\langle \hat{H}, r, s' \rangle) \right]$$

6: **end for**

7: **if** $\arg\max_{\theta_A \in \Theta_A} \in \Theta_A^{\text{Random}}$ **then**

8:     $\pi_P(H^t, r^*) = 1$, where $r^* = \text{RANDOM KM}(k, l, k_c)$

9: **else**

10:     Distribute $\pi_P(H^t, \cdot)$ uniformly over $\arg\max_r E_{s^t}^r(H^t)$

11: **end if**

3.1   HBA Types (Advertiser's Strategies)

In this section, we define the hypothesised type space Θ_A of the advertiser. The first two strategies don't involve an adaptive component, so they are in a sense naive. However, there are works [17] that suggest that this kind of bidding can often be found in real world auctions. The rest of the strategies of the advertiser that we specify, are well studied learning models that involve distinct learning and strategy components. This set is chosen to capture the diversity of potential types of behaviour of the unknown adversary.

We also present the best response strategies of the publisher to each of the respective strategies of the advertiser, under the assumption that all the private information of the advertiser is known by the publisher. These best response strategies consist a set of *offline optimal* benchmarks, that will serve as an upper bound on the revenue of our method, which assumes no private knowledge other than the type space of the advertiser.

Greedy Strategy. Advertiser's greedy policy, given his action space $\mathcal{A}_A = \{0, \ldots, v_{\max}\}$, is to always to bid his maximum value for the impression.

$$\pi_A^{\text{greedy}}(a_A = v_{\max}) = 1$$

One can see that publisher's best response policy is to simply match his maximum value and offer it as a reserve price in every turn of the game.

$$\pi_P^*(a_P = v_{\max}|\pi_A^{\text{greedy}}(a_A \in \mathcal{A}_A)) = 1$$

Random Strategy. In the second strategy, the advertiser places a bid i.i.d. from a fixed distribution over his value vector. We use several random distributions, such as the uniform, the normal, the logistic, the log-normal and the exponential.

The best response strategy against a random advertiser, with the distribution $\mathcal{D}$ over $\mathbf{v}$ known by the publisher, is the reserve price that maximises the publisher's expected revenue.

$$\pi_P^*(a_P = r^* | \pi_A^{\text{random}}(b \sim_{\mathcal{D}} \mathbf{v})) = 1, r^* = \arg\max_{r \in \mathbf{v}} r T_{\mathcal{D}}(r)$$

where $T_{\mathcal{D}}(r)$ denotes the tail probability of the distribution $\mathcal{D}$ for the value r.

Learn Then Bid Strategy. The next adaptive bidding strategy of the advertiser is given in the work of Ghosh et al. [12], where the advertiser chooses to opt out for a specified number of m out of n rounds in order to observe the prices of the reserve and then, based on his observations, decides between the price P_m^* that guarantees, in expectation, the target fraction f of impressions he sets, and the price Z_m^* satisfying the maximum amount he wants to spend.

The best response strategy against an advertiser playing Learn Then Bid strategy, with the parameters of the Learn Then Bid algorithm known by the publisher, is the following.

1. Find the maximum value of the price that satisfies the target of his campaign reach, times the probability of him playing that price.

$$P_m^* = \arg\max P_m \frac{fn}{(n-m)\mathcal{P}_m(P_m)}$$

2. Find the maximum value of the price that satisfies the advertiser's target spent.

$$Z_m^* = \sup\left\{z : \mathcal{P}_m(z) \geq \frac{fn}{n-m}\right\}$$

$\mathcal{P}_m(x)$ is the estimated distribution of the market from the advertiser after the learning phase. The optimal policy for the publisher is to exhaust the advertiser's budget, by deterministically selecting the maximum of those two prices.

$$\pi_P^*(a_P = \max\{P_m^*, Z_m^*\} | \pi_A^{\text{Learn Then Bid}}(a_A \in \mathcal{A}_A)) = 1$$

Multi Armed Bandits Strategy. Another strategy we use for the advertiser is the well known UCB algorithm [6]. We implement it using a ϵ-greedy action selection policy. Publisher's optimal policy is to offer the maximum value, as a reserve price in every turn of the game.

$$\pi_P^*(a_P = v_{\max}, \cdot | \pi_A^{\text{UCB}}(a_A \in \mathcal{A}_A)) = 1$$

It is not hard to see that any traditional *no-regret* strategy is easily manipulable, therefore inadequate for this interactive problem setting, something also highlighted in [5].

Q-learning Strategy. The last algorithm in the type space of the advertiser is the well known Q-learning algorithm [20]. We implement it using a soft-max action selection policy. The states for the advertiser are the different levels of his remaining budget and the action space is defined by his value vector.

Publisher's optimal policy, similar to when he faces a random distribution, is finding the reserve price that maximises his expected revenue w.r.t. the Boltzmann distribution produced by the Q-values, which dictates the soft-max action selection.

$$\pi_P^*(r^*|\pi_A^{\text{Q-learning}}(s, r \in \mathcal{A}_A)) = 1$$

where,

$$r^* = \arg \max_{r \in \mathcal{A}_P} r \sum_{r'=r}^{v_{max}} \frac{e^{Q(s,r')/\tau}}{\sum_{a'_A \in \mathcal{A}_A} e^{Q(s,a'_A)/\tau}}.$$

3.2 HBA Beliefs and Best Responses

Over the recently introduced types, HBA maintains and updates beliefs, that will determine action selection at each step. In step 2 of Algorithm 1, HBA keeps a posterior belief over types, $Pr(\theta_A^*|H^t)$ by keeping track of the sequence of actions of the opposing player and calculating the probability that these actions come from a specific type. Afterwards, in step 5, it computes the Q-values of every possible action at this state $Q_s^r(\hat{H})$ and, following this, it calculates the expected revenue $E_s^r(\hat{H})$ of every action based on the posterior over types it maintains and on the Q-values it has just computed. In the final step 7, depending on whether it recognises a stochastic opponent, $\theta_A \in \Theta_A^{\text{Random}}$, or a deterministic one, HBA decides between calling a procedure designed specifically for random opponents (discussed in Sect. 3.4) and playing a single price as a best response.

By exchanging between iterations of these two procedures, the posterior belief calculation and the Q-values computation followed by a single expectation maximisation step, HBA succeeds in modelling in a dynamic fashion the opposing agent, while also plays optimally, in expectation, against her at each step.

3.3 HBA Censored

As mentioned earlier, accommodating censored observations requires estimating actions that can be used for belief updating. There are two specific values that are needed for such an estimate. The first one concerns the probability of the last observed action, conditional to a player being of a specific type. This probability is used to calculate the posterior of the opposing agent's type, according to the Bayes rule. In our setting, where the observations are censored, we estimate this value by using structural characteristics of the distribution he plays. For instance, let $\mathcal{Q}$ be the distribution associated with the Q-values for a Q-Learning advertiser. If the publisher sells the impression at price r, he doesn't observe the bid, but he can update the probability, conditional on his opponent's type:

$$Pr(a_A^{t-1}|\theta_A^{\text{Q-learning}}) = T_{\mathcal{Q}}(r)$$

The second one concerns the computation of the HBA's own Q-values in the expectation maximisation step. Recall from Algorithm 1 that we compute the Q-values by computing every possible outcome in expectation,

$$Q_s^v(\hat{H}) \leftarrow Q_s^v(\hat{H}) + \alpha[v + \gamma \max_{r'} Q_{s'}^{v'}(\hat{H}) - Q_s^v(\hat{H})],$$

where v denotes the utility of the previous step:

$$v = \begin{cases} v, & \text{if } v \leq r \text{ and the impression was sold,} \\ 0, & \text{if } v \geq r \text{ and the impression was not sold,} \\ vT_Q(v), & \text{otherwise.} \end{cases}$$

Again here we utilise the tail probability of the distribution in order to estimate the required Q-values from the censored observations.

With this, we achieve performance close to the offline optimal metric, against advertising strategies that play a single price or that are converging to a price, i.e. Greedy, Learn Then Bid or UCB.

From the defined type space Θ_A we see cases where the output of our opponent's algorithm is randomised, either according to a fixed random distribution or a dynamic one, in the case of the Q-learning algorithm. It is known from the theory underpinning HBA [2] that this case requires different treatment.

3.4   KM Estimator for Stochastic Opponents

We now present an approach based on the Kaplan-Meier estimator [14], for estimating distributions from censored samples. When HBA recognises a randomised opponent, we let this algorithm decide both the query values and the optimal reserve price. KM estimator is a powerful tool for approximating distributions based on censored samples and has found use in both e-commerce [4] and financial applications [11].

The Random KM algorithm uses a few simple ideas from random sampling and the Kaplan-Meier estimator. We start by scanning the support of the distribution for potential candidate values and for every candidate we query for a sufficient number of times, in order to have a good estimation.

Essentially the Random KM algorithm works in two steps. In the first step, it queries for k steps over all the support of the opponent's possible values and makes a loose estimation of each value's tail probability by calculating the fraction of *right censored* observations ($R^t(x)$, times the impression gets sold and the value $x \leq r$ is less or equal than the reserve price), to the sum of right and *left censored* observations.

In the second step it isolates the candidate values which are the most probable to generate the most revenue, and queries each of them for a number of k_c steps, which Kaplan-Meier dictates, in order to get a precise approximation of their tail probabilities. Using the estimated tail probabilities, it calculates the price that maximises the expected payoff and returns it.

Algorithm 2. Random Querying - KM

Input: Distribution q to make CDF queries
Output: Optimal reserve price r_q^*
1: **procedure** RANDOM KM(k, l, k_c)
2:    **for** $t = 1, \ldots, k$ **do**
3:       Set the reserve price r^t uniformly at random.
4:       **if** $b^t \geq r^t$ **then**
5:          $\forall x$ such that $x \leq r^t : R^t(x) \leftarrow R^t(x) + 1$
6:       **else**
7:          $\forall x$ such that $x \geq r^t : L^t(x) \leftarrow L^t(x) + 1$
8:       **end if**
9:    **end for**
10:    $\forall x \in$ support $: \hat{T}_q(x) = \frac{R^k(x)}{R^k(x) + L^k(x)}$
11:    Compute $r_q^* = \arg\max_{r \in \text{support}} r\hat{T}_q(r)$
12:    Create the list candidates $= [r_q^* - l, \ldots, r_q^* + l]$
13:    **for all** $c \in$ candidates **do**
14:       Set the reserve price $r^t = c$ for k_c steps.
15:       Keep counter $R_c^t = \sum_{\tau=1}^{t} \mathbb{1}_{\{b^\tau \geq r^\tau\}}$
16:       Update:
$$\hat{T}_q(c) = \frac{R_c^{k_c}}{R_c^{k_c} + L_c^{k_c}}$$
17:    **end for**
18:    $r_q^* = \arg\max_{r \in \text{candidates}} r\hat{T}_q(r)$
19: **end procedure**

   In Fig. 1 we can see how these two steps are implemented in practice. The green area denotes the estimation of the revenue function, denoted by the blue area, during the first step (random querying). Similarly, the red area denotes the approximation of the expected revenue after implementing the KM estimator on the second step, for a selected number of candidate values. The precise approximation KM provides us, allows for optimal, in expectation, action selection against stochastic opponents.

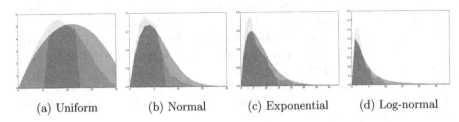

(a) Uniform   (b) Normal   (c) Exponential   (d) Log-normal

Fig. 1. Random KM estimation of the empirical payoff functions of various random distributions. The loose estimation is the random querying step (in green), followed by the precise KM step (in red). (Color figure online)

4   Experimental Results

4.1   Agents in the Type Space Θ_A

The AdX domain is of significant commercial importance. However, this also means that obtaining real world data from live auctions is difficult. We conduct empirical studies using a domain that captures many aspects of this domain. The domain is that of the *Trading Agents Competition (TAC) AdX '15* [18], which simulate an Ad Exchange game. We use our own implementation of these specifications. In particular we have a setting of 60 days, with 1000 impression opportunities each day, specified daily Budget B and Campaign Reach C_R for the advertiser and defined advertiser's type space: Θ_A = [Random, Greedy, LTB, UCB, Q-learn].

We use three basic benchmarks for the evaluation of our HBA-KM algorithm.

1. The Offline Optimal algorithm that knows the true type θ_A of the advertiser *a priori* and decides the optimal policy π_P^* as his best response.
2. The Q-learning algorithm, a well known reinforcement learning technique.
3. The UCB algorithm, a MAB technique.

The Offline Optimal algorithm realises the best response strategies, discussed in Sect. 3.1, and serves as an upper bound for our method. We also choose a Q-learning agent and a UCB-based online learning agent as our baselines, since, given the stochastic game formulation of the problem, one may hope to solve it using techniques from reinforcement learning. Q-learning with soft-max, and UCB with ϵ-greedy action selection, are two of the simplest algorithms for reinforcement learning, giving good results in a wide spectrum of applications.

We use different parameters for each of the strategies in the advertiser's type space. For the comparative evaluation of our algorithm, against the specified benchmarks, we consider metrics, such as the *revenue* of our algorithm and the *standard deviation* to quantify our agent's payoff variation between consecutive games.

The parameter settings for our experiments were chosen for the opposing agents, in a way to demonstrate that our results hold, for every way one could distribute the probability mass over the value vector for the impressions ($\mathbf{v}$ = $[0, 1]$), according to a specific random distribution or an adaptive strategy. The exact parameters that were used for all the simulations follow.

1. For the *randomised* strategies: $\mathcal{U}\{a, b\}$ for the uniform distribution, with $a = 0$ and $b = \{0.5, \ldots, 1\}$. $\mathcal{N}(\mu, \sigma^2)$ for the normal distribution, with $\mu = \{0.25, \ldots, 0.65\}$ and $\sigma^2 = \{2 \times 10^{-6}, \ldots, 6 \times 10^{-6}\}$. $\ln \mathcal{N}(\mu, \sigma^2)$ for the log-normal distribution, with $\mu = \{-7, \ldots, -5.4\}$ and $\sigma = \{0.5, \ldots, 1\}$. $f(x; \beta)$ for the exponential distribution, with $\beta = \{1/900, \ldots, 1/500\}$.
2. For the *deterministic* and *adaptive* strategies: For the *Greedy* strategy, we set the bid to be $v_{max} = \{0.5, \ldots, 1\}$. For the *Learn-Then-Bid* strategy, we set the exploration length to be $m = \{20, \ldots, 200\}$ and the target fraction $f = \{0.3, \ldots, 0.7\}$. For the *UCB* strategy, we set the exploration

step $k = \{20, \ldots, 200\}$ and the exploitation with probability $1 - \epsilon$, where $\epsilon = \{0.01, \ldots, 0.30\}$. For the *Q-learn* strategy we set the learning rate to be $\alpha = \{0.1, \ldots, 0.3\}$, the discount factor $\gamma = \{0.80, \ldots, 0.99\}$ and the temperature of the soft-max selection policy $\tau = \{100, \ldots, 1000\}$

For every opposing strategy, we consider the cartesian product of the parameters we specified and the results that follow are averaged over every possible run using these parameters, across 100 simulations for each individual opponent.

In Fig. 2, we can see the HBA-KM's performance against the deterministic and adaptive strategies of the advertiser. The performance is close to the offline optimal benchmark and outperforms the other two baselines.

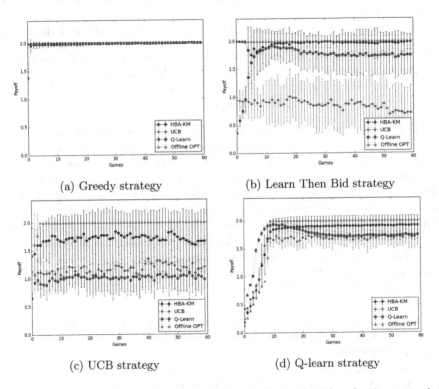

Fig. 2. Revenue comparison and one standard deviation against the adaptive strategies.

In Fig. 3, we can see the HBA-KM's performance against the randomised strategies of the advertiser. The performance of HBA-KM approximates the optimal offline benchmark, on every single occasion based on the distribution approximation that the subroutine Random KM performs.

As another metric we consider the average competitive ratio of each algorithm, when compared to the *online optimal* one. The online optimal algorithm is the one with the best case cost; imagine that the publisher knows *a priori*

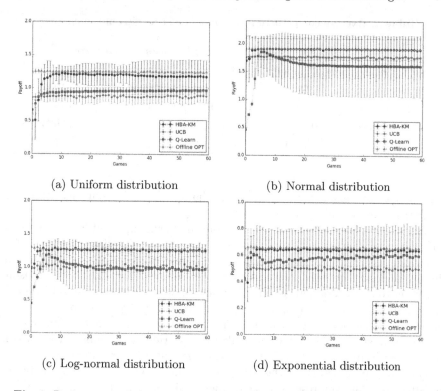

(a) Uniform distribution

(b) Normal distribution

(c) Log-normal distribution

(d) Exponential distribution

Fig. 3. Revenue comparison and standard deviation against random strategies.

the sequence of bids the advertiser is going to play. Then he attains the online optimal policy by greedily selecting to sell the impression at the highest possible cost, up until the budget of the advertiser is exhausted. So for a bid b and budget B, we have:

$$\pi_P(b, B) = \max_{r \in \{0,1\}} r \ \ s.t. \ r \leq b, \ r \leq B$$

The competitive ratio is defined as $\rho_{\text{ALG1}} = \frac{\text{ALG1}}{\text{ONLINE-OPT}}$ and Table 1 summarises the results over all the opposing strategies, adaptive and randomised respectively. The significant drop on the competitive ratio of every benchmark we witness when we move from facing adaptive strategies to randomised is expected, as the optimal online algorithm has knowledge of the exact sequence of bids, something powerful against truly random opponents.

4.2 Neural Network Agent

So far our experiments only included opposing agents that are in the hypothesised type space Θ_A of our own algorithm. Unfortunately, this is not always the case in real life scenarios, as an agent in this marketplace should be able to face opposing strategies he cannot expect, or model explicitly, in real time. This is the question we sought to answer in the second part of our experiments;

Table 1. Average competitive ratio, across all strategies and simulations

	Against adaptive		Against randomised	
Algorithm name	Competitive ratio	Std	Competitive ratio	Std
OFFLINE-OPT	0.9721	0.0064	0.7657	0.1009
HBA-KM	0.9245	0.2073	0.7434	0.1585
Q-LEARN	0.7976	0.1650	0.6165	0.2673
UCB	0.7004	0.2334	0.6218	0.1751

what happens when such an agent enters the market and is our algorithm still competitive against him?

We choose a Neural Network agent as our unknown opponent for two reasons. The first reason is that a NN does not belong to the hypothesised type space Θ_A of our own agent, therefore our type space should be descriptive enough to be able to model adequately such unknown agents on the fly. The second reason, consistent to the second part of our experiments, where we use a Neural Network trained in a mixture of the opposing publishers, is that we want to capture the inherent heterogeneity an agent faces in this market, where his opponents are trained against a variety of pricing algorithms. Here we simulate the Neural Network with up to 4 Hidden Layers and train it at each arriving impression.

Our exact parameters for the simulation follow: Two input layers, the bid of the advertiser and the reserve price of the publisher. 1 up to 4 Hidden Layers. One output layer, the bidder's immediate payoff. Each node is fully-connected with nodes of next layer and we use a standard sigmoidal threshold function. We train online for every instance of the first day of simulations and for 100 impressions for each subsequent day. The network is trained until convergence at the end of each simulation day. The optimisation step is using a Hill Climber approach.

NN Trained on a Single Opposing Agent. In the first set of experiments, we train the neural network using samples from his current opponent. The two input layers of the network are the bid and the reserve price for each impression that arrives and the single output is the revenue of the advertiser. We run the simulations using a neural network with 1 up to 4 hidden layers. Table 2 summarises the competitive ratio of each algorithm against this agent.

Table 2. Average competitive ratio against a NN agent, across all simulations

Algorithm name	Competitive ratio	Std
HBA-KM	0.9423	0.2697
Q-LEARN	0.8699	0.3470
UCB	0.8469	0.4119

NN Trained on a Mixture of Opposing Agents. In the second set of experiments, we train the neural network in a mixture of the opposing publishing agents. Specifically, for subsequent chunks of 100 impressions, the NN agent is trained with samples from the HBA-KM, the UCB and the Q-LEARN respectively, throughout the first day of the simulations (1000 impressions). The reasoning behind this type of training is that in real world auctions we should expect to face algorithms trained in a variety of scenarios and, as such, we will not be able to model these agents explicitly. Table 3 shows that against this opposing network, our algorithm stays highly competitive, even compared to the online optimal benchmark through the competitive ratio.

Table 3. Average competitive ratio against a mixed Neural Network agent, across all simulations

Algorithm name	Competitive ratio	Std
HBA-KM	0.9501	0.2222
Q-LEARN	0.7957	0.4336
UCB	0.8630	0.3897

5 Discussion and Conclusions

In this paper, we address the learning problem faced by the publisher in an ad exchange, an interaction that is both practically significant and scientifically challenging. We propose the use of a novel methodology for learning in multiagent interactions, showing how this enables us to achieve substantial empirical improvements in simulations involving the TAC AdX domain.

Although we have not performed the theoretical analysis of these extensions, we conjecture that HBA-KM best response actions will always converge to an approximately optimal policy against either stochastic or deterministic opponents, within this posted price auction mechanism. This is based on the observation that the challenge is twofold, with each individual piece having known properties. The Random KM estimator can approximate successfully any given distribution, or a specific family of random distributions, while HBA converges to the correct beliefs over his hypothesised type space Θ_A.

A useful future direction would be to expand this to the case where there are multiple advertisers and a publisher interacting with each other in the ad exchange market. The main question here becomes whether there is a way (a) to model explicitly every single one of your opponents, or (b) to model the market price, i.e. the price ρ^t that an agent $A \in \mathbf{A}$ faces in each step t of the auction and is derived from the joint actions of every other agent in the auction $\mathbf{A} - \{A\}$. An algorithm that answers successfully either of those questions, would come a long way to us understanding the implicit interactions between different agent types, and will probably have other implications in situations where the modelling of your opponent, or teammate, on the fly is the core of the problem, such as the *ad hoc teams* challenge [19].

References

1. Albrecht, S.V., Ramamoorthy, S.: A game-theoretic model and best-response learning method for ad hoc coordination in multiagent systems. In: Proceedings of the 2013 International Conference on Autonomous Agents and Multi-Agent Systems, pp. 1155–1156. IFAAMAS (2013)
2. Albrecht, S.V., Ramamoorthy, S.: On convergence and optimality of best-response learning with policy types in multiagent systems. In: Proceedings of the 30th Conference on Uncertainty in Artificial Intelligence, pp. 12–21 (2014)
3. Albrecht, S., Crandall, J., Ramamoorthy, S.: Belief and truth in hypothesised behaviours. Artif. Intell. **235**, 63–94 (2016). https://doi.org/10.1016/j.artint.2016.02.004
4. Amin, K., Kearns, M., Key, P., Schwaighofer, A.: Budget optimization for sponsored search: censored learning in MDPs. In: Proceedings of the 28th Conference on Uncertainty in Artificial Intelligence, pp. 54–63. AUAI Press (2012)
5. Amin, K., Rostamizadeh, A., Syed, U.: Learning prices for repeated auctions with strategic buyers. In: Advances in Neural Information Processing Systems, pp. 1169–1177 (2013)
6. Auer, P., Cesa-Bianchi, N., Fischer, P.: Finite-time analysis of the multiarmed bandit problem. Mach. Learn. **47**(2–3), 235–256 (2002)
7. Barrett, S., Stone, P., Kraus, S.: Empirical evaluation of ad hoc teamwork in the pursuit domain. In: Proceedings of the 10th International Conference on Autonomous Agents and Multiagent Systems, pp. 567–574. IFAAMAS (2011)
8. Cesa-Bianchi, N., Gentile, C., Mansour, Y.: Regret minimization for reserve prices in second-price auctions. In: Proceedings of the 24th Annual ACM-SIAM Symposium on Discrete Algorithms, pp. 1190–1204. SIAM (2013)
9. Cole, R., Roughgarden, T.: The sample complexity of revenue maximization. In: Proceedings of the 46th Annual ACM Symposium on Theory of Computing, pp. 243–252. ACM (2014)
10. Insights from buyers and sellers on the RTB opportunity. Forrester Consulting, commissioned by Google, White Paper (2011)
11. Ganchev, K., Nevmyvaka, Y., Kearns, M., Vaughan, J.W.: Censored exploration and the dark pool problem. In: Proceedings of the 25th Conference on Uncertainty in Artificial Intelligence, pp. 185–194 (2009)
12. Ghosh, A., Rubinstein, B.I., Vassilvitskii, S., Zinkevich, M.: Adaptive bidding for display advertising. In: Proceedings of the 18th International Conference on World Wide Web, pp. 251–260. ACM (2009)
13. Huang, Z., Mansour, Y., Roughgarden, T.: Making the most of your samples. In: Proceedings of the 16th ACM conference on Economics and Computation, pp. 45–60. ACM (2015)
14. Kaplan, E.L., Meier, P.: Nonparametric estimation from incomplete observations. J. Am. Stat. Assoc. **53**, 457–481 (1958)
15. Mohri, M., Medina, A.M.: Learning theory and algorithms for revenue optimization in second-price auctions with reserve. In: Proceedings of the 31st International Conference on Machine Learning, pp. 262–270 (2014)

16. Muthukrishnan, S.: Ad exchanges: research issues. In: Leonardi, S. (ed.) WINE 2009. LNCS, vol. 5929, pp. 1–12. Springer, Heidelberg (2009). https://doi.org/10.1007/978-3-642-10841-9_1
17. Pin, F., Key, P.: Stochastic variability in sponsored search auctions: observations and models. In: Proceedings of the 12th ACM Conference on Electronic Commerce, pp. 61–70. ACM (2011)
18. Schain, M., Mansour, Y.: Ad exchange – proposal for a new trading agent competition game. In: David, E., Kiekintveld, C., Robu, V., Shehory, O., Stein, S. (eds.) AMEC/TADA -2012. LNBIP, vol. 136, pp. 133–145. Springer, Heidelberg (2013). https://doi.org/10.1007/978-3-642-40864-9_10
19. Stone, P., Kaminka, G.A., Kraus, S., Rosenschein, J.S., et al.: Ad hoc autonomous agent teams: collaboration without pre-coordination. In: AAAI (2010)
20. Watkins, C.J., Dayan, P.: Q-learning. Mach. Learn. **8**(3–4), 279–292 (1992)

Towards Online Electric Vehicle Scheduling for Mobility-On-Demand Schemes

Ioannis Gkourtzounis[1], Emmanouil S. Rigas[2([⊠])], and Nick Bassiliades[2]

[1] Department of Computing, The University of Northampton,
Northampton NN15PH, UK
`ioannisgk@live.com`
[2] Department of Informatics, Aristotle University of Thessaloniki,
54124 Thessaloniki, Greece
`{erigas,nbassili}@csd.auth.gr`

Abstract. We study a setting where electric vehicles (EVs) can be hired to drive from pick-up to drop-off stations in a mobility-on-demand (MoD) scheme. Each point in the MoD scheme is equipped with battery charge facility to cope with the EVs' limited range. Customer-agents announce their trip requests over time, and the goal for the system is to maximize the number of them that are serviced. In this vein, we propose two scheduling algorithms for assigning EVs to agents. The first one is efficient for short term reservations, while the second for both short and long term ones. While evaluating our algorithms in a setting using real data on MoD locations, we observe that the long term algorithm achieves on average 2.08% higher customer satisfaction and 2.87% higher vehicle utilization compared to the short term one for 120 trip requests, but with 17.8% higher execution time. Moreover, we propose a software package that allows for efficient management of a MoD scheme from the side of a company, and easy trip requests for customers.

Keywords: Electric vehicles · Mobility on demand · Scheduling ·
Demand response · Software

1 Introduction

In a world where over 60% of the total population will be living in, or around cities, the current personal transportation model is not viable as it is based almost entirely on privately owned internal combustion engined vehicles. These vehicles cause high air and sound pollution, and suffer from low utilization rates [1]. One of the key elements of the vision of future Smart Cities is the development of Mobility-on-Demand (MoD) systems, especially ones using fleets of Electric Vehicles (EVs) [2]. Such vehicles emit no tailpipe pollutants and, once powered by electricity produced from renewable sources, they can play an important role towards the transition to a new and sustainable transportation era.

© Springer Nature Switzerland AG 2019
M. Slavkovik (Ed.): EUMAS 2018, LNAI 11450, pp. 94–108, 2019.
https://doi.org/10.1007/978-3-030-14174-5_7

Most of the deployed MoD schemes use normal cars. However, EVs present new challenges for MoD schemes. For example, EVs have a limited range that requires them to charge regularly their battery when they stop. Moreover, if such MoD schemes are to become popular, it will be important to ensure that charging capacity is managed and scheduled to allow for the maximum number of consumer requests to be serviced across a large geographical area. In this context, Pavone et al. have developed mathematical programming-based rebalancing mechanisms for deciding on the relocation of vehicles to restore imbalances across a MoD network, either using robotic autonomous driving vehicles [3], or human drivers [4], while Smith et al. [5] use mathematical programming to optimally route such rebalancing drivers. Moreover, Carpenter et al. [6] propose solutions for the optimal sizing of shared vehicle pools. However, in all these works internal combustion engine-based vehicles are assumed and hence do not account for the limited range of EVs and how to balance the number of pending requests at specific nodes across the network while serving the maximum number of users. In contrast, [7] consider on-demand car rental systems for public transportation. To address the unbalanced demand across stations and maximise the operator's revenue, they adjust the prices between origin and destination stations depending on their current occupancy, probabilistic information about the customers' valuations and estimated relocation costs. Using real data from an existing on-demand mobility system in a French city, they show that their mechanisms achieve an up to 64% increase in revenue for the operator and at the same time up to 36% fewer relocations. In addition, Rigas et al. [8] use mathematical programming techniques and heuristic algorithms to schedule EVs in a MoD scheme taking into consideration the limited range of EVs and the need to charge their batteries. They goal of the system is to maximize serviced customers.

In this paper, we step upon the work of Rigas et al. [8], and we solve the problem of assigning EVs to customers online. In so doing, we propose two scheduling algorithms for the EV-to-customer assignment problem aiming to maximize the number of serviced customers. The first one is shown to be efficient for short term bookings, while the second for both short and long term ones. Both algorithms are evaluated in a setting using real data regarding MoD locations in Bristol, UK. Moreover, we propose a software package which consists of a web platform that supports the efficient monitoring and management of a MoD scheme from the side of a company, and a mobile application for easy trip requests for customers.

The rest of the paper is organized as follows: Sect. 2 presents a mathematical formulation of the problem, while Sect. 3 describes the scheduling algorithms. Section 4 presents the software package for the management of the MoD scheme and Sect. 5 contains the evaluation. Finally, Sect. 6 concludes and presents ideas for future work.

2   Problem Formulation

We study a MoD setting where customer-agents $k \in K$, announce their intentions to drive between pairs of locations over time. Each time a new request is received by the MoD company (we assume a single MoD company to exist), it applies an algorithm that schedules an available EV, if such an EV exists, to drive across the set of requested locations. In assigning EVs to trips, the MoD company aims to maximize the number of agents that will be serviced. We assume that EVs have a limited driving range which requires them to have their battery charged at the stops that form part of the MoD scheme.

In more detail, we denote a set of EVs $a \in A$ and a set of locations $l \in L$ which are pick-up and drop-off stations, where each $l \in L$ has a maximum capacity $c_l \in N$. We consider a set of discrete time points $T \subset \mathbb{R}, t \in T$, where time is global for the system and the same for all EVs. Moreover, we have a set of tasks $i \in \Delta$ where each task is denoted by a tuple $p_i = <l_i^{start}, l_i^{end}, t_i^{start}, \tau_i, b_i>$. l_i^{start} and l_i^{end} are the start and end locations of the task, t_i is the starting time point of the task, while τ_i is its travel time (each task has also an end time $t_i^{end} = t_i^{start} + \tau_i$), and b_i is the energy cost of the task. Each agent has a valuation $v_k(i) = 1$ for executing the requested task i and $v_k(i') = 0$ for any other task $i' \neq i$. Note that a task is a trip taking place at a particular point in time. Also, note that one-way rental is assumed, and therefore, start and end locations of a task are always different. Moreover, we assume that customers drive the cars between start and end locations without stopping or parking them during the trip. One-way rental introduces significant flexibility for users, but management complexities (e.g., complex decision making in choosing which customers to service, and high importance of the initial location of EVs) [9]. Henceforth, index a stands for EVs, l for locations, t for time points and i for tasks.

Each EV a has a current location at time point t, denoted as $l_{a,t}$, and this location changes only each time a executes one task. Here, we assume that at time point $t = 0$ all EVs are at their initial locations $l_{a,t=0}^{initial} \in L$, and that their operation starts at time point $t \geq 1$. Moreover, each a has a current battery level $b_{a,t} \in N$, a consumption rate con_a and therefore, a current driving range in terms of time $\tau_{a,t} = [b_{a,t}/con_a] \in N$. Now, for a task i to be accomplished, at least one EV a must be at location l_i^{start} at time point t_i, having enough energy to execute the task (i.e., $b_{a,t} > b_i$). We also define binary variable $prk_{t,a,l} \in \{0,1\}$ to capture the location where each EV is parked at each time point, binary variable $\epsilon_{a,i,t} \in \{0,1\}$ denoting whether EV a is executing task i at time t, and binary variable $\delta_i \in \{0,1\}$ denoting whether task i is executed or not. At any t, each EV should either be parked at exactly one location, or travelling between exactly one pair of locations. In the next section we present the decision making algorithms.

3   Scheduling Algorithms

In this section, we study the scheduling of EVs to customers, based on their requests and EV availability across the set of stations. In this vein we develop two decision making algorithms, one for short-term bookings and another for long-term ones.

3.1   Short Mode Algorithm

The Short mode algorithm (see Algorithm 1) receives as input trip requests from agents to drive an EV between stations l_i^{start} and l_i^{end} starting at time point t_i^{start} (t^{cur} defines the current time point). Based on these data, the algorithm calculates the duration τ_i of the task and the energy demand b_i (line 1). In this way the tuple p_i that describes a task i is completed. Next, the vehicles that are currently at the start station or are travelling to it and will arrive in $t < t_i^{start}$ (i.e., $\forall a : prk_{t_i^{start}-1,a,l_i^{start}} = 1$) are added to the set $candidateEV \subseteq A$ (lines 2–4). If $candidateEV = \emptyset$, then task i cannot be executed. Otherwise, we check if the candidate vehicles have future routes and we add the vehicles without future routes, to set $betterCandidateEV \subseteq candidateEV$ (lines 5–10).

Algorithm 1. EVs Scheduling Algorithm - Short mode.

Require: i and A and L and T and t^{cur} and $\forall a, b_{a,t^{cur}}$ and $\forall a,t,l \ \epsilon_{a,t,l}$ and $\forall a,t,l$
   and $prk_{a,t,l}$
1: Calculate b_i and τ_i
2: **for** $\forall a \in A$ **do**
3:    **if** $prk_{t_i^{start}-1,a,l_i^{start}} = 1$ **then**
4:       Assign a to $candidateEV$
5: **if** $candidateEV = \emptyset$ **then**
6:    $\delta_i = 0$
7: **else**
8:    **for** $\forall a \in candidateEV$ **do**
9:       **if** $\forall i, t > t_i^{start} + \tau_i, \ \epsilon_{a,i,t} = 0$ **then**
10:          Assign a to $betterCandidateEV$
11: **if** $betterCandidateEV \neq \emptyset$ **then**
12:    **for** $\forall a \in betterCandidateEV$ **do**
13:       Calculate battery charge $b_{a,t_i^{start}-1}$
14:       **if** $b_{a,t_i^{start}-1} - b_i > 0$ **then**
15:          Assign a to $bestCandidateEV$
16:    Sort $bestCandidateEV$ based on remaining energy after executing task i
17:    Assign first vehicle a to task i and set $\delta_i = 1$
18: **else**
19:    $\delta_i = 0$
   return δ_i, ϵ, prk

If $betterCandidateEV = \emptyset$, then the task is not executed. Otherwise, we calculate and set the future charge (we use the term "charge" for "the current state of charge level") for each vehicle by subtracting the energy cost of the task b_i from the charge level of each vehicle, $b_{a,t_i^{start}-1}$. We sort the vehicles by future charge in descending order and if the future charge of a vehicle is greater than zero, we add it to set $bestCandidateEV \subseteq betterCandidateEV$. We assume that for $\forall a,t,l : prk_{a,t,l} = 1$ a is charging its battery, unless it is fully charged. Now, the $bestCandidateEV$ contains all the vehicles that are suitable to execute task i. The first vehicle in the list has the maximum charge level, so it is the best candidate vehicle. Thus, we assign it to task i (i.e., $\forall t : t >= t_i^{start}$ and $t <$ $t_i^{start} + \tau_i, \epsilon_{a,i,t} = 1$ and $\delta_i = 1$). If $bestCandidateEV = \emptyset$, the task is not executed (lines 11–19).

In summary, the Short mode algorithm gets the vehicles that currently are, or finish in the start station before start time of the task and adds the vehicles without future routes to a list. We calculate the charge cost for the request trip, get the vehicles with enough charge and assign the best vehicle to the new route.[1]

The Short mode algorithm has a relatively simple implementation, but its biggest drawback is that if a vehicle is assigned for a route that starts for example, after 6 h, this vehicle can not be used for another route for this period of time. One way to tackle this problem is to restrict the minutes in the future, that users are allowed to request a trip.

3.2   Long Mode Algorithm

A restriction that does not allow users to request vehicles at any time during the day, may not be a viable solution. In order to overcome this problem, we developed and implemented the Long mode algorithm that applies a one-step look ahead technique to reschedule EVs to other tasks in order to increase vehicle utilization and customer satisfaction.

The Long mode algorithm (see Algorithm 2) receives trip requests and the $candidateEV$ set is populated in the same way as in the short mode algorithm (lines 1–4). Next, we get the vehicles without future routes and the vehicles with one future route and add them to $betterCandidateEV \subseteq candidateEV$ and $betterCandidateEVWithLateRoutes \subseteq candidateEV$ respectively (lines 5–12). We calculate the future charge $b_{i,t_i^{start}}$ of the EVs in the same way as described in the previous section and sort them by future charge in descending order. We subtract the task's energy cost b_i from $b_{i,t_i^{start}}$ and if the value is greater than zero, we add it to the $bestCandidateEV$. Now, the $bestCandidateEV$ contains all the vehicles that are suitable for the requested trip. The first vehicle in the list has the maximum charge level, so it is the best candidate vehicle. Thus, we assign this vehicle to execute task i. If there are no vehicles in the $bestCandidateEV$, the first part of the algorithm which is actually the same to the short mode one, ends (lines 13–19).

At this point, there are no vehicles without future routes and with enough charge level to execute task i. Thus, we need to find a candidate vehicle a' with a future task i', assign this task to a substitute vehicle a and then assign the vehicle a' to task i (i.e., $a \rightarrow i'$, $a' \rightarrow i$). First, we check the task of the vehicles in the $betterCandidateEVWithLateRoutes$. If this task starts later than the start time plus the trip duration of the user request, we need to find a substitute vehicle. We add the vehicles that are currently in the start station and the vehicles that finish at the start station before the start time of task i', to the $substituteEV \subseteq betterCandidateEVWithLateRoutes$. Note that $t_{i'}^{start} > t_i^{start}$ and for this reason $candidateEV \subseteq substituteEV$ (lines 20–25).

---

[1] A flowchart for the Short mode is available at https://goo.gl/dxpfer.

Algorithm 2. EVs Scheduling Algorithm - Long mode.

Require: i and A and L and T and t^{cur} and $\forall a, b_{a,t^{cur}}$ and $\forall a,t,l$ $\epsilon_{a,t,l}$ and $\forall a,t,l$
and $prk_{a,t,l}$

1: Calculate b_i and τ_i
2: **for** $\forall a \in A$ **do**
3:     **if** $prk_{t_i^{start}-1,a,l_i^{start}} = 1$ **then**
4:         Assign a to $candidateEV$
5: **if** $candidateEV = \emptyset$ **then**
6:     $\delta_i = 0$
7: **else**
8:     **for** $\forall a \in candidateEV$ **do**
9:         **if** $\forall i, t > t_i^{start} + \tau_i, \epsilon_{a,i,t} = 0$ **then**
10:            Assign a to $betterCandidateEV$
11:        **else if** $\forall i, \sum_{t:t>t_i^{start}+\tau_i}(|\epsilon_{a,i,t} - \epsilon_{a,i,t-1}| = 2)$ **then**
12:            Assign a to $betterCandidateEVWithLateRoutes$
13: **if** $betterCandidateEV \neq \emptyset$ **then**
14:    **for** $\forall a \in betterCandidateEV$ **do**
15:        Calculate battery charge $b_{a,t_i^{start}-1}$
16:        **if** $b_{a,t_i^{start}-1} - b_i > 0$ **then**
17:            Assign a to $bestCandidateEV$
18:    Sort $bestCandidateEV$ based on remaining energy after executing task i
19:    Assign first vehicle a to task i and set $\delta_i = 1$
20: **else**
21:    **for** $\forall a' \in betterCandidateEVWithLateRoutes$ **do**
22:        For the task i', vehicle a' is assigned to:
23:        **for** $\forall a \in A$ **do**
24:            **if** $prk_{t_{i'}^{start}-1,a,l_{i'}^{start}} = 1$ **then**
25:                Assign a to $substituteEV$
26:        **for** $\forall a \in substituteEV$ **do**
27:            **if** $\forall i, t > t_{i'}^{start} + \tau_{i'}, \epsilon_{a,i,t} = 0$ **then**
28:                Assign a to $substitutesWithoutRoute$
29:        **if** $substitutesWithoutRoute \neq \emptyset$ **then**
30:            **for** $\forall a \in substitutesWithoutRoute$ **do**
31:                Calculate battery charge $b_{a,t_{i'}^{start}-1}$
32:                **if** $b_{a,t_{i'}^{start}-1} - b_{i'} > 0$ **then**
33:                    Assign a to $bestCandidatesubstituteEV$
34:            Sort $bestCandidatesubstituteEV$ based on remaining energy after executing task i'
35:            Assign a to i' and a' to i and set $\delta_i = 1$
36:            Break for loop
37:        **else**
38:            $\delta_i = 0$
    return δ_i, ϵ, prk

We add the substitute EVs without future routes to the *substitutes WithoutRoute* $\subseteq$ *substituteEV* and we sort them by future charge in descending order. If the substitute vehicle of the best candidate vehicle has enough charge for the future trip of the candidate vehicle, we assign the substitute vehicle to the candidate vehicle's future route and EV $a \in betterCandidateEVWithLateRoutes$ is assigned to task i and $\delta_i = 1$ (lines 26–38).

In summary the Long mode algorithm gets the vehicles in the start station without future routes and those with one future route and adds them to lists. If there are vehicles with enough charge in the first list, we assign the best vehicle to the new route, else we search for a substitute vehicle to replace the future route of a vehicle in the second list. We get substitute vehicles without future routes and with enough charge, and we assign the best substitute to the future route of the best vehicle in the second list. Now, the best vehicle in the second list is no longer "locked/assigned" and we assign it to the new route.[2] The differences between the two algorithms are summarized in Fig. 1.

In the next section, we present a software package that integrates the previously described algorithms and provides an efficient user interface for MoD companies to manage their fleet, and customers to request trips.

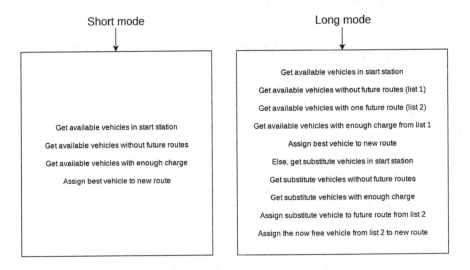

Fig. 1. Comparing Short mode and Long mode algorithms functionalities.

4   MOD Software Package

Apart from the scheduling algorithms, we designed and developed a fully functional software package (see Fig. 2), for a MoD company to monitor the state and locations of their EVs, and for the customers to make trip requests and bookings.[3]

---

[2] A flowchart for the Long mode is available at https://goo.gl/zFdWvh (part 1) and https://goo.gl/WZHrRc (part 2).

[3] A video demo is available at https://youtu.be/flyixErIE-A.

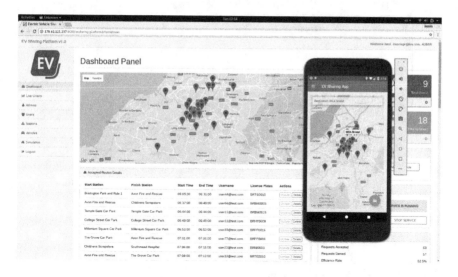

Fig. 2. Web platform and mobile application screenshot.

The web platform is installed on a server and it consists of the web pages presented to the administrators and a MySQL database to manage all the data. Administrators interact with the platform using web pages designed for specific functionalities. Thus, the interface allows them to login, upload a map with stations, upload vehicles' details and manage administrator and user accounts, stations, vehicles and routes. All operations submitted through the interface of the pages are implemented as transactions in the MySQL database. From the customer perspective, the mobile application lets users register, login, access their profile, their request history, select stations and request a vehicle for a trip. Creating an account and accessing account and station details, require access to the MySQL database on the server (Fig. 3).

Information exchange with the database is done via a RESTful web service that authenticates users and allows them to access, create and edit data on the database. Messages for vehicle requests use WebSockets, so the server listens to a specific port for incoming TCP messages. Those messages are encrypted and sent from the mobile device to the server through a TCP connection, making the communication safe and reliable. Then, the platform applies one of the algorithms presented in the previous section to accept or deny the user request. The status of the request (accepted or denied) will be sent back to the mobile application as an encrypted message through the TCP connection and the user can see if his request is accepted by the system.

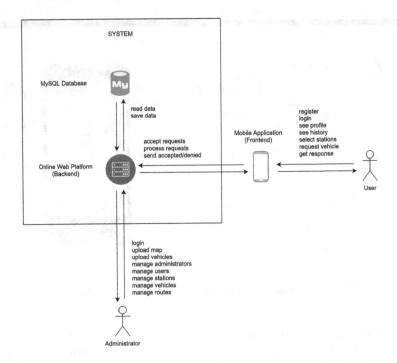

Fig. 3. System architecture with web platform, database and mobile app.

4.1   Web Platform

The pages of the web platform[4] let the administrators perform the functionalities related to managing user accounts, stations, vehicles and routes, as discussed earlier. The login page allows administrators to login with their username and password. On successful login, they are redirected to the Dashboard Panel. The Dashboard Panel shows route data and contains buttons to update, delete or add a new route. The stations are shown as pins on a Google map and buttons allow administrators to start or stop the service.

The Manage Admins and Manage Users pages show administrator and user data respectively, and contain buttons to update their existing details, delete them or add new administrators and users. The Manage Stations and Manage Vehicles pages offer the same functionalities, but they also allow administrators to upload stations and vehicle details from files in XML format. For each object there are two secondary pages, the Add New page and the Update page to provide secondary functionalities, such as adding and updating objects.

The Graphical User Interface (GUI) of the web platform consists of elements that are similar and shared between the pages. The navigation options link to the most important pages: Dashboard, Live Charts, Admins, Users, Stations, Vehicles, Simulation and Logout.

---

[4] https://github.com/ioannisgk/evsharing-platform3-release.

4.2   Mobile Application

The Android mobile application[5] consists of a set of screens: The Main Login screen allows users to login with their username and password. If they do not have an account, they can click the Register button to create a new account with their details. On successful login, they are redirected to My Profile screen. The Profile screen shows their details and contains buttons to update them and to move on to the next screens, such as the Open Map and User History.

The Open Map screen initiates the first step for requesting a vehicle and the map is shown with the stations as pins. Users can select the start and finish stations and on the second step, on the Request screen they can select a specific time and send the request to the platform. The platform will process all current data and decide whether to accept or deny the request. The mobile application gets the status of the request from the server and saves it to the device. The User History screen contains the history of requests and the Settings screen allows users to change the application settings.

The GUI of the mobile application consists of elements that are similar and shared between the screens. The navigation options link to the most important screens: My Profile, Open Map, Request Vehicle, User History, Settings and Logout. The navigation menu is hidden when the user touches the rest area of the application.

5   Testing and Evaluation

In this section, we present the evaluation of the scheduling algorithms. In so doing, we use real data on locations of pick-up and drop-off stations and real-istic data regarding trip requests. We evaluate the algorithms based on a set of criteria: (1) the maximum number of route requests in a day, (2) the maximum number of minutes that a user is allowed to request a vehicle in the future, (3) the start times density factor, that essentially means how "close" or how "further away" in time, the start times of the route requests are, and (4) the number of available EVs. We generated test cases with different configurations and assigned them to the Short mode and Long mode algorithms.

Since we aim to maximize the number of customers being served in the MoD scheme, we apply all those different criteria in order to better evaluate the system under different conditions. The criteria 1 and 4 deal with the number of requests and the number of vehicles respectively, while criteria 3 and 4 focus on different requests times, as the problem is also directly associated with when the customers post their requests for routes.

Charging stations coordinates were collected for Bristol, UK as it is a strong candidate city to host a MoD scheme. We selected 27 stations that are also EV charging stations today[6] and categorized them in 5 traffic levels in order to resemble realistic travel conditions. We assume that the stations nearest to the

---

[5] https://github.com/ioannisgk/evsharing-app-release.

[6] Charging stations data were collected from https://goo.gl/pWXFm6.

center of the city have a higher number of requests, so the initial locations of the vehicles are at the stations closer to the city center, with a 100% charge level.

We developed a "route requests" generation tool that generates requests with random request times, start stations and finish stations. This tool helped us simulate a large number of vehicles requests from users with many different settings. We generated more than 400 test cases with different trip requests and used them as input to the MoD software. The platform now uses both the Short mode and Long mode algorithms to determine whether to accept or deny these requests. The algorithms also need to calculate the travel time and energy consumption of an EV during a trip, so we first calculate the average theoretical speed of the vehicle, depending on a base speed and the traffic level of the stations. Then, we compute the distance between the stations[7] and the trip duration in minutes. Energy consumption can be calculated by multiplying the duration with a base charge cost per minute.

With a total of 54 EVs, 60 and 120 maximum requests per day, we ran simulations with 60–600 min between the time the request is communicated to the MoD company and the start time of the trip, and different values of the requests allocation density within a day. The last factor shows how "concentrated" or close, the start times of the requests are, and we tested with 5 different density factors (i.e., the lower the value the more uniformly distributed the requests are).

First, we tested with 60 maximum requests per day. The Short mode produced an efficiency rate (the percentage of the requests that get accepted by the system and become routes) from 76.00%–86.67% and the Long mode from 77.00%–87.67% and it performed better with an average gain of 1.40%. When testing with 120 maximum requests per day, the Short mode produced efficiency rate from 57.17%–78.17% and the Long mode from 57.67%–80.83% and it performed better with an average gain of 2.08%. Since the more trip requests are accepted by the system the more customers are satisfied, we can directly connect the average gain of the efficiency rate of an algorithm, with the customer satisfaction levels. In other words, the Long mode algorithm achieves on average 2.08% higher agent satisfaction.

We also recorded the total minutes travelled by EVs and found that the gain of the Long mode algorithm translates to 8.88 more minutes of travelling time for every 60 requests, and 32.52 more minutes for every 120 requests for routes. In terms of vehicle utilization, the Long mode achieves a 2.87% higher average rate than the Short mode in 120 requests per day. However, the Short mode algorithm due to its simpler design achieves 17.8% lower execution times.

Looking closer at the charts in Figs. 4, 5 and 6, we can observe the following: (1) the Long mode algorithm performs better than the Short mode in all tests, (2) the Long mode offers more gain in the efficiency of the system when the number of requests increase, (3) when the time between the start times of the requests increases, the efficiency increases, (4) the overall efficiency seems not to be influenced by the requests allocation density within the day, (5) when the number of requests within the day increases, the efficiency of the system drops.

---

[7] Distance calculation using latitude and longitude https://goo.gl/3bDKuT.

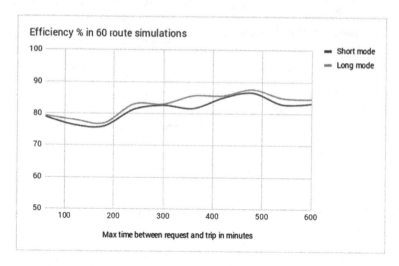

Fig. 4. Efficiency rate and max time between request and trip, when the system handles 60 requests and it is equipped with 54 EVs.

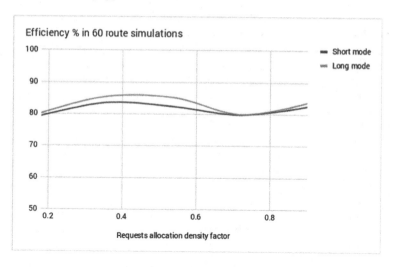

Fig. 5. Efficiency rate and requests allocation density, when the system handles 60 requests and it is equipped with 54 EVs.

Finally, we tested the system with different numbers of vehicles, as this is a very important factor for any MoD scheme company due to the high cost of EVs. The chart in Fig. 7 shows that the average gain of the Long mode algorithm is higher when using more vehicles and when the number of requests per day increase. When using 18, 27, 36, 45 and 54 EVs in 60 requests per day, the gain is 1.17%, 1.27%, 1.70%, 1.47% and 1.40% respectively. In the case of 120 requests per day, we have 1.83%, 1.93%, 2.27%, 2.05% and 2.08% gain. We conclude that

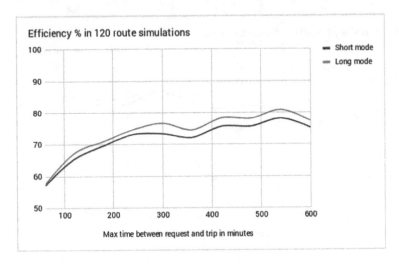

Fig. 6. Efficiency rate and max time between request and trip, when the system handles 120 requests and it is equipped with 54 EVs.

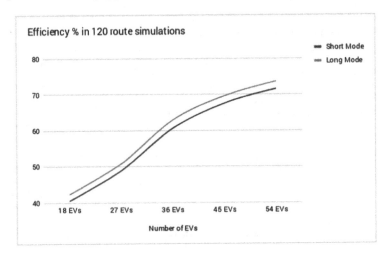

Fig. 7. Efficiency rate graph when the system handles 120 requests and the system is equipped with 18, 27, 36, 45 and 54 EVs.

the optimal number of EVs for our use case of 27 stations in Bristol, is 36 vehicles as it offers a slightly better increase in performance.

The Long mode algorithm has the advantage of looking ahead for vehicles that can substitute a current candidate vehicle with enough charge and a future route. That vehicle would be characterized as unavailable in Short mode, but now we can make one swap with the help of the substitute vehicle, and make the current candidate one, available for trips. Due to the fact that the Long mode incorporates only one such swap, the gains are relatively small in percentage when compared with the results of the Short mode. We can see that this is confirmed by the observations in the charts as discussed earlier.

In Fig. 5 the overall efficiency shows some small fluctuations and it is not influenced significantly by the requests allocation density within the day, mainly because of the low number of requests. Moreover, if we look closer in the chart at Fig. 7, we will see that as the number of vehicles increases, the gain of the efficiency rate of the Long mode algorithm is almost constant. This is to be expected, because the current version of the Long mode incorporates only one swap functionality and the number of requests (120) is relatively too small to significantly affect the gain. A future version of the Long mode with more swaps would increase the gain when the vehicles increase.

For our test strategy on the software package, we used black-box test design techniques that included equivalence partitioning and boundary value analysis for proper input validation testing. The system was evaluated when the user scenarios were executed according to the test cases generated by our "route requests" generation tool, which simulates user requests for trips. Taking all tests into account, we can say that our software solution is thoroughly tested and achieves its aim and objectives.

6   Conclusions and Future Work

In this paper, we studied a setting where electric vehicles (EVs) can be hired to drive from pick-up to drop-off stations in a mobility-on-demand (MoD) scheme. In this vein, we proposed two scheduling algorithms for assigning EVs to trips. The first one is efficient for short term reservations, while the second for both short and long term ones. While evaluating our algorithms in a setting using real data on MoD locations, we observe that the long term algorithm achieves on average 2.08% higher customer satisfaction and 2.87% higher vehicle utilization compared to the short term one in 120 requests per day.

Moreover, we designed, implemented and tested a system that consists of a web platform and a mobile application. The web platform allows administrators to manage users, stations, vehicles, routes, and accepts or denies vehicle requests using our algorithms. The mobile application lets users register, login, see the available stations in their area and send vehicle requests for trips to the web platform. The platform executes our algorithms in order to decide whether to accept or deny them.

As future work we aim to apply machine learning techniques in order to efficiently predict future customers' demand. Moreover, we aim to use load balancing techniques to enhance the charging procedure of the EVs. In addition, given that EVs hold large batteries we consider using them as temporal energy storage devices in a vehicle-to-grid domain. Finally, we aim to study ways to manage the uncertainty in future trip execution caused by unexpected circumstances such as traffic congestion and accidents.

References

1. Tomic, J., Kempton, W.: Using fleets of electric-drive vehicles for grid support. J. Power Sources **168**, 459–468 (2007)
2. Mitchell, W., Borroni-Bird, C., Burns, L.: Reinventing the Automobile: Personal Urban Mobility for the 21st Century. The MIT Press, Cambridge (2010)
3. Pavone, M., Smith, S., Frazzoli, E., Rus, D.: Robotic load balancing for mobility-on-demand systems. Int. J. Robot. Res. **31**, 839–854 (2012)
4. Pavone, M., L. Smith, S., Frazzoli, E., Rus, D.: Load balancing for mobility-on-demand systems. In: Robotics: Science and Systems (2011)
5. Smith, S., Pavone, M., Schwager, M., Frazzoli, E., Rus, D.: Rebalancing the rebalancers: optimally routing vehicles and drivers in mobility-on-demand systems. In: 2013 American Control Conference, pp. 2362–2367 (2013)
6. Carpenter, T., Keshav, S., Wong, J.: Sizing finite-population vehicle pools. IEEE Trans. Intell. Transp. Syst. **15**, 1134–1144 (2014)
7. Drwal, M., Gerding, E., Stein, S., Hayakawa, K., Kitaoka, H.: Adaptive pricing mechanisms for on-demand mobility. In: Proceedings of the 16th International Conference on Autonomous Agents and Multiagent Systems, pp. 1017–1025 . International Foundation for Autonomous Agents and Multiagent Systems (2017)
8. Rigas, E., Ramchurn, S., Bassiliades, N.: Algorithms for electric vehicle scheduling in large-scale mobility-on-demand schemes. Artif. Intell. **262**, 248–278 (2018)
9. Barth, M., Shaheen, S.: Shared-use vehicle systems: framework for classifying car-sharing, station cars, and combined approaches. Transp. Res. Rec. J. Transp. Res. Board **1791**, 105–112 (2002)

Two-Sided Markets: Mapping Social Welfare to Gain from Trade

Rica Gonen[1(⊠)] and Ozi Egri[2]

[1] Department of Management and Economics,
The Open University of Israel, The Dorothy de Rothschild Campus,
1 University Road, 43107 Raanana, Israel
ricagonen@gmail.com
[2] Department of Mathematics and Computer Science,
The Open University of Israel, The Dorothy de Rothschild Campus,
1 University Road, 43107 Raanana, Israel
ozieg@hotmail.com

Abstract. Though the definition of gain from trade extends the definition of social welfare from auctions to markets, from a mathematical point of view the additional dimension added by gain from trade makes it much more difficult to design a gain from trade maximizing mechanism. This paper provides a means of understanding when a market designer can choose the easier path of maximizing social welfare rather than maximizing gain from trade.

We provide and prove the first formula to convert a social welfare approximation bound to a gain from trade approximation bound that maintains the original order of approximation. This makes it possible to compare algorithms that approximate gain from trade with those that approximate social welfare. We evaluate the performance of our formula by using it to convert known social welfare approximation solutions to gain from trade approximation solutions. The performance of all known two-sided markets solutions (that implement truthfulness, IR, BB, and approximate efficiency) are benchmarked by both their theoretical approximation bound and their performance in practice. Surprisingly, we found that some social welfare solutions achieve a better gain from trade than other solutions designed to approximate gain from trade.

1 Introduction

In recent years the established research on one-sided markets in economics and computer science was extended to two-sided markets where both buying and selling agents are strategic (e.g. [10] and many citations within). A key difficulty that arises when moving from one-sided to two-sided markets is handling the additional intricacies that arise when approximating gain from trade (acronym GFT). The need to approximate gain from trade stems from Myerson and Satterthwaite's impossibility result [13]. [13] states that no two-sided market can simultaneously satisfy the economically desirable properties (of truthful reporting, participation without a loss and not running a deficit) while maintaining

© Springer Nature Switzerland AG 2019
M. Slavkovik (Ed.): EUMAS 2018, LNAI 11450, pp. 109–126, 2019.
https://doi.org/10.1007/978-3-030-14174-5_8

efficiency. The two-sided market literature largely chooses to approximating efficiency in order to maintain the other economic properties.

Given the intricacies involved in approximating gain from trade the literature often chooses to approximate social welfare (acronym SWF) [1–3,5,7,10,16]. Social welfare is the parallel of gain from trade in a one-sided market setting. In one-sided markets, where only the buying agents are strategic and there is a single selling agent, efficiency is measured by maximizing the sum of the buying agents' valuations, i.e., social welfare. Social welfare in one-sided markets extends to two-sided markets by summing the buying agents' valuations and subtracting the selling agents' costs, i.e., gain from trade [11,13]. Much of the literature on two-sided markets provides solutions that approximates social welfare efficiency maximization as opposed to approximating gain from trade efficiency, i.e., maximizing the sum of the buying agents' valuations plus the sum of the non sold commodities' costs held by selling agents at the end of trade [1–3,5,7,10,16].

One would expect that notionally gain from trade would be similar to social welfare in two-sided markets as gain from trade extends the definition of social welfare from auctions (one-sided markets) to two-sided markets[1]. Despite their conceptual similarity, it is much more complex to design a gain from trade maximizing approximation mechanism than a social welfare maximizing approximation mechanism. For example consider a buying agent with value $10 for a commodity and a selling agent with cost $7 for the commodity. If they trade the social welfare is $10, while the gain from trade is $3. It is easy to see that any mechanism that maximizes social welfare also maximizes gain from trade. However, the two objectives are rather different in approximation. In the example above, if the buying agent and the selling agent do not trade, the mechanism achieves a social welfare of $7 which is 70% of the optimal social welfare, however it achieves 0 gain from trade which is not within any constant factor of the optimal gain from trade. It can be observed that any constant factor approximation of a mechanism's gain from trade is necessarily a constant factor approximation of the mechanism's social welfare, however the other direction does not hold. Thus, gain from trade is a more difficult objective to approximate. Even so, gain from trade is an important market concept that accurately captures the value of the market to both sides; buyers and sellers (see [11,13] as an example). This paper provides a means of understanding when a designer can take the easier path of designing a market that approximates the maximization of social welfare instead of gain from trade.

Two-sided market research is motivated by numerous applications such as web advertising and securities trading, and indeed the literature contains multiple two-sided market designs. Some solutions present two-sided markets with a single commodity and unit-demand [1,3,4,6,7,11,12,14,16] while others are

---

[1] Illustratively, the broad explanation of maximizing social welfare in a two-sided market is that sellers who place a relatively higher cost on a given commodity should end up retaining that commodity while the broad explanation of maximizing gain from trade is that sellers who place a relatively lower cost on a given commodity should end up selling that commodity.

combinatorial markets with multiple commodities and demand for bundles [2,5,10,15]. Some of the two-sided markets are offline, i.e., optimize given all agents' bids in advance [2,4–7,11,12,14,15] while others are online, i.e., optimize as agents' bids arrive [1,3,10,16]. Lastly both deterministic [3,10,11,16] and randomized [1–7,12,14,15] solutions exist. The above literature seeks to maintain the desirable economic properties of truthfulness (agents dominant strategy is to report their true valuation/cost), individual rationality (IR - no agent should end up with a negative utility if the agent?s true valuation/cost is submitted to the mechanism) and budget balance (BB - the price paid by the buying agents is at least as high as the price received by the selling agents, i.e., the market does not run a deficit), while keeping as much as possible of the trade efficiency. However, the existing theoretical tools do not allow a designer to compare the efficiency of all the available solutions as some approximate social welfare [1,2,5,7,10] while others approximate gain from trade [4,6,8,9,11,12,14,15].

We provide and prove the first formula to convert a social welfare approximation bound to a gain from trade approximation bound that maintains the original order of approximation (under natural conditions on percentage of commodities switching hands). This makes it possible to compare solutions that approximate gain from trade with those that approximate social welfare. The conversion formula applies to the most general setting of two sided markets, i.e. each agent can buy or sell bundles of multiple distinct commodities which may have a different number of identical units from each. This is the most general combinatorial market setting and the conversion bound does not require any restrictions on the valuation functions. Indeed the formula can convert the bounds of single unit-single demand markets as well as combinatorial markets (see Sect. 4). Moreover the formula does not change the mechanism's allocation nor the computed prices and therefore the economic properties remain the same.

We evaluate the performance of our formula by using it to convert social welfare approximating solutions in the literature to gain from trade approximating solutions. We compare the performance of all known two-sided market solutions (that implement truthfulness, IR, BB, and approximate efficiency)[2] according to the theoretical approximation bound as well as in practice. With respect to the comparisons of theoretical bounds, we show that the converted bounds perform well even when our conditions are not met. More specifically, the converted bounds are guaranteed to maintain the competitiveness order of the original social welfare bound (roughly speaking) only when more commodities switched hands in market than not. However, the converted bounds perform well even when most commodities did not change hands and were not sold.

We also implement and run the various algorithms in practice using synthetic data to evaluate their relative performance at maximizing gain from trade. These results are compared to the converted theoretical bounds. Surprisingly, in the practical runs we found that some of the social welfare solutions achieve better

---

[2] [9] and [8] are not presenting a classic two-sided market but rather a multi-side market as they have mediators in their model. Since we do not know of a similar model for maximizing SWF we decided not to include their result in our comparisons.

gain from trade than solutions that were designed to approximated gain from trade. This even happens in cases where the social welfare solution was intended for a combinatorial market settings and the gain from trade solution was intended for single commodity and unit-demand settings.

Another interesting aspect of our conversion formula is that it can be used to indicate, without an actual practical run, the practical performance of a social welfare maximizing two-sided market compared to a gain from trade maximizing two-sided market. By converting the social welfare bound of a two-sided social welfare maximizing algorithm to a gain from trade maximizing bound and comparing it with another gain from trade maximizing two-sided market algorithm we found that one can estimate the practical performance of the two algorithms with respect to gain from trade maximization.

In summary, this paper's contributions are threefold. First, we provide the first means of comparing the performance of previously uncomparable solutions. Second, the experimental tests show that our conversion formula can be used to predict the practical performance difference between the compared mechanisms. Third, we show that for combinatorial two-sided markets it is better to use the known social welfare maximizing solutions than to use a solution that directly maximizes gain from trade.

2   Preliminaries

There are multiple commodities and each one comes in a number of units. Let A be the total supply of any kind of commodities in the market, i.e., the number of all units of all commodities in the market. There are l agents interested in selling commodities. These agents may also be interested in selling multiple units of each of their commodities. There are n agents who are interested in buying commodities. These agents may also be interested in buying multiple units of these commodities. An allocation in a two-sided market can be represented as a pair of vectors $(X, Y) = ((X_1, \ldots, X_n), (Y_1, \ldots, Y_l))$ such that sum of elements in the union of $X_1, \ldots, X_n, Y_1, \ldots Y_l$ is A, and $X_1, \ldots X_n, Y_1, \ldots, Y_l$ are mutually non-intersecting. Each buying agent i, $1 \leq i \leq n$ has a valuation function v_i that assigns a non-negative value to each allocation X_i. Each selling agent t, $1 \leq t \leq l$ has a bundle of commodities S_t that he initially owns and a cost function c_t that assigns a positive cost for each allocation Y_t. The auctioneer's goal in the one-sided auction is to partition the commodities by allocating each buying agent i, X_i, so as to maximize $\sum_{i=1}^{n} v_i(X_i)$. This goal is referred to as maximizing social welfare (SWF) (or efficiency).

In a two-sided market the market maker's goal is to change hands and partition the commodities by allocating each buying agent i, X_i and each selling agent t, Y_t, so as to maximize $\sum_{i=1}^{n} v_i(X_i) - \sum_{t=1}^{l} c_t(Y_t)$. This goal is referred to as maximizing gain from trade (acronym GFT) (efficiency).

As discussed in Sect. 1 much of the literature on two-sided markets provides solutions that approximates social welfare efficiency maximization as opposed to approximating gain from trade efficiency. In a two-sided market SWF means maximizing the sum of the buying agents' valuations plus the sum of the unsold commodities' costs [1–3, 5, 7, 10, 16]. The motivation behind this extension is that SWF accounts for all agents that end up with commodities at the end of the trade.

Let (X^o, Y^o) be the pair of vectors containing the optimal allocation in the two-sided market. Let (X, Y) be the pair of vectors containing the two-sided market algorithm's allocation solution. Let $V_{ALG} = \sum_{i=1}^{n} v_i(X_i)$ and let $V_{OPT} = \sum_{i=1}^{n} v_i(X^o_i)$ be the two-sided market algorithm's solution and the optimal SWF maximization solution computed only using the buying agents, and without accounting for the unsold commodities. Let $C_{ALG} = \sum_{t=1}^{l} c_t(Y_t)$ and let $C_{OPT} = \sum_{t=1}^{l} c_t(Y^o_t)$ be the two-sided market algorithm's solution and optimal's SWF minimization solution for the selling agents. Let $G_{ALG} = \sum_{t=1}^{l}(c_t(S_t) - c_t(Y_t))$ and let $G_{OPT} = \sum_{t=1}^{l}(c_t(S_t) - c_t(Y^o_t))$ be the two-sided market's solution and the SWF maximization solution computed using only the unsold commodities.

Let $W_{ALG} = V_{ALG} + G_{ALG}$ and let $W_{OPT} = V_{OPT} + G_{OPT}$.
Let $\gamma = \frac{V_{OPT}}{C_{OPT}}$, let $\delta = \frac{G_{OPT}}{C_{OPT}}$ and let $\mu \geq \frac{W_{OPT}}{W_{ALG}}$.

3   Converting Social Welfare to Gain from Trade

In this section we show how to convert a SWF maximization approximation bound in two-sided markets into a GFT maximization approximation ratio guarantee.

In the following theorem we assume non trivial market mechanisms, i.e. mechanisms where at least one trade occurs where the seller has a positive cost for that trade and the optimal GFT is strictly positive. That is $\gamma > 1$, $C_{OPT} > 0$ and $\mu > 0$. Let $H = \left(\frac{2\gamma + (-\mu + 2)\delta - \mu - \frac{\mu W_{ALG}}{C_{OPT}}}{\mu(\gamma - 1)} \right)$.

Theorem 1. *Any two-sided market mechanism, such that $\gamma > 1, C_{OPT} > 0$, that maximizes SWF[3] within a factor of $\mu > 1$, i.e., $W_{ALG} \geq \frac{1}{\mu} W_{OPT}$ is H-competitive with respect to the optimal GFT, i.e., $(V_{ALG} - C_{ALG}) \geq \frac{1}{H}(V_{OPT} - C_{OPT})$. Moreover if $\delta \leq 1$ and $W_{ALG} > G_{OPT} + C_{OPT}$[4] then $0 < H \leq 1$ and $\frac{1}{H}$ approximation factor maintains the original $\frac{1}{\mu}$ order of approximation.*

Hence, for $W_{ALG} = \frac{1}{\mu} W_{OPT}$ the competitive ratio is $V_{ALG} - C_{ALG} \geq \left(\frac{\gamma + \delta}{\mu(\gamma - 1)} - \frac{\delta + 1}{\gamma - 1} \right) (V_{OPT} - C_{OPT})$.

Intuitively the formula's outcome GFT bound maintains the μ order of approximation only in settings where the optimal solution SWF from sold commodities is at least as high as SWF from unsold commodities, i.e. δ less or equal

---

[3] Maximizes SWF of buying agents and remaining commodities.

[4] Note that the requirement for $W_{ALG} > G_{OPT} + C_{OPT}$ is trivial in the context of two-sided markets where the SWF resulting from unallocated commodities is included as the algorithm can at least gain the SWF resulting from not allocating any commodities.

1. The performance condition is easier to understand when one considers a market where most trade can not occur and most commodities are left unsold. In such a market the cost of the unsold commodities will contribute to the SWF sum while the GFT will be unboundedly low including only the value of the few sold commodities in the few trades that will occur. Furthermore the larger the ratio of buyers' optimal SWF to sellers' optimal SWF, i.e., γ, is with respect to the converted SWF bound, the closer the converted approximation is to μ. Similarly to the above intuition, a market with a higher γ has a high GFT "potential" as there are high values of sold commodities compared to their costs.

Proof. The proof of Theorem 1 is composed of Lemmas 1 and 2.

Lemma 1 that shows that the two-sided market mechanism is $\left(\frac{\gamma + \delta - \frac{\mu G_{ALG}}{C_{OPT}}}{\mu \gamma} \right)$-competitive with respect to the buying agents' optimal SWF, i.e., $V_{ALG} \geq \left(\frac{\gamma + \delta - \frac{\mu G_{ALG}}{C_{OPT}}}{\mu \gamma} \right) V_{OPT}$.

Lemma 2 that shows that the two-sided market mechanism is $\left(1 + \delta(1 - \frac{1}{\mu}) - \frac{\gamma}{\mu} + \frac{V_{ALG}}{C_{OPT}} \right)$-competitive with respect to the selling agents' optimal SWF, i.e., $C_{ALG} \leq \left(1 + \delta(1 - \frac{1}{\mu}) - \frac{\gamma}{\mu} + \frac{V_{ALG}}{C_{OPT}} \right) C_{OPT}$.

For simplicity of exposition, let $\alpha = \frac{\mu \gamma}{\gamma + \delta - \frac{\mu G_{ALG}}{C_{OPT}}}$ and let $\beta = 1 + \delta(1 - \frac{1}{\mu}) - \frac{\gamma}{\mu} + \frac{V_{ALG}}{C_{OPT}}$.

Combining the two Lemmas we have that

$$V_{ALG} - C_{ALG} \geq \frac{1}{\alpha} V_{OPT} - \beta C_{OPT} = \left[\frac{\gamma}{\alpha} - \beta \right] C_{OPT} \tag{1}$$

$$= \left[\frac{\frac{\gamma}{\alpha} - \beta}{\gamma - 1} \right] (V_{OPT} - C_{OPT}) \tag{2}$$

$$= \left(\frac{\gamma + \delta - \frac{\mu G_{ALG}}{C_{OPT}} - \mu - \delta(\mu - 1) + \gamma - \frac{\mu V_{ALG}}{C_{OPT}}}{\mu (\gamma - 1)} \right) (V_{OPT} - C_{OPT}) \tag{3}$$

$$= \left(\frac{2\gamma + (-\mu + 2)\delta - \mu - \frac{\mu W_{ALG}}{C_{OPT}}}{\mu (\gamma - 1)} \right) (V_{OPT} - C_{OPT})$$

Equalities (1) and (2) follow since $\gamma = \frac{V_{OPT}}{C_{OPT}}$. By substituting α and β in equality (2) we achieve equality (3).

It remains to show that the competitive ratio claimed in Theorem 1 is greater than zero and less or equal to one. Since $\mu > 0$ and $\gamma > 1$, in order for the competitive ratio to be greater than zero we assume that

$$2\gamma + (-\mu + 2)\delta - \mu - \frac{\mu W_{ALG}}{C_{OPT}} > 0 \Rightarrow$$

$$2\frac{V_{OPT}}{C_{OPT}} - \mu \frac{G_{OPT}}{C_{OPT}} + \frac{2G_{OPT}}{C_{OPT}} - \mu - \frac{\mu W_{ALG}}{C_{OPT}} > 0 \Rightarrow$$

$$\frac{2V_{OPT} - \mu G_{OPT} + 2G_{OPT} - \mu C_{OPT} - \mu W_{ALG}}{C_{OPT}} > 0$$

Since $C_{OPT} > 0$ we only need to assume that

$$2V_{OPT} - \mu G_{OPT} + 2G_{OPT} - \mu C_{OPT} - \mu W_{ALG} > 0 \Rightarrow$$
$$2(V_{OPT} + G_{OPT}) > \mu(G_{OPT} + C_{OPT} + W_{ALG}) \Rightarrow$$
$$2W_{OPT} > \mu(G_{OPT} + C_{OPT}) + W_{OPT} \Rightarrow$$
$$W_{OPT} > \mu(G_{OPT} + C_{OPT}) \Rightarrow W_{ALG} > G_{OPT} + C_{OPT}$$

Note that the requirement for $W_{ALG} > G_{OPT} + C_{OPT}$ is trivial in the context of two-sided markets where the social welfare resulting from unallocated commodities is included as the algorithm can at least gain the social welfare resulting from not allocating any commodities.

For the competitive ratio to be less or equal 1 we need to assume that

$$2\gamma + (-\mu + 2)\delta - \mu - \frac{\mu W_{ALG}}{C_{OPT}} \leq \mu(\gamma - 1) \Rightarrow$$

$$(-\mu + 2)\delta - \mu - \frac{\mu W_{ALG}}{C_{OPT}} \leq \mu(\gamma - 1) - 2\gamma \Rightarrow$$

$$(-\mu + 2)\delta - \frac{\mu W_{ALG}}{C_{OPT}} \leq \mu\gamma - 2\gamma \Rightarrow$$

$$(2 - \mu)\delta - \frac{\mu W_{ALG}}{C_{OPT}} \leq \gamma(\mu - 2)$$

If $\mu \geq 2$ any positive γ will satisfy the above condition. Since we assume non trivial market where $\gamma > 1$ then in this case no additional assumption is needed. If $\mu < 2$ then it has to hold that

$$(2 - \mu)(\delta + \gamma) < \frac{\mu W_{ALG}}{C_{OPT}} = \gamma + \delta \Rightarrow$$

$$2 - \mu < 1 \Rightarrow \mu > 1$$

Note that the above requirement for $\mu > 1$ is natural since $\mu \geq \frac{W_{OPT}}{W_{ALG}}$ and is an approximation factor of a combinatorial problem.

In the case where $W_{ALG} = \frac{1}{\mu}W_{OPT}$ the expression $\mu\frac{W_{ALG}}{C_{OPT}}$ can be simplify to $\gamma + \delta$ and therefore

$$= \frac{2\gamma + (-\mu + 2)\delta - \mu - \gamma - \delta}{\mu(\gamma - 1)}(V_{OPT} - C_{OPT})$$

$$= \frac{\gamma + \delta(-\mu + 1) - \mu}{\mu(\gamma - 1)}(V_{OPT} - C_{OPT})$$

$$= \left(\frac{\gamma + \delta}{\mu(\gamma - 1)} - \frac{\mu(\delta + 1)}{\mu(\gamma - 1)}\right)(V_{OPT} - C_{OPT})$$

$$= \left(\frac{\gamma + \delta}{\mu(\gamma - 1)} - \frac{\delta + 1}{\gamma - 1}\right)(V_{OPT} - C_{OPT})$$

Lemma 1. *The two-sided market mechanism is* $\frac{\gamma+\delta-\frac{\mu G_{ALG}}{C_{OPT}}}{\mu\gamma}$*-competitive with respect to the buying agents' optimal social welfare, i.e.,* $V_{ALG} \geq \frac{1}{\alpha}V_{OPT}$.

Proof (Proof of Lemma 1). From μ definition we know that $\frac{W_{ALG}}{W_{OPT}} \geq \frac{1}{\mu}$ or in other words that $\frac{V_{ALG}+G_{ALG}}{V_{OPT}+G_{OPT}} \geq \frac{1}{\mu}$. Therefore $V_{ALG} \geq \frac{V_{OPT}+G_{OPT}}{\mu} - G_{ALG}$, dividing by V_{OPT} we get that $\frac{V_{ALG}}{V_{OPT}} = \frac{1}{\alpha} \geq \frac{V_{OPT}+G_{OPT}}{\mu V_{OPT}} - \frac{G_{ALG}}{V_{OPT}} = \frac{1}{\mu} + \frac{G_{OPT}-\mu G_{ALG}}{\mu V_{OPT}}$.
By multiplying $\frac{G_{OPT}-\mu G_{ALG}}{\mu V_{OPT}}$ numerator and denominator by $\frac{1}{C_{OPT}}$ we get that
$= \frac{1}{\mu} + \frac{\frac{G_{OPT}}{C_{OPT}}-\frac{\mu G_{ALG}}{C_{OPT}}}{\frac{\mu V_{OPT}}{C_{OPT}}} = \frac{1}{\mu} + \frac{\delta-\frac{\mu G_{ALG}}{C_{OPT}}}{\mu\gamma} = \frac{\gamma+\delta-\frac{\mu G_{ALG}}{C_{OPT}}}{\mu\gamma}$.

Lemma 2. *The two-sided market mechanism is* $1 + \delta(1 - \frac{1}{\mu}) - \frac{\gamma}{\mu} + \frac{V_{ALG}}{C_{OPT}}$*-competitive with respect to the selling agents' optimal social welfare, i.e.,* $C_{ALG} \leq \beta C_{OPT}$.

Proof (Proof of Lemma 2). From μ definition we know that $\frac{W_{ALG}}{W_{OPT}} \geq \frac{1}{\mu}$ or in other words that $\frac{V_{ALG}+G_{ALG}}{V_{OPT}+G_{OPT}} \geq \frac{1}{\mu}$. Therefore $G_{ALG} \geq \frac{V_{OPT}+G_{OPT}}{\mu} - V_{ALG}$, dividing by G_{OPT} we get that

$$\frac{G_{ALG}}{G_{OPT}} =$$

$$\tilde{\beta} \geq \frac{V_{OPT}}{\mu G_{OPT}} + \frac{1}{\mu} - \frac{V_{ALG}}{G_{OPT}} = \frac{1}{\mu} + \frac{V_{OPT} - \mu V_{ALG}}{\mu G_{OPT}} \tag{4}$$

Now we need to convert $\tilde{\beta}$ the approximation ratio of $\frac{G_{ALG}}{G_{OPT}}$ to β the approximation ratio of $\frac{C_{ALG}}{C_{OPT}}$. Or in other words we need to convert the social welfare approximation resulting from the unallocated selling agents' commodities to social welfare approximation resulting from allocated selling agents at the market.
We know from (4) that $1 - \left(1 - \frac{1}{\mu} - \left(\frac{V_{OPT}-\mu V_{ALG}}{\mu G_{OPT}}\right)\right) \leq \tilde{\beta} \leq 1 + \varepsilon$ holds, then we can say for β that

$$1 - \delta\varepsilon \leq \beta \leq 1 + \delta\left(1 - \frac{1}{\mu} - \left(\frac{V_{OPT} - \mu V_{ALG}}{\mu G_{OPT}}\right)\right) \tag{5}$$

It follows from (5) that $\beta \leq 1 + \delta(1 - \frac{1}{\mu}) - \left(\frac{V_{OPT}-\mu V_{ALG}}{\mu C_{OPT}}\right) = 1 + \delta(1 - \frac{1}{\mu}) - \frac{\gamma}{\mu} + \frac{V_{ALG}}{C_{OPT}}$. It is easy to see that in order to keep $\tilde{\beta}$'s approximation quality one needs to assume that $\delta \leq 1$.

4   Experimental Results

We used simulations to empirically study the performance of our conversion formula. We investigated the questions, if and when a mechanism designer can replace the use of gain-from-trade maximizing algorithm with a social-welfare

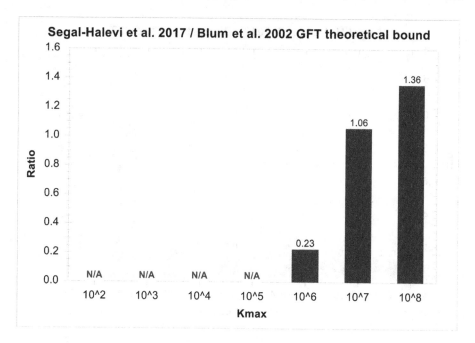

Fig. 1. Segal-Halevi et al. 2017's theoretical bound on gain from trade competitive ratio vs. Blum et al. 2002's converted theoretical bound on gain from trade competitive ratio. When under 10^6 units are traded Segal-Halevi et al. 2017's theoretical bound results in a negative value. For very large markets where over 10^8 units are traded one might consider using Segal-Halevi et al. 2017's solution if no online aspect is required from the market.

maximizing algorithm given the convergence formula. Our simulations involved two types of empirical evaluations. The first evaluation type reflected by Figs. 1, 3, 6 and 7 studies the converted theoretical gain from trade (acronym GFT) approximation bounds on the simulated data as a function of δ and $\frac{1}{K_{\max}}$, where $\frac{1}{K_{\max}}$ is the maximal demand/supply number of units of any commodity by any agent. The second evaluation type reflected by Figs. 2, 4, 5 and 8 shows the actual GFT approximation achieved by the benchmarked algorithms' runs on the same simulated data, shown as a function of δ and $\frac{1}{K_{\max}}$. In order to compute the actual GFT approximation achieved by the benchmarked algorithms we implemented the algorithms in [1–3,5,7,10–12,14–16].

Inputs were generated based on various random distributions and we found minimal to no qualitative difference between distributions. In the figures the uniform distribution was used in the following manner: Agents' costs and values bids were selected as uniformly random independent values between 1 and 10^5. The supply/demand of different commodities the agents hold/desire were also selected as uniformly random independent values between 1 and 5000. For each supplied/demanded commodity we selected the number of units as uni-

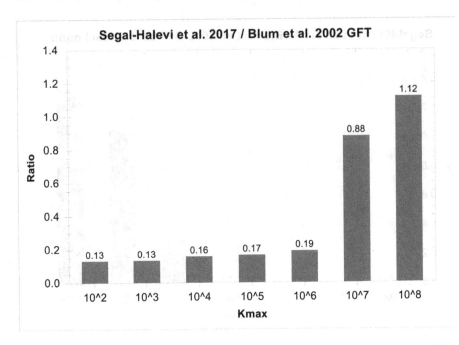

Fig. 2. Segal-Halevi et al. 2017's practical gain from trade competitive ratio vs. Blum et al. 2002's practical gain from trade competitive ratio. Similar to Fig. 1 it can be seen that for very large markets where over 10^8 units are traded one might consider using Segal-Halevi et al. 2017's solution if no online aspect is required from the market.

formly random independent values between 1 and 10^6. All parameters showing in the plots such as δ, K_{max} are histograms based on the instances generated for empirical evaluation.

The literature we compare makes different assumptions and valuation distribution limitations under which their theoretical bounds are guaranteed. In our experiments, for all figures, when comparing two algorithms we construct markets that maintain the distribution assumptions of both algorithms by choosing markets that fulfill the most restrictive assumptions needed by the compared algorithms. It is important to note that some algorithms are more restrictive than others in which case the worst case of one may fall beyond the restriction of the other. For every comparison presented in the paper we also performed a comparison based on the less restrictive algorithm of the two. However, the results were not significantly different and therefore those figures were not included.

The results presented in Figs. 5, 7, 6, and 8 were averaged over 30,000 trials and reflect millions of sellers, buyers and units of each of 5000 commodities. The comparisons were performed on markets with $K_{max} = 1600$. While we could compute the theoretical bounds of all the two-sided markets in the literature using the above magnitude we could not do so with [15]'s theoretical bound as

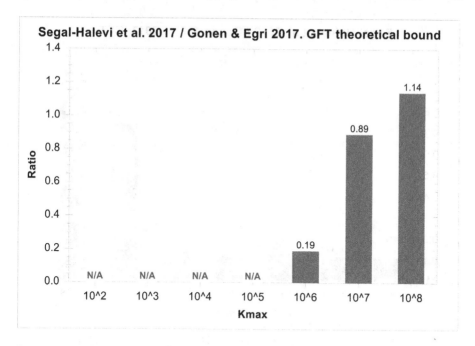

Fig. 3. Segal-Halevi et al. 2017's theoretical bound on GFT competitive ratio vs. Gonen & Egri 2017's converted theoretical bound on GFT competitive ratio. When under 10^6 units are traded Segal-Halevi et al. 2017's theoretical bound results in a negative value. For very large markets, where over 10^8 units are traded, one might consider using Segal-Halevi et al. 2017's solution if no online aspect is required from the market.

it gives a negative (N/A) bound in these cases[5]. Therefore for the comparisons with [15]'s algorithm and bound, presented in Figs. 2, 1, 4, and 3 we used a different setting. Figures 2, 1, 4, and 3 were averaged over 15,000 trails per each column due to their very large size. These markets support millions of sellers, buyers and units but can only run on 4 commodities. The upper bound for a single trade is 0.3, $\delta = 0.5$ and every agent supplies/demands a single unit.

We first empirically evaluate the performance of our formula for converting from a SWF approximation bound to a GFT approximation bound by applying the formula to the known two-sided market mechanisms that provide a bound for the SWF approximation maximization [1,2,5,7,10]. We convert the above five bounds to a GFT approximation bound and compare the resulting bounds with the known two-sided market mechanisms that provide a direct bound on their GFT approximation maximization [11,12,14,15]. The figures comparing the theoretical GFT bound with the formula converted theoretical SWF bound illustrates the formula guaranty, i.e., even when the formula is used in some

---

[5] [15]'s theoretical bound is negative unless markets are very large as the bound is not tight and the algorithm only performs well on very large markets.

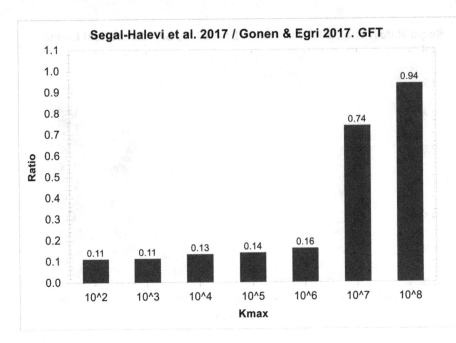

Fig. 4. Segal-Halevi et al. 2017's practical GFT competitive ratio vs. Gonen & Egri 2017's practical GFT competitive ratio. It seems that in practice Gonen & Egri 2017 performs better than Segal-Halevi et al. 2017 despite the fact that Gonen & Egri 2017 provides an online solution as opposed to Segal-Halevi et al. 2017's offline solution and Gonen & Egri 2017 are designed to maximize SWF as opposed to Segal-Halevi et al. 2017's which is designed to maximize GFT.

worst-case scenarios it is better for a designer to consider SWF maximizing algorithm over a GFT one.

Conversion of the bounds is accomplished by simply computing the values of γ and δ for the simulated data, after which the SWF bound μ of the SWF maximizing algorithms [1, 2, 5, 7, 10] is plugged into the formula of Theorem 1 and compared with [11, 12, 14, 15] algorithms' bound on GFT. We show that though the converted bounds are guaranteed to maintain the competitiveness order of the original SWF bounds only if $\delta \leq 1$, the converted bounds perform well even when δ is as large as 5 and most commodities did not change hands and were not sold (see Figs. 6 and 7). This result holds across all converted bounds whether they bound a single-unit single-commodity setting or a combinatorial market setting.

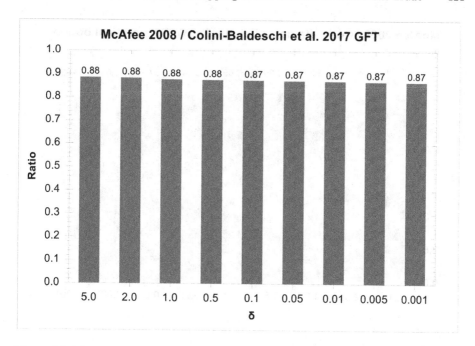

Fig. 5. McAfee 2008's practical GFT competitive ratio vs. Colini-Baldeschi et al. 2017's practical GFT competitive ratio. It appears that in practice Colini-Baldeschi et al. 2017 performs better than McAfee 2008 despite the fact that Colini-Baldeschi et al. 2017 provides a combinatorial solution as opposed to McAfee 2008's single commodity single-unit demand solution and Colini-Baldeschi et al. 2017 is designed to maximize SWF as opposed to McAfee 2008's which is designed to maximize GFT.

It is important to note that for conducting Figs. 1, 3, 6 and 7 one does not need to compute γ directly from an algorithm's run. A bound on γ can be concluded without running an algorithm to compute GFT. This results from much of the current literature assuming a bound on the maximum valuation bid and minimum cost bid from which one can conclude a bound on V_{OPT} and C_{OPT}.

In addition to the theoretical bound comparison we empirically compare the various algorithms in practice using synthetic data to evaluate their relative performance at maximizing GFT. These results are compared to the converted theoretical bounds. Surprisingly, in the practical runs, we found that some of the SWF solutions achieve better GFT than other solutions designed to approximated GFT. More specifically we found that Colini-Baldeschi et al. 2017 [5] that originally approximates SWF achieves a higher GFT than McAfee 2008 [12], which approximates GFT though Colini-Baldeschi et al. 2017, is intended

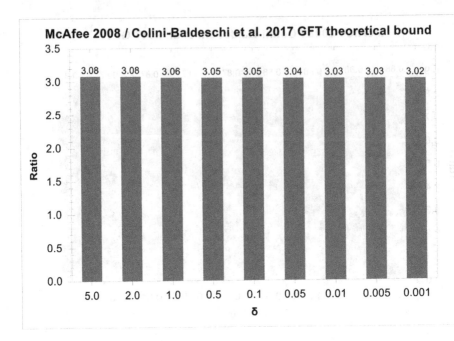

Fig. 6. McAfee 2008's theoretical bound on gain from trade competitive ratio vs. Colini-Baldeschi et al. 2017's converted theoretical bound on gain from trade competitive ratio. The theoretical bounds indicate that McAfee 2008 should perform better than Colini-Baldeschi et al. 2017. However the figure also indicates that the gap in performance might not be very large in particular if one wishes to design a combinatorial market (as Colini-Baldeschi et al. 2017) as opposed to a single-commodity unit-demand mechanism (as McAfee 2008 provides). Another interesting aspect of the figure is the effect of δ on the converted bound. One can see that the relative difference in the converted theoretical bound between a market where most commodities are sold ($\delta = 0.001$) to a market where most commodities are not sold ($\delta = 5$) is less than 2%.

for combinatorial market settings and McAfee 2008 is intended for single commodity and unit-demand settings (see Fig. 5). Another example is the work by Blum et al. [1] that originally approximates SWF in the single-commodity unit-demand setting and performs better at GFT approximation in practice than Segal-Halevi et al. [15] in settings where Segal-Halevi et al.'s algorithm runs instances with unit-demand (see Fig. 2). The above observation is particularly interesting given the fact that Blum et al. is an online algorithm (i.e., computes the SWF optimization function on an ongoing input steam) while Segal-Halevi

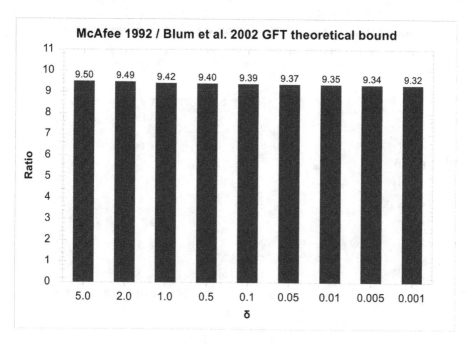

Fig. 7. McAfee 1992's theoretical bound on GFT competitive ratio vs. Blum et al. 2002's converted theoretical bound on GFT competitive ratio. Based on the theoretical bounds McAfee 1992 should perform better than Blum et al. 2002. However, the figure also indicates another interesting aspect which is the effect of δ on the converted bound. One can see that the relative difference in the converted theoretical bound between a market where most commodities are sold ($\delta = 0.001$) to a market where most commodities are not sold ($\delta = 5$) is less than 2%.

et al. is an offline algorithm (that computes the GFT optimization function given all input agents' bids in advance). Gonen and Egri [10] is similarly interesting in that it originally approximates SWF in an online combinatorial market environment and in practice performs better at approximating GFT than Segal-Halevi et al. [15]. This is achieved despite the fact that Segal-Halevi et al. is an offline algorithm (see Fig. 4).

Another interesting observation is that the practical runs appear to show that for combinatorial markets one is better off using the known social-welfare maximizing solutions [5, 10] than using the known GFT maximizing solution [15], even if the designer is interested in maximizing GFT. This observation was made from Figs. 8 and 4 comparing [15] to [5] and [10] which show the social-welfare maximizing solutions consistently outperforming the GFT solutions.

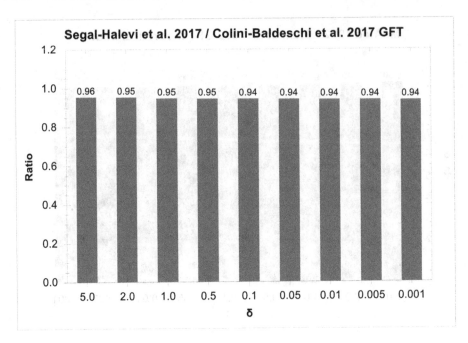

Fig. 8. Segal-Halevi et al. 2017's practical gain from trade competitive ratio vs. Colini-Baldeschi et al. 2017's practical gain from trade competitive ratio. In practice Colini-Baldeschi et al. 2017 performs better than Segal-Halevi et al. 2017 despite the fact that Colini-Baldeschi et al. 2017 is geared to maximize social welfare as opposed to Segal-Halevi et al. 2017's result which is geared to maximize gain from trade.

An interesting aspect of our conversion formula is that it can be used to estimate, without an actual practical run, the practical performance of a SWF maximizing two-sided market compared to a GFT maximizing two-sided market. By converting the SWF bound of a two-sided SWF maximizing algorithm to a GFT maximizing bound and comparing it with another GFT maximizing two-sided market algorithm we found that one can evaluate the practical performance of the two algorithms with respect to GFT maximization. For example see Fig. 3 showing that in the worst case analysis running Gonen and Egri will result in almost the same performance as running Segal-Halevi et al. and indeed in practice (see Fig. 4) Gonen and Egri performs slightly better. We found similar prediction in the case of Blum et al. with respect to Segal-Halevi et al. See Figs. 1 and 2.

A detailed table summarizing the comparative practical performance of the known social-welfare maximizing two-sided markets against the known GFT maximizing two-sided market can be found in Fig. 9.

SWF \ GFT	Segal-Halevi et al. 2017 Off/Comb/Rand				
Segal-Halevi et al. 2017 Off/Comb/Rand	1.45[*]	0.14[*]	3.34	2.24	0.16[*]
Segal-Halevi et al. 2016 Off/Sngl/Detr	16.92	2.27	3.61	2.42	2.21
McAfee 2008 Off/Sngl/Rand	14.54	1.95	3.10	2.08	1.90
McAfee 1992 Off/Sngl/Detr	16.95	2.27	3.61	2.42	2.21
GFT \ **SWF**	**Blumrosen & Dobzinski 2014** Off/Comb/Rand	**Colini-Baldeschi et al. 2016** Off/Sngl/Rand	**Berdin et al. 2007** On/Sngl/Rand	**Wurman et al. 1998** On/Sngl/Rand	**Gonen & Egri 2017** On/Comb/Detr

Off : offline , **On**: online
Comb: combinatorial market , **Sngl** : Single good market
Rand: algorithm is random , **Detr** : deterministic algorithm
[*] Were run in markets where Segal-Halevi et al. 2017 has a valid theoretical bound.

Fig. 9. This table demonstrates the practical performance of most known two-side market mechanisms with respect to GFT. The results suggest that it is better for a two-sided market designer who wishes to maximize GFT to use algorithms that directly maximize GFT. However, if the designer wishes to design a GFT maximizing two-sided combinatorial market than he/she should consider using a SWF maximizing algorithm as opposed to an algorithm that directly maximizes GFT.

5   Conclusion and Discussion

In this paper we provided and proved the first formula to convert a SWF approximation bound for two-sided markets into a bound on a GFT approximation that maintains the original order of approximation. This conversion makes it possible to compare solutions that approximate GFT with those that approximate SWF. We evaluate the performance of our conversion formula by using it to convert SWF approximation solutions in the literature to GFT approximation solutions.

The experimental results showed that our conversion formula can be used to estimate (without an actual practical run) the practical performance (in GFT) of a social-welfare maximizing two-sided market compared to a (directly computed) GFT maximizing two-sided market.

We found that in some cases the solutions designed for SWF maximization perform better at maximizing GFT than algorithms designed for directly maximizing GFT. This is true in particular in the case of combinatorial markets where most known social-welfare maximizing solutions consistently perform better at maximizing GFT than the known combinatorial GFT (directly) optimizing solution.

We would like to see our conversion formula used in future work in multi-sided market design as a means for evaluation and comparison of new solutions to existing ones.

References

1. Blum, A., Sandholm, T., Zinkevich, M.: Online algorithms for market clearing. In: SODA, pp. 971–980 (2002)
2. Blumrosen, L., Dobzinski, S.: Reallocation mechanisms. In: EC, pp. 617–640 (2014)
3. Bredin, J., Parkes, D., Duong, Q.: Chain: a dynamic double auction framework for matching patient agents. J. Artif. Intell. Res. **30**, 133–179 (2007)
4. Brustle, J., Cai, Y., Wu, F., Zhao, M.: Approximating gains from trade in two-sided markets via simple mechanisms. In: Proceedings of the 18th ACM Conference on Economics and Computation, EC, pp. 589–590 (2017)
5. Colini-Baldeschi, R., Goldberg, P., de Keijzer, B., Leonardi, S., Roughgarden, T., Turchetta, S.: Approximately efficient two-sided combinatorial auctions. In: EC, pp. 591–608 (2017)
6. Colini-Baldeschi, R., Goldberg, P., de Keijzer, B., Leonardi, S., Turchetta, S.: Fixed price approximability of the optimal gain from trade. In: Devanur, N.R., Lu, P. (eds.) WINE 2017. LNCS, vol. 10660, pp. 146–160. Springer, Cham (2017). https://doi.org/10.1007/978-3-319-71924-5_11
7. Colini-Baldeschi, R., de Keijzer, B., Leonardi, S., Turchetta, S.: Approximately efficient double auctions with strong budget balance. In: SODA, pp. 1424–1443 (2016)
8. Feldman, M., Frim, G., Gonen, R.: Multi-sided advertising markets: dynamic mechanisms and incremental user compensations. In: Bushnell, L., Poovendran, R., Başar, T. (eds.) GameSec 2018. LNCS, vol. 11199, pp. 227–247. Springer, Cham (2018). https://doi.org/10.1007/978-3-030-01554-1_13
9. Feldman, M., Gonen, R.: Removal and threshold pricing: truthful two-sided markets with multi-dimensional participants. In: Deng, X. (ed.) SAGT 2018. LNCS, vol. 11059, pp. 163–175. Springer, Cham (2018). https://doi.org/10.1007/978-3-319-99660-8_15
10. Gonen, R., Egri, O.: DYCOM: a dynamic truthful budget balanced double-sided combinatorial market. In: Proceedings of the 16th Conference on Autonomous Agents and MultiAgent Systems, AAMAS 2017, São Paulo, Brazil, 8–12 May 2017, pp. 1556–1558 (2017)
11. McAfee, R.P.: A dominant strategy double auction. J. Econ. Theory **56**, 434–450 (1992)
12. McAfee, R.P.: The gains from trade under fixed price mechanisms. Appl. Econ. Res. Bull. **1**, 1–10 (2008)
13. Myerson, R.B., Satterthwaite, M.A.: Efficient mechanisms for bilateral trading. J. Econ. Theory **29**, 265–281 (1983)
14. Segal-Halevi, E., Hassidim, A., Aumann, Y.: SBBA: a strongly-budget-balanced double-auction mechanism. In: Gairing, M., Savani, R. (eds.) SAGT 2016. LNCS, vol. 9928, pp. 260–272. Springer, Heidelberg (2016). https://doi.org/10.1007/978-3-662-53354-3_21
15. Segal-Halevi, E., Hassidim, A., Aumann, Y.: MUDA: a truthful multi-unit double-auction mechanism. In: Proceedings of AAAI (2018)
16. Wurman, P., Walsh, W., Wellman, M.: Flexible double auctions for electronic commerce: theory and implementation. Decis. Support Syst. **24**, 17–27 (1998)

Affective Decision-Making in Ultimatum Game: Responder Case

Jitka Homolová$^{(\boxtimes)}$, Anastasija Černecka, Tatiana V. Guy, and Miroslav Kárný

The Czech Academy of Sciences, Institute of Information Theory and Automation,
Adaptive System Department, P.O. Box 18, 182 08 Prague 8, Czech Republic
{jkratoch,guy,school}@utia.cas.cz, chernetskaan@gmail.com

Abstract. The paper focuses on *prescriptive* affective decision making in Ultimatum Game (UG). It describes preliminary results on incorporating emotional aspects into *normative* decision making. One of the players (responder) is modelled via Markov decision process. The responder's reward function is the weighted combination of two components: economic and emotional. The first component represents pure monetary profit while the second one reflects overall emotional state of the responder. The proposed model is tested on simulated data.

Keywords: Decision making · Emotions in economic game ·
Markov decision process · Ultimatum Game

1 Introduction

Empirical evidence repeatedly confirms that emotions and mood significantly influence human decision making. Systematic deviation of human decisions from the economically plausible (rational) predictions has motivated broad research in this direction. Though many interesting experimental results and observations have been obtained, [3,15] the standard economic theory still considers a purely rational decision maker. Neither the role of the moods in human strategic decision making is fully understood yet, nor clear vision of how the this human feature could be incorporated into normative decision making is made.

Extensive experimental work, see for instance [7,9,10,14], revealed different patterns of affective human behaviour, that can make normative DM theory more human-like. However these descriptive patterns cannot be directly used in normative decision making. The present paper makes a step towards challenging this state, formalises some of the patterns reported [3,11] and use them in prescriptive manner. Unlike game theory approach we consider the distributed, responder-centric setup. The main aim thus was to test the normative optimised responder model enriched by descriptive findings.

Supported by GA16-09848S, LTC18075 and EU-COST Action CA16228.

© Springer Nature Switzerland AG 2019
M. Slavkovik (Ed.): EUMAS 2018, LNAI 11450, pp. 127–139, 2019.
https://doi.org/10.1007/978-3-030-14174-5_9

We consider *expected emotions*, i.e. emotions not available at the moment of deciding but occurring as additional outcomes associated with different decisions made. The expected emotions can be naturally incorporated into the reward function. Then the expected reward will depend on the predicted emotional state of the decision maker. Inclusion of the expected emotional state in the reward seems to be consistent with the empirical evidence that human decision maker aims on improving his emotional state [4,10].

There is extensive research on how decision results affect human feelings, for instance, [12], though influence of emotions on decisions is more important for practical applications but less studies. Papers [6] and [14] focus on fairness and include emotions only *indirectly* in fairness elements. The work [5] contains an emotional parameter which reflects status and reciprocity between adversaries. In utility function, the parameter multiplies monetary profit of an opponent so that utility increases with increasing emotion. The emotional parameter is to be determined experimentally and could be positive as well as negative. Although this model deals with so called *emotional state function*, emotions included are very specific and related only to relationship between the players.

A model that is more closely related to emotions was introduced by [17]. Their model is based on two psychological theories. This approach has good foundation, nevertheless, it has some limitations (in particular, limited ability to predict human players) [9]. Another shortcoming of the approach is that initial player's emotions are not considered. All mentioned approaches give many important insights into human decision making and provide solution of many subtasks, but none of them systematically solves affective DM.

In this paper we include an emotional component to the reward function of an optimizing responder and, based on empirical evidence, proposed a model of emotion development. The emotional state of the responder is influenced by the course of the game and vice versa. In our study we introduce 5 emotional states: 1 is the least positive and 5 is the most positive influence. The proposed model does not distinguish different qualitative types of emotions but only discrete states of them from the worst to the best reflecting their effect on DM. These states are the resulting emotional state given by a combination of different basic emotions, which thus reflects the final current emotional state of the player.

The paper outline is as follows. Section 2 introduces necessary definitions and mathematical apparatus. Section 3 describes the UG game as MDP, mainly the model of the responder, introduces the reward function and the model of the responder's emotional state development, as well as derives the corresponding DM strategy based on MDP and dynamic programming formalism. Section 4 describes the experimental part of this paper including the detailed description of all individual types of responders. The paper is concluded by a summary of findings and description of the main open problems in Sect. 5.

2 Preliminaries

2.1 General Notation

x_t — value of random variable x at discrete time t;

$p_t(x|y)$ — conditional probability density function of a random variable x conditioned on random variable y, known at discrete time t;

$E[x], E[x|y]$ — expectation of random variable x and conditional expectation of x conditioned on random variable y;

$\delta(x, y)$ — Kronecker delta function that equals 1 for $x = y$ and 0 otherwise.

2.2 Ultimatum Game Rules

The Ultimatum Game (UG) [8] is an economic game for two players: *proposer* and *responder*. The proposer offers how to divide an amount of money among players. The responder can either accept or reject the offer. If the responder accepts the amount is divided in accordance with the offer, otherwise both players get nothing. The DM aim of each player is to maximize his/her own profit.

This game simulates many common trade situations in real life: a seller of goods and a buyer have roles of the proposer and the responder respectively. The seller proposes the price and the buyer decides whether the deal will take place.

2.3 Markov Decision Process

The paper describes the responder in UG via Markov decision process (MDP) [13] where the proposer plays role of the system. A decision maker chooses action a_t at decision epoch $t \in \mathbf{T}$ based on observed state s_t that evolves probabilistically and depends on state s_{t-1} and action a_{t-1} only[1].

At each decision epoch t, the system stands at state s_t. Then action a_t is realized and the new system state s_{t+1} is determined stochastically by transition probability $p_t(s_{t+1}|a_t, s_t)$. After this transition, the decision-maker gains reward $r_t(s_{t+1}, a_t, s_t)$. Its value indicates a degree of reaching his/her DM preferences.

Formally, MDP is determined by the following definition. It assumes that state s_{t+1} is fully observable after choosing action a_t.

Definition 1 *(Markov decision process). The discrete-time* **Markov decision process** *is defined as a 5-tuple* $\{\mathbf{T}, \mathbf{S}, \mathbf{A}, p, r\}$*, where* $\mathbf{T}$ *denotes a discrete finite set of decision epochs;* $\mathbf{T} = \{1, 2, ..., N\}$*,* $N \in \mathbb{N}$*,* $\mathbf{S}$ *is a discrete finite set of states,* $\mathbf{S} = \cup_{t \in \mathbf{T}} \mathbf{S}_t$*,* $\mathbf{S}_t$ *is a set of possible states and* $s_t \in \mathbf{S}_t \subset \mathbf{S}$ *is the state at decision epoch* $t \in \mathbf{T}$*;* $\mathbf{A}$ *denotes a discrete finite set of actions,* $\mathbf{A} = \cup_{t \in \mathbf{T}} \mathbf{A}_t$ *is a set of possible actions and* $a_t \in \mathbf{A}_t \subset \mathbf{A}$ *is the action chosen at decision epoch* $t \in \mathbf{T}$*. The function* p_t *represents a transition probability function known at decision epoch* t*. The probability function* $p_t(s_{t+1}|a_t, s_t)$ *is a probability that action* $a_t \in \mathbf{A}_t$ *changes state* $s_t \in \mathbf{S}_t$ *into state* $s_{t+1} \in \mathbf{S}_{t+1}$ *satisfying the*

---

[1] This is so-called *Markov assumption*.

condition $\sum_{s_{t+1} \in \mathbf{S}_{t+1}} p_t(s_{t+1}|a_t, s_t) = 1, \forall t \in \mathbf{T}, \forall s_t \in \mathbf{S}_t, \forall a_t \in \mathbf{A}_t.$ *Finally,*
$r_t : \mathbf{S}_{t+1} \times \mathbf{A}_t \times \mathbf{S}_t \longrightarrow \mathbb{R}$ *stands for a reward function that is received after taking action a_t and transiting from state s_t into state s_{t+1}, $r_t = r_t(s_{t+1}, a_t, s_t)$.*

Further on, time-invariant reward function $r_t(s_{t+1}, a_t, s_t) = r(s_{t+1}, a_t, s_t)$ is considered. The solution to MDP (**Optimal DM policy**) is a sequence of optimal DM rules $\left\{ p_t^{opt}(a_\tau|s_\tau) \right\}_{\tau=t}^{t+T-1}$ that maximizes the expected total reward:

$$\pi_{t,T}^{opt} \in \arg\max_{\pi_{t,T} \in \Pi_t} E \left[\sum_{\tau=t}^{t+T-1} r(s_{\tau+1}, a_\tau, s_\tau) \mid s_t \right]. \tag{1}$$

The *expected reward function* is defined as:

$$E_t \left[r(s_{t+1}, a_t, s_t)|s_t \right] = \sum_{s_{t+1} \in \mathbf{S}, a_t \in \mathbf{A}} r(s_{t+1}, a_t, s_t) \cdot p_t(s_{t+1}, a_t|s_t), \tag{2}$$

where

$$p_t(s_{t+1}, a_t|s_t) = p_t(s_{t+1}|s_t, a_t) \cdot p_t(a_t|s_t). \tag{3}$$

To compute an optimal policy (1), the first factor in (3) is needed. In the inspected UG, it is a model of the proposer.

3   Ultimatum Game as Markov Decision Process: Responder's Strategy

The paper primary aims at incorporating responder's *expected emotional states* [11] into the reward function in order to reflect empirical patterns [3] how emotions influence responder's economic behaviour. The responder intends not only to maximize his/her monetary profit, but to retain the best achievable emotional state or to improve it at least.

Let us emphasise that qualitative characteristics of emotional state (joy, anger, relief, ...) have no importance here. Let us consider emotional state appearing after decision made and reward is obtained. If the original responder's preferences or expectations are fulfilled the emotional state becomes more positive (increases), otherwise it becomes more negative (decreases). So we can say that the emotional state reflects degree of plausibility of the game outcomes for the responder. To formalize that we introduce a completely ordered set of possible emotional states $m_t \in \mathbf{M} = \{m_{min}, ..., m_{max}\} \in \mathbb{N}$, where m_{min} denotes the least positive state and m_{max} is the most positive emotional state.

3.1   Model of the Responder

The considered UG is treated as an N-round repeated game and decision epochs $t \in \mathbf{T} = \{1, 2, ..., N\}$ correspond to these game rounds. There is an constant amount of money $q \in \mathbb{N}$ to be divided in each round of the game. The proposer forms the responder's system and as such is modelled.

Definition 2 *(UG as MDP of the responder). The MDP for proposed UG is modeled by* $\{\mathbf{T}, \mathbf{S}, \mathbf{A}, p, r\}$, *see Definition 1, where* $\mathbf{A} = \{1, 2\}$ *is a set of all possible actions of the responder, where* $a_t = 1$ *is the rejection and* $a_t = 2$ *the acceptance of the actual proposer's offer;* $s_t = (o_t, m_t) \in \mathbb{S} = \mathbf{O} \times \mathbf{M} \subset \mathbb{N}^2$ *is a state of the system in round* $t \in \mathbf{T}$; $o_t \in \mathbf{O} = \{o_{min}, ..., o_{max}\} \in \mathbb{N}$ *denotes a proposer's offer,* $0 < o_{min}, o_{max} < q$; $m_t \in \mathbf{M} = \{m_{min}, ..., m_{max}\} \in \mathbb{N}$ *represents an emotional state of the responder. The corresponding reward of the responder is defined by*

$$r(s_{t+1}, a_t, s_t) = (1 - \omega) \cdot (a_t - 1) \cdot o_t + \omega \cdot \frac{q - o_t}{q - 1} \cdot m_{t+1}, \qquad (4)$$

where $\omega \in [0, 1]$ *is a weight reflecting balance between importance of emotional and economic components in the responder's reward.*

The first part of the reward function formula describes an economic profit of the game round. The total economic profit of the responder after $t \in \mathbf{T}$ rounds is a sum of the actual profits at each round, i.e. $z_R(t) = \sum_{i=1}^{t} z_i = \sum_{i=1}^{t} (a_i - 1) \cdot o_i$.

Model of the Emotional State

The second part of the reward (4) is a responder's "emotional" profit where m_{t+1} is a deterministic, dynamically changing emotional state. It depends on action a_t and proposer's offer o_t as follows:

$$m_{t+1}(a_t, o_t, m_t) = \begin{cases} \min\{m_t + \chi_t(a_t, o_t, m_t), m_{max}\}, & \chi_t(a_t, o_t, m_t) \geq 0, \\ \max\{m_t + \chi_t(a_t, o_t, m_t), m_{min}\}, & \chi_t(a_t, o_t, m_t) < 0, \end{cases} \qquad (5)$$

where

$$\chi_t(a_t, o_t, m_t) = \begin{cases} -2, & \text{if } a_t = 2 \wedge o_t \in [o_{min}, p_o) \wedge m_t \in [p_m, m_{max}] \\ -1, & \text{if } a_t = 1 \\ 0, & \text{if } a_t = 2 \wedge o_t \in [o_{min}, p_o) \wedge m_t \in [m_{min}, p_m] \\ 1, & \text{if } a_t = 2 \wedge o_t \in [p_o, o_{max}] \end{cases} \qquad (6)$$

Parameters $p_o \in (o_{min}, o_{max})$ and $p_m \in (m_{min}, m_{max})$ are specific to a given responder. They represent threshold values of the offers and the emotional states. Changing the parameters causes changing the responder's behaviour.

The emotional evolution scenario is as follows. The responder's emotional state (5) is supposed to be dependent on offer o_t and current emotional state m_t. Any rejection implies decrease the emotional state, $\chi = -1$ in (6).

Three cases are possible when the responder accepts the offer $o_t \in [o_{min}, o_{max}]$. If the offer is higher than the responder's 'personal' threshold $p_o \in (o_{min}, o_{max})$, the responder's emotional state increases, $\chi = 1$ in (6). This thresholds determine the "emotional stability point".

If the offer is low $o_t < p_o$, the change of the responder's emotional state depends on the current value m_t. Once the current state is too low, $m_t \leq p_m$, it will remain the same, $\chi = 0$ in (6). Parameter $p_m \in (m_{min}, m_{max})$ determines the proposer's "emotional tolerance point". If the current emotional state is more positive, $m_t \geq p_m$, it naturally worsens, $\chi = -2$ in (6) with the acceptance of the low-valued offer.

The emotional model has been partially inspired by the results reported in [15] and [3]. In particular that *the intensity of negative (positive) emotions is negatively (positively) related to the offer*; and that *the probability of rejection depends positively on the intensity of bad emotions*.

3.2   Dynamic Programming in UG

The responder chooses action $a_t \in A$ based on the *randomized DM rule* $p_t(a_t|s_t)$ in each decision epoch $t \in \mathbf{T}$. The DM rule is a non-negative function representing probability of action a_t in state $s_t \in \mathbf{S}$. The responder searches for an optimal *DM policy* $\pi_{t,T}$ (a sequence of DM rules mapping states to actions) maximizing expected reward (4) over some horizon T through the following algorithm:

$$
a_\tau^{opt} \in \arg\max_{a_\tau \in \{1,2\}} E\Big[r(m_{\tau+1}, a_\tau, o_\tau) + V_{k-1}^{opt}(o_{\tau+1}, m_{\tau+1}) \mid a_\tau, o_\tau, m_\tau\Big]
$$

$$
= \arg\max_{a_\tau \in \{1,2\}} E\Big[[(1-\omega)\cdot(a_\tau - 1)\cdot o_\tau + \omega \cdot \frac{q - o_t}{q-1} \cdot m_{\tau+1}(a_\tau, o_\tau, m_\tau)]
$$

$$
+ V_{k-1}^{opt}(o_{\tau+1}, m_{\tau+1}) \mid a_\tau, o_\tau, m_\tau\Big], \tag{7}
$$

where

$$
V_k^{opt}(o_\tau, m_\tau) = E\Big[r(m_{\tau+1}, a_\tau^{opt}, o_\tau) + V_{k-1}^{opt}(o_{\tau+1}, m_{\tau+1}) \mid a_\tau^{opt}, o_\tau, m_\tau\Big]
$$

$$
= E\Big[[(1-\omega)\cdot(a_\tau^{opt} - 1)\cdot o_\tau + \omega \cdot \frac{q - o_t}{q-1} \cdot m_{\tau+1}(a_\tau^{opt}, o_\tau, m_\tau)]
$$

$$
+ V_{k-1}^{opt}(o_{\tau+1}, m_{\tau+1}) \mid a_\tau^{opt}, o_\tau, m_\tau\Big], \tag{8}
$$

$$
V_0(o_{t+T}, m_{t+T}) = V_0^{opt}(o_{t+T}, m_{t+T}) = 0, k = t + T - \tau.
$$

Finally, the decision policy over horizon T is built by a sequence of resulting decision rules $\forall \tau \in \{t, ..., k\}$:

$$
p_\tau^{opt}(a_\tau | m_\tau, o_\tau) = \delta(a_\tau, a_\tau^{opt}(m_\tau, o_\tau)). \tag{9}
$$

4   Illustrative Experiments

The following sections describe the performed experiments, initialize necessary constants and values and summarize the results obtained. The simulations were performed by using MATLAB® software [1].

4.1 Models of Proposer

To compute an optimal policy of the responder (1) we need the proposer's model (the first factor in (3)). Several types of proposers considered in the paper belong to the two main groups: so-called "open loop" proposers and "closed loop" proposers. The proposer belongs to a particular group depending on whether they take into account the responder's feedback, i.e. acceptance/rejection decision.

Open Loop. These proposers do not respect the responder's previous action, i.e. $p(o_{t+1}|a_t, o_t) = p(o_{t+1}|o_t)$, $\forall t \in \mathbf{T}$, so the fixed testing sequence of the offers can be generated off-time. This significantly simplifies computations, however, a feedback from the responder is neglected while the next offer is generated. The following algorithm models an open-loop proposer:

$$p_t(o_{t+1}|a_t, o_t) \propto \begin{cases} \exp\left(-\dfrac{\left(o_{t+1}-(o_t+p)\right)^2}{2\sigma^2}\right), & \forall o_{t+1} \in \mathbf{O}, \text{if } o_t + p \in [b_l, b_u], \\ \exp\left(-\dfrac{\left(o_{t+1}-o_t\right)^2}{2\sigma^2}\right), & \forall o_{t+1} \in \mathbf{O}, \text{otherwise}, \end{cases}$$

$$(10)$$

where o_t is an offer in game round t and p is a random variable from set $\{-1, 1\}$. Parameters b_l and b_u are lower and upper bounds. The constant σ is a standard deviation of normal distribution.

We distinguish the three subtypes of the proposer: "greedy", "neutral" and "generous", each subtype being specified by parameters b_l and b_u. The "greedy" proposer suggests very low and unfair offers. The "neutral" proposer's offers are more fair and offers of the "generous" proposer are very "generous" to the responder but unfair to himself/herself.

Closed Loop. The more realistic proposer follows the most simple and intuitive algorithm: if the previous round is successful, the proposer tries to make more money in the next round by decreasing the offer. In case of rejection, the proposer increases the offer to make it successful.

Virtually, this proposer increases his/her next offer o_{t+1} by one if previous offer o_t has been accepted, and decreases the next offer by one otherwise:

$$p_t(o_{t+1}|a_t, o_t) \propto \begin{cases} \exp\left(-\dfrac{\left(o_{t+1}-(o_t+1)\right)^2}{2\sigma^2}\right), & \forall o_{t+1} \in \mathbf{O}, \text{if } a_t = 1 \\ \exp\left(-\dfrac{\left(o_{t+1}-(o_t-1)\right)^2}{2\sigma^2}\right), & \forall o_{t+1} \in \mathbf{O}, \text{if } a_t = 2. \end{cases}$$

$$(11)$$

The described model is useful for preliminary testing, however, it does not correspond human thinking as it does not discriminate extremely low or high offers and does not respect fairness as humans do [2].

4.2 Experiment Setup

For each game, the number of game rounds is preset to $N = 30$ and the amount to split is $q = 10$ CZK. The proposer's offers vary from 1 to 9, $o_t \in \mathbf{O} = \{1, 2, ..., 9\}$, and the emotional states from 1 to 5, $m_t \in \mathbf{M} = \{1, 2, 3, 4, 5\}$, where 1−worst and 5−best. In relation to Sect. 3, it holds $o_{min} = 1$, $o_{max} = 9$, $m_{min} = 1$ and $m_{max} = 5$. The proposer is non-optimizing with a pre-defined DM algorithm, while the responder uses T-step optimization with time horizon $T = 10$ as described in Subsect. 3.2. For all experiments, the personal parameters of the emotional state model (5) are set to $p_m = 3$ and $p_o = 6$. The parameter σ used in models of the proposer (see Subsect. 4.1) is always set to $\sigma = 1$.

The simulations are carried out for two different types of the proposer, see Subsect. 4.1. The "open loop" proposer is additionally represented by three subtypes differing in lower and upper bounds of the chosen offers. The specific values are: $b_l = 1$ and $b_u = 3$ for "greedy" proposer, $b_l = 3$ and $b_u = 6$ for the "neutral" proposer and $b_l = 6$ and $b_u = 9$ for the "generous" one.

Each experiment is performed for five different values of the weight $\omega \in \{0, 0.2, 0.4, 0.6, 0.8\}$, combined with different initial emotional states $m_1 \in \mathbf{M} = \{1, 2, 3, 4, 5\}$. Terminal emotional state m_N, total profits of players $z_R(N)$ and success rate of the game (i.e. percentage of successful rounds) are monitored.

4.3 Results

The results for each combination of the proposer type, weight ω and initial emotional state m_1 were calculated as average of results from 100 Monte Carlo simulations.

Open Loop Results. The proposer's offers were sampled from the transition probability (10). The used bounds b_l, b_u corresponded to individual types of the proposers, see Sect. 4.2. The sequences of offers generated were fixed for all performed simulations.

Greedy Proposer. The "greedy" proposer makes extremely unfair offers. It could be expected that the responder would reject the majority of them. However, the responder mostly accepts. The average results, see Table 1, show "unfair" profits corresponding to the offers. The best results correspond to purely rational responder, i.e. without emotional component in the reward (4). Adding the emotional component worsens monetary profit but improves the final emotional state. Noticeably this is valid for all initial moods except of $m_1 = 1$. In the last case the responder accepts everything and does not change this strategy with growing ω.

Neutral Proposer. The offers of the "neutral" proposer are more fair than the offers of the "greedy" proposer. It could be expected that the success rates would be generally higher in these simulations. However, such trend is not visible in our results, see Table 2. The small values of the weight $\omega = (0; 0.2)$ imply the

Table 1. Game with the optimizing responder and the "greedy" proposer

Weight	0	0.2	0.4	0.6	0.8	1
Initial mood $m_1 = 1$						
Final mood of responder	1	1	1	1	1	1
Profit of responder (in CZK)	86	86	86	86	86	0
Profit of proposer (in CZK)	214	214	214	214	214	0
Success Rate (in %)	100	100	100	100	100	0
Initial mood $m_1 = 3$						
Final mood of responder	1	1.76	2	2	2	2
Profit of responder (in CZK)	86	84.48	84	84	84	84
Profit of proposer (in CZK)	214	207.92	206	206	206	206
Success Rate (in %)	100	97.47	96.67	96.67	96.67	96.67
Initial mood $m_1 = 5$						
Final mood of responder	1	1.75	2	2	2	2
Profit of responder (in CZK)	86	84.5	84	84	82	80
Profit of proposer (in CZK)	214	208	206	206	198	190
Success Rate (in %)	100	97.5	96.67	96.67	93.33	90

responder strategy to be close to the purely rational one (accept every offer). Increasing ω changes this and decreases total profit of the responder. The terminal emotional state is equal to 5 for any initial setting. This phenomenon is due to the fair offers sufficient to maximize the responder's emotional state during 30 game rounds. Therefore the influence of the initial emotional state on the success rate is not noticeable with the "neutral" proposer. This confirms again the correspondence of the previous simulations for the "greedy" proposer [16] because the offers are fair in that case.

Generous Proposer. The offers of this proposer are very high, so the game of 30 game rounds improves any initial emotional state to its maximum value 5, see Table 3. The success rates are the highest from all open-loop experiments as it can be expected.

The results show the clear positive correlation between the monitored quantities and the weight ω as in the previous two tests with "greedy" and "neutral" proposers. The success rate and the total economic profits depend on the weight. A dependence of the monitored variables on the initial emotional state is not noticeable on the overall results.

Closed Loop Proposer. The proposer dynamically updating his strategy is considered, see Sect. 4.1. The offers of this proposer *stochastically* depend on the previous offers and the responder's decisions, i.e. the newly generated offer respects whether the previous offer was accepted or rejected. The transition probabilities are computed by the model (11).

Table 2. Game with the optimizing responder and the "neutral" proposer

Weight	0	0.2	0.4	0.6	0.8	1
Initial mood $m_1 = 1$						
Final mood of responder	5	5	5	5	5	5
Profit of responder (in CZK)	172	172	169.98	166.3	153.9	136
Profit of proposer (in CZK)	128	128	125.22	120.3	105.9	84
Success Rate (in %)	100	100	98.4	95.53	86.6	73.33
Initial mood $m_1 = 3$						
Final mood of responder	5	5	5	5	5	5
Profit of responder (in CZK)	172	172	170.22	164.65	148.85	136
Profit of proposer (in CZK)	128	128	125.48	118.75	100.85	84
Success Rate (in %)	100	100	98.57	94.47	83.23	73.33
Initial mood $m_1 = 5$						
Final mood of responder	5	5	5	5	5	5
Profit of responder (in CZK)	172	172	170.13	166.35	148.95	136
Profit of proposer (in CZK)	128	128	125.27	120.45	100.95	84
Success Rate (in %)	100	100	98.47	95.60	83.30	73.33

Table 3. Game with the optimizing responder and the "generous" proposer

Weight	0	0.2	0.4	0.6	0.8	1
Initial mood $m_1 = 1$						
Final mood of responder	5	5	5	5	5	5
Profit of responder (in CZK)	205	205	205	202.7	195	180
Profit of proposer (in CZK)	95	95	95	92.7	85	70
Success Rate (in %)	100	100	100	98.47	93.33	83.33
Initial mood $m_1 = 3$						
Final mood of responder	5	5	5	5	5	5
Profit of responder (in CZK)	205	205	205	202.3	195	180
Profit of proposer (in CZK)	95	95	95	92.3	85	70
Success Rate (in %)	100	100	100	98.2	93.33	83.33
Initial mood $m_1 = 5$						
Final mood of responder	5	5	5	5	5	5
Profit of responder (in CZK)	205	205	205	202.3	195	180
Profit of proposer (in CZK)	95	95	95	92.3	85	70
Success Rate (in %)	100	100	100	98.2	93.33	83.33

Again the best results correspond to the purely economical responder. Increasing ω causes the emotional component of the reward function (4) becomes more significant than the economic component. The responder's behavior is than influenced by the emotional state function (6) which values decrease with each rejection. Importantly that starting from $\omega = 0.2$ the average profit and final emotional state grow with growing *omega*. Then they start to decrease for high values ω. The acceptance rate shows opposite tendency. This behavior implies existence of 'optimal' combination of the economic and emotional components in the reward function (4). This optimal combination provides the best results.

Dependence on the initial emotional state is not so significant. There is slight decrease of the acceptance once the initial emotional state is high. The results obtained partially confirm the results of research [16]: the correlation between initial emotional state and success rate exists only in the case of unfair offers (Table 4).

Table 4. Game with the optimizing responder and the closed-loop proposer

Weight	0	0.2	0.4	0.6	0.8	1
Initial mood $m_1 = 1$						
Final mood of responder	3.22	2.85	3.54	3.58	2.22	2.03
Profit of responder (in CZK)	120.1	107.91	104.54	104.93	66.98	60.2
Profit of proposer (in CZK)	36.9	21.99	32.46	42.27	188.32	213.3
Success Rate (in %)	52.33	43.3	45.67	49.07	85.1	91.17
Initial mood $m_1 = 3$						
Final mood of responder	2.94	2.75	3.61	3.19	2.24	2.27
Profit of responder (in CZK)	113.62	106.49	105.93	95.06	68.89	67.7
Profit of proposer (in CZK)	36.68	21.61	34.37	46.14	178.61	193
Success Rate (in %)	50.1	42.7	46.77	47.07	82.5	86.9
Initial mood $m_1 = 5$						
Final mood of responder	2.76	2.85	3.42	3.5	2.45	2.26
Profit of responder (in CZK)	113.77	107.33	104.6	101.44	77.92	76.64
Profit of proposer (in CZK)	37.03	21.67	36	45.56	140.68	167.76
Success Rate (in %)	50.27	43	46.87	49	72.87	81.47

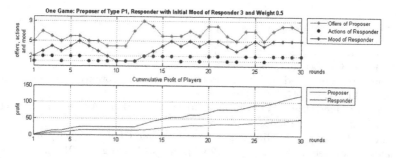

Fig. 1. One game with proposer of type P1

5   Conclusions

The paper focuses on *prescriptive* affective decision making in Ultimatum Game (UG). It describes preliminary results on explicit incorporating expected emotion into *normative* decision making. The responder is modelled via Markov decision process with the reward function be the weighted combination of two components: economic and emotional. The T-step-ahead optimizing policy for the responder in UG has been developed and tested on simulated data. For simulation several types of the proposer were developed and implemented.

Two main groups of experiments imitated two different types of trading: (i) *open-loop* when the decision maker does not influence the price and decides only on buying/not buying (this models, for instance, the case of trading with future) and (ii) *closed-loop* when the decision maker indirectly influences the price (this models, for instance, trading with goods and services).

Affective DM strategy of the responder worsens the economic results, but improves emotional state. The proposed model of the emotional development is very intuitive and reasonably corresponds with the reported empirical evidence [3,4,15]. Based on the results obtained the number of emotional states can be further extended to 12 levels, see [18].

The proposed reward function includes the parameter ω balancing the monetary profit and emotional state. The results show the existence of "optimal" weight balancing the compromised between emotional and monetary profit in the reward function.

Inevitable direction of further research is creating a *learning* adaptive responder. Currently, the responder obtains the model of the proposer. This can be interpreted as that the responder already knows usual behavior of the co-player. However, this is not a realistic assumption and the learning ability should be added. This can be done by using Bayesian learning suitable for a very short learning periods.

Acknowledgement. We would like to thank Eliška Zugarová for comments that greatly influenced the manuscript. The authors express their gratitude to anonymous reviewers for the valuable suggestions.

References

1. MATLAB version 7.5.0 (R2007b): The MathWorks Inc., Natick, Massachusetts, USA (2010)
2. Binmore, K.G.: Game Theory and the Social Contract: Just Playing, vol. 2. The MIT Press, Cambridge (1998)
3. Bosman, R., van Winden, F.: Emotional hazard in a power-to-take experiment. Econ. J. **112**(474), 147–169 (2002)
4. Capra, M.: Mood-driven behavior in strategic interactions. Am. Econ. Rev. **94**(2), 367–372 (2004)
5. Cox, J.C., Friedman, D., Gjerstad, S.: A tractale model of reciprocity and fairness. Games Econ. Behav. **59**(1), 17–45 (2007). https://doi.org/10.1016/j.geb.2006.05.001

6. Fehr, E., Schmidt, K.M.: A theory of fairness, competition, and cooperation. Q. J. Econ. **114**(3), 817–868 (1999). https://doi.org/10.1162/003355399556151

7. Grecucci, A., Giorgetta, C., van't Wout, M., Bonini, N., Sanfey, A.G.: Reappraising the ultimatum: an FMRI study of emotion regulation and decision making. Cereb. Cortex **23**, 399–410 (2013)

8. Guth, W., Schmittberger, R., Schwarze, B.: An experimental analysis of ultimatum bargaining. J. Econ. Behav. Org. **3**(4), 367–388 (1982)

9. Haselhuhn, M.P., Mellers, B.A.: Emotions and cooperation in economic games. Elsevier **23**(1), 24–33 (2005). https://doi.org/10.1016/j.cogbrainres.2005.01.005

10. Livet, P.: Rational choice, neuroeconomy and mixed emotions. Philos. Trans. R. Soc. B **365**, 259–269 (2010)

11. Loewenstein, G., Lerner, J.S.: The role of affect in decision making. In: Handbook of Affective Sciences, pp. 619–642. Oxford University Press (2003). (Chap. 31)

12. McGraw, A.P., Larsen, J.T., Kahneman, D., Schkade, D.: Comparing gains and losses. Psychol. Sci. **21**(10), 1438–1445 (2010)

13. Puterman, M.L.: Markov Decision Processes. Wiley, Hoboken (1994)

14. Rabin, M.: Incorporating fairness into game theory and economics. Am. Econ. Rev. **83**(5), 1281–1302 (1993)

15. Sanfey, A., Rilling, J., Aronson, J., Nystrom, L., Cohen, J.: The neural basis of economic decision-making in the ultimatum game. Science **300**, 1755–1758 (2003)

16. Srivastava, J., Espinoza, F., Fedorikhin, A.: Coupling and decoupling of unfairness and anger in ultimatum bargaining. J. Behav. Decis. Making **22**, 475–489 (2008). https://doi.org/10.1002/bdm.631

17. Tamarit, I., Sanchez, A.: Emotions and strategic behavior: the case of the ultimatum game. PloS One **11**(7) (July 2016). https://doi.org/10.1371/journal.pone.0158733

18. Woodruffe-Peacock, C., Turnbull, G.M., Johnson, M.A., Elahi, N., Preston, G.C.: The quick mood scale: development of a simple mood assessment scale for clinical pharmacology studies. Hum. Psychopharmatology Clin. Exp. **13**(1), 53–58 (1998)

Implementing Argumentation-Enabled Empathic Agents

Timotheus Kampik$^{(\boxtimes)}$ ⓘ, Juan Carlos Nieves ⓘ, and Helena Lindgren ⓘ

Umeå University, 901 87 Umeå, Sweden
{tkampik,jcnieves,helena}@cs.umu.se

Abstract. In a previous publication, we introduced the core concepts of *empathic agents* as agents that use a combination of utility-based and rule-based approaches to resolve conflicts when interacting with other agents in their environment. In this work, we implement proof-of-concept prototypes of empathic agents with the multi-agent systems development framework *Jason* and apply argumentation theory to extend the previously introduced concepts to account for inconsistencies between the beliefs of different agents. We then analyze the feasibility of different admissible set-based argumentation semantics to resolve these inconsistencies. As a result of the analysis, we identify the *maximal ideal extension* as the most feasible argumentation semantics for the problem in focus.

Keywords: Agent architectures ·
Agent-oriented software engineering · Argumentation

1 Introduction

Complex information systems that act with high degrees of autonomy and automation are ubiquitous in human society and shape day-to-day life. This leads to societal challenges of increasing frequency and impact, in particular when systems are primarily optimized towards simple metrics like views and clicks and are then used by malevolent actors for deceptive purposes, for example, to manipulate political opinions. Emerging research highlights these challenges and suggests the development of new multi-agent system concepts to address the problem [12,17]. In a recent publication, we outlined the basic concepts of empathic agents [13], based on an established definition of empathy as the ability to "simulate another's situated psychological states, while maintaining clear self–other differentiation" [8][1]. Our empathic agent is utility-based but additionally uses acceptability rules to avoid conflicts with agents that act in its environment. Thereby, we attempt to emulate empathic human behavior, which

---

[1] We based our empathic agent on a rationality-oriented definition of empathy, to avoid the technical ambiguity definitions that focus on emotional empathy imply. A comprehensive discussion of definitions of empathy is beyond the scope of this work.

© Springer Nature Switzerland AG 2019
M. Slavkovik (Ed.): EUMAS 2018, LNAI 11450, pp. 140–155, 2019.
https://doi.org/10.1007/978-3-030-14174-5_10

considers rules and societal norms, as well as the goals and intentions of others. We consider our agent a research contribution to the "synergistic combination of modelling methods" for "autonomous agents modelling other agents" as outlined as an open research problem in the discussion of a survey by Albrecht and Stone [1].

This paper addresses the following research questions:

- How can proof-of-concept prototypes of empathic agents be implemented with a multi-agent systems development framework?
- How can abstract argumentation (see: Dung [9]) be used to extend the agents to be able to resolve scenarios in which the beliefs of different agents are initially inconsistent?
- Which admissible set-based argumentation semantics are most feasible for resolving these inconsistencies?

To illustrate our work in a context that can be considered of societal relevance, we introduce the following example scenario: A *persuader* agent can select exactly one item from a list of persuasive messages (*ads*) it can display to a *mitigator* agent. Based on the impact the message will have on a particular end-user the mitigator represents, the mitigator will either accept the message and allow the end-user to consume the persuader's service offering (and the message), or terminate its session with the persuader. The persuader's messages are considered *empathic agent actions*; the mitigator does not *act* in the sense of the empathic agent core framework (see: Sect. 2). If the mitigator accepts the action proposal of the persuader, the environment will pay different utility rewards (or punishments) to persuader and mitigator. The goal of the persuader is to select a message that is utility-optimal for itself, considering that the mitigator has to accept it. We suggest that the scenario, albeit simple, reflects a potential type of real-world use case for future empathic agents of greater maturity. Improving the end-user experience of systems that are traditionally primarily optimized towards one simple metric can be considered a societally beneficial use case for empathic agents: (self-)regulation could motivate advertisement-financed application providers to either implement both persuader and mitigator to create a more purposeful user experience, or to open up their APIs to third-party clients that try to defend the interests of the end-users.

The rest of this paper is organized as follows: First, we elaborate on the core concepts of the empathic agent (Sect. 2), taking into consideration the belief-desire-intention (BDI) architecture in which the concepts are to be implemented. Then, we describe the implementation of the concepts with the multi-agent systems development framework Jason [6] (Sect. 3). Next, we apply argumentation theory and extend the core concepts and their implementation with Jason to handle inconsistencies between the beliefs of different agents (Sect. 4). To clarify our choice of argumentation semantics, we then analyze the applicability of different admissible set-based semantics (Sect. 5). Subsequently, we examine other work related to argumentation with Jason, briefly discuss the work in the context of existing negotiation approaches, discuss limitations of the implemented agents, and outline possible steps towards more powerful empathic agents (Sect. 6), before we conclude the paper (Sect. 7).

2   Empathic Agent Concepts

In this section, we elaborate on the empathic agent core concepts as initially sketched out in [13].

A set of interacting empathic agents can be described as follows:

- In an environment, n empathic agents $\{A_0, ..., A_n\}$ are acting.
- Each agent $A_i(0 \leq i \leq n)$ can execute a finite set of actions $Acts_{A_i} = \{Act_{A_{i_0}}, ..., Act_{A_{i_m}}\}$. All agents execute all actions simultaneously at one specific point in time; (inter)actions in continuous time or over a series of discrete time steps are beyond the scope of the basic concept.
- In its *belief base*, each agents has a utility function that assigns a numeric value to all possible action combinations: $u_{A_i} : Acts_{A_0} \times ... \times Acts_{A_n} \rightarrow \{-\infty, \mathbb{R}, +\infty\}$.
- Each agent strives to determine a set of actions it should execute that is considered *acceptable* by all agents. For this, an agent A_i determines the sets of actions that maximize its own utility function (arg max u_{A_i})[2]: If there is no own best set of actions $acts_{am_{i,k}} \in$ arg max u_{A_i} that matches a set of best actions of all of the other agents, a conflict of interests exists. Hence, a conflict of interest can be determined by the following boolean function:

$$c(\arg \max u_{A_0}, ..., \arg \max u_{A_i}, ..., \arg \max u_{A_n}) =$$
$$true, if :$$
$$\nexists \, acts_{am_{0,l}}, ..., acts_{am_{i,k}}, ..., acts_{am_{n,m}}$$
$$in \arg \max u_{A_0} \times ... \times \arg \max u_{A_i} \times ... \times \arg \max u_{A_n} :$$
$$\{acts_{am_{0,l}} \cap ... \cap acts_{am_{i,k}} \cap ... \cap acts_{am_{n,m}}\} \neq \{\}$$
$$else : false$$

- In addition to its utility mapping, each agent A_i has a set of acceptability rules $Accs_{A_i}$ in its belief base. An acceptability rule $Acc_{A_{i_j}} \in Accs_{A_i}$ is a boolean function that takes as it input the to-be-executed actions and returns a boolean value[3]: $Acts_{A_0} \times ... \times Acts_{A_n} \rightarrow \{true, false\}$.
- If the conflict determination function does not determine a conflict, the agents can go ahead and execute their optimal actions. If the conflict determination function returns *true* (determines a conflict), each agent applies the acceptability rules to check whether the actions it would ideally execute are in fact not acceptable or whether they can *potentially* be executed nevertheless. Here, the agents can employ different game-theoretical approaches (for example: minimizing the maximal loss) for ensuring that the execution of conflicting, but acceptable action sets by multiple agents does not lead to utility outcomes that are "bad" (in particular: worse than executing the actions that

---

[2] Note that arg max u_{A_i} returns a set of sets.
[3] Note that a single acceptability rule does not necessarily consider all to-be-executed actions, i.e. it might ignore some of its input arguments.

maximize combined utility) for all parties[4]. If the actions are not acceptable or are acceptable but chosen to not be executed, each agent can either choose to maximize the shared utility of all agents (arg $\max(u_{A_0} \times ... \times u_{A_n})$) or to select the next best action set and execute the conflict and acceptability check for them. We refer to agent variants that employ the former approach as *lazy* empathic agents, and agents that use the latter as *full* empathic agents. Lazy empathic agents save computation time by not iterating through action sets that do not optimize the ideally possible individual utility but provide better individual utility than the action sets that maximize combined utility. However, they might not find "good" solutions that are represented by action sets they do not cover[5].

Listing 1 describes the empathic agent base algorithm (*lazy* variant) in a two-agent scenario.

Listing 1. Empathic agent base algorithm

1. Determine actions that maximize own utility (arg $\max u_{A\,self}$).
2. Check for conflict with actions that maximize utility of other agent (arg $\max u_{A\,other}$).
3. If conflicts appear (apply conflict determination function c):
    Check if actions that maximize own utility are acceptable despite conflicts.
    if conflicts are acceptable
    (apply acceptability rules as function $Acts_{self} \times Acts_{other} \rightarrow \{true, false\}$):
        Execute (arg $\max u_{A\,self}$) $\cap Acts_{self}$.
    else: Execute (arg $\max(u_{A\,self} \times u_{A\,other})$) $\cap Acts_{self}$.
    else: Execute (arg $\max u_{A\,self}$)$\cap Acts_{self}$ (or (arg $\max(u_{A\,self} \times u_{A\,other})$))$\cap Acts_{self}$).

3   Implementation of an Empathic Agent with Jason

To provide proof-of-concept prototypes of empathic agents, we implemented a simple empathic agent example scenario with Jason, a development framework that is well-established in the multi-agent systems community[6].

The agents implement the scenario type we explained in the introduction as follows:

---

[4] As the simple examples we implement in this paper feature only one *acting* agent (a second agent is merely approving or disapproving of the actions), such game-theoretical considerations are beyond scope. Hence, we will not elaborate further on them.

[5] For now, we assume all agents in a given scenario have the same implementation variant. Empathic agents that are capable to effectively interact with empathic agents of other implementation variants or with non-empathic agents are–although interesting–beyond scope.

[6] The implementation of our empathic agents with Jason (including the Jason extension we introduce below, as well as a technical report that documents the implementation) is available at https://github.com/TimKam/empathic-jason.

1. The persuader agent has a set of utility mappings–(*utility value, unique action name*)-tuples–in its belief base[7]:

         revenue(3, "Show vodka ad").
         revenue(1, "Show university ad").
              . . .

   The mappings above specify that the persuader can potentially receive three utility units for showing a vodka advertisement, or one utility unit for showing a university advertisement. The persuader communicates these utility mappings to the mitigator.

2. The mitigator has its own utility mappings:

         benefit(-100, "Show vodka ad").
         benefit(10, "Show university ad").
              . . .

   The mitigator's utility mappings are labelled *benefit*, in contrast to the persuader's *revenue* label. While this should reflect the difference in impact the actions have on persuader and mitigator (or rather: on the end-user the mitigator is proxying), it is important to note that we consider the utility of the two different mappings *comparable*. The mitigator responds to the persuader's announcement by sending back its own mapping. In addition to the utility mappings, both persuader and mitigator have a set of *acceptability rules* in their belief base[8] :

         acceptable(
             "Show university ad",
             "Show community college ad").
         acceptable(
             "Show community college ad",
             "Show university ad").
              . . .

   The *intended meaning* the empathic agents infer from these rules is that both agents agree that *Show university ad* is always acceptable if the preferred action of the mitigator is *Show community college ad* and vice versa. After sending the response, the mitigator determines the action it thinks should be executed (the *expected* action), using the algorithm as described in Sect. 2. In the current implementation, the agents are *lazy* empathic agents, i.e. they will only consider the actions with maximal individual and maximum shared utility (as explained above)[9].

3. The persuader determines its action proposal in the same way and announces it to the mitigator.

4. When the response is received and the own expected action is determined, the mitigator compares the received action with the determined action. If

---

[7] The mappings are end-user specific. In a scenario with multiple end-users, the persuader would have one set of mappings per user.

[8] Note that in Jason terminology, acceptability rules are *beliefs* and not *rules*.

[9] If at any step of the decision process, several actions could be picked because they provide the same utility, the agents will always pick the first one in the corresponding list to reach a deterministic result.

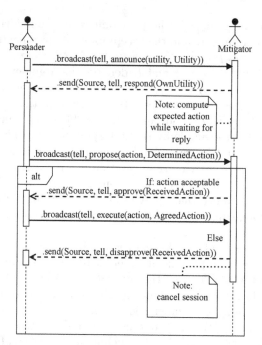

Fig. 1. Empathic agent interaction: sequence diagrams, basic example

the actions are identical, the mitigator sends its approval of the action proposal to the persuader. If the actions are inconsistent, the mitigator sends a disapproval message and cancels the session.
5. If the persuader receives the approval from the mitigator, it executes the agreed-upon action.

Figure 1 shows a sequence diagram of the basic empathic agent example we implemented with Jason. The message labels show the internal actions the agents use for communicating with each other.

4 Argumentation for Empathic Agents: Reaching Consensus in Case of Inconsistent Beliefs

In real-world scenarios, agents will often have inconsistent beliefs because of an only partially observable environment or subjective interpretations of environment information. Then, further communication is necessary for the agents to reach consensus on what actions to execute and to prevent mismatches between the different agents' decision making processes that lead to poor results for all involved agents. In this section, we show how abstract argumentation can be used to synchronize inconsistent agents *acceptability rules* in the agents' belief bases. I.e., we improve the *autonomy* for the agent from the perspective of moral philosophy in that we enhance ability of an empathic agent to "impose the [...] moral law on [itself]" [7].

Abstract argumentation, as initially introduced by Dung formalizes a theory on the exchange of arguments and the acceptability of arguments within a given context (*argumentation framework*) as a tuple $\langle A, R \rangle$ where A is a finite set of arguments (for example: $\{a, b, c\}$ and R denotes the arguments' binary attack relations (for example: $\{(a, b), (b, c)\}$, i.e. *a attacks b and b attacks c*). Dung introduces different properties of argument sets in argumentation frameworks, such as *acceptability, admissibility,* and *completeness* [9]. Dung's initial theory has since been extended, for example by Bench-Capon, who introduced value-based argumentation to account for values and value preferences in argumentation [3]. In multi-agent systems research, argumentation is a frequently applied concept, "both for automating individual agent reasoning, as well as multiagent interaction" [16]. It can be assumed that applying (abstract) argumentation to enhance the consensus-finding capabilities of the empathic agents is an approach worth investigating.

In the previous example implementation, which primarily implements the empathic agent core concepts, inconsistencies between acceptability rules are irreconcilable (and would lead to the mitigator's rejection of the persuader's action proposal). To address this limitation, we use an abstract argumentation approach based on Dung's argumentation framework [9] and its *maximal ideal extension* definition [10]. We enable our agents to launch attacks at acceptability rules. An attack is a tuple (a, b), with a being the identifier of an action or of another attack and b being the identifier of the attack itself. Each attack is added to a list of arguments, with the rule it attacks as its target. Attacked rules are added as arguments without attack targets. Additional arguments can be used to launch attacks on attacks. The argumentation framework is resolved by determining its *maximal ideal extension*[10]. The instructions in Listing 2 describe the algorithm the argumentation-enabled empathic agents use for belief synchronization. The algorithm is executed if the agents initially cannot agree on a set of actions, which would lead to a cancellation of the interaction using only the basic approach as introduced above. We extended Jason to implement an argumentation function as an internal action. The function calls the argumentation framework solver of the *Tweety Libraries for Logical Aspects of Artificial Intelligence and Knowledge Representation* [18], which we wrapped into a web service with a RESTful HTTP interface. In our example implementation, we extend our agents as follows:

1. The persuader starts with some acceptability rules in its belief base that are not existent in the belief set of the mitigator:
   ```
   acceptable("Show vodka ad", "Show university ad").
   ```
   . . .

   In the provided example, the persuader believes (for unknown reasons) that the action *Show vodka ad* is acceptable even if the action *Show university ad* provides greater utility to the mitigator.
2. When the mitigator assesses the action proposal, it detects that its own expected action (*Show university ad*) is inconsistent with the persuader's

---

[10] Note that we compare different argumentation semantics in Sect. 5.

expected action (*Show vodka ad*). It sends a disapproval message to the persuader that includes an attack on the acceptability rules of the *Show vodka ad* action:

```
attack("Show vodka ad", "Alcoholic").
```

3. The persuader then constructs an argumentation framework from acceptability rules and attacks. With the help of the Jason argumentation extension, it determines all preferred extensions of the argumentation framework and removes the arguments (acceptability rules and attacks) that are not part of any preferred extension. Subsequently, the persuader updates its belief base accordingly, re-determines the action proposal, and sends the proposal to the mitigator, who then re-evaluates it (and ideally accepts it). Potentially, the persuader could also launch attacks on the mitigator's attacks and so on, until consensus is reached or the session is aborted[11].

Listing 2. Argumentation-enabled belief synchronization

1. **Mitigator agent** (After having received not acceptable action proposal): Request acceptability rules from mitigator.
2. **Persuader agent**: Send acceptability rules.
3. **Mitigator agent** (Upon receiving acceptability rules):
   (a) Consider acceptability rules as initial arguments $\{acc_0, ..., acc_n\}$ in argumentation framework $AF = \langle \{acc_0, ..., acc_n\}, \{\} \rangle$. Resolve AF by determining maximal ideal extension.
   (b) Decide whether:
       new attacks should and can be launched,
       or AF can be accepted as-is,
       or session should be cancelled.
   (c) If attacks should be launched:
       Add additional set of arguments to AF that attack acceptability rules that are inconsistent with own beliefs. Send updated AF to persuader (go to 4a).
       else if AF can be accepted as-is:
       Accept and resolve AF.
       else:
       Cancel session.
4. **Persuader agent** (Upon receiving AF):
   (a) Generate/update own argumentation framework AF. Resolve AF by determining maximal ideal extension.
   (b) Decide whether:
       new attacks should and can be launched,
       or AF can be accepted as-is,
       or session should be cancelled.
   (c) If attacks should be launched:
       Add additional set of arguments to AF that attack acceptability rules that are inconsistent with own beliefs. Send updated AF to mitigator (go to 3c).
       else if AF can be accepted as-is:
       Accept and resolve AF.
       else:
       Cancel session.

---

[11] However, the provided example code implements only one argumentation cycle.

Figure 2 shows a sequence diagram of the argumentation-capable empathic agent example we implemented with Jason.

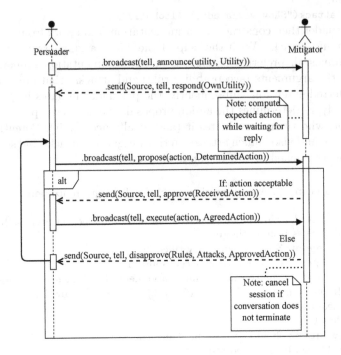

Fig. 2. Empathic agent interaction: sequence diagrams, argumentation example

5   Argumentation Semantics Analysis

In this section, we explain why our argumentation-enabled empathic agents use argumentation semantics that select the *maximal ideal* extension. The purpose of the required argumentation reasoner is to remove all arguments from an argumentation framework that have been attacked by arguments that in turn have not been successfully attacked. I.e., the naive requirements are as follows:

1. Given a set of arguments S, the semantics must exclude all successfully attacked arguments from S, and only these.
2. An argument in S is considered successfully attacked if it is attacked by any argument in S that is not successfully attacked itself.

According to Dung, a "preferred extension of an argumentation framework AF is a maximal (with respect to set inclusion) admissible set of AF". In this context, "a conflict-free set of arguments S is admissible iff each argument in S is acceptable with respect to S", given an "argument $A \in AR$ is acceptable with respect to a set S of arguments iff for each argument $B \in AR$: if B attacks A then B is attacked by S" [9]. As Dung's definition of preferred extensions is congruent with the stipulated requirements, it seems reasonable for the empathic agent implementations to resolve argumentation frameworks by using the corresponding argumentation semantics.

However, if there are arguments that attack each other directly or that form any other circular structure with an even number of nodes, multiple preferred extensions exist. For example, given the arguments $\{a, b\}$ and the attacks $\{(a, b), (b, a)\}$ both $\{a\}$ and $\{b\}$ are preferred extensions.

This behavior implies ambiguity, which prevents the empathic agents from reaching a deterministic resolution of their belief inconsistency. As this is to be avoided, we introduce an additional requirement:

3. The semantics must determine exactly one set of arguments by excluding ambiguous arguments.

Dung defines the grounded extension of an argumentation framework as "the least (with respect to set inclusion) complete extension" and further stipulates that an "admissible set S of arguments is called a complete extension iff each argument, which is acceptable with respect to S, belongs to S" [9]. In the above example, the grounded extension is $\{\}$. As determining the grounded extension avoids ambiguity, using grounded semantics seems to be a reasonable approach at first glance. However, as Dung et al. show in a later work, grounded semantics are *sceptical* beyond the stipulated requirements, in that they exclude some arguments that are not successfully attacked [10][12]. In the same work, the authors define a set of arguments X as *ideal* "iff X is admissible and it is contained in every preferred set of arguments" and show that the maximal ideal set is "a proper subset of all preferred sets" [10]. Hence, we adopt maximal ideal semantics for determining the set of acceptable arguments of our empathic agents.

The following two running examples highlight the relevance of the distinction between preferred, maximal ideal, and grounded semantics in the context of our empathic agents.

Example 1. Example 1 highlights the advantage of grounded and maximal ideal semantics over preferred and complete semantics for the use case in focus. The to-be-solved empathic agent scenario is as follows:

- Acceptability rules persuader: $\{(Show\ steak\ ad, *)\}$ (a_1)[13]
- Acceptability rules mitigator: $\{\}$.

To resolve the acceptability rule inconsistency, the agents exchange the following arguments:

1. Mitigator: $\{(a_1, Steak\ ad\ too\ unhealthy)\}$ (b_1)

---

[12] In Example 2, we illustrate an empathic agent argumentation scenario, in which grounded semantics are overly strict.

[13] For the sake of simplicity, we use a wild card (*) to denote that the acceptability rule applies no matter which preference the mitigator agent has. Note that this syntax is not supported by our implementation.

2. Persuader: $\{(b_1, \text{Steak ad healthy enough})\}$ (c_1)
3. Mitigator: $\{(c_1, b_1)\}$

As a result, we construct the following argumentation framework:

$$AF_1 = \langle \{a_1, b_1, c_1\}, \{(b_1, a_1)(c_1, b_1), (b_1, c_1)\} \rangle$$

Applying preferred, complete, grounded, and maximal ideal semantics to solve the framework yields the following results:

- Preferred: $\{b_1\}, \{a_1, c_1\}$
- Complete: $\{b_1\}, \{a_1, c_1\}, \{\}$
- Grounded: $\{\}$
- Maximal ideal: $\{\}$

As can be seen, applying preferred or complete semantics does not allow for a clear decision on whether the acceptability rule should be discarded or not, whereas both grounded and maximal ideal semantics are addressing this problem by discarding all arguments that are ambiguous in complete or ideal semantics, respectively. Figure 3 contains a graphical overview of how the different argumentation semantics solve the argumentation framework of the first example.

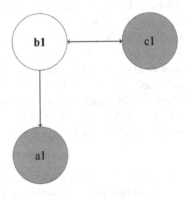

Fig. 3. Example 1: argumentation semantics visualization: preferred (bold and grey, respectively). Grounded and maximal ideal extensions are $\{\}$. The preferred argument sets are also complete. $\{\}$ is an additional complete set of arguments.

Example 2. Example 2 highlights the advantage of maximal ideal over grounded semantics for the use case in focus. We start with the following empathic agent scenario:

- Acceptability rules persuader: $(\text{Show steak ad}, *)$ (a_2)
- Acceptability rules mitigator: $\{\}$

To resolve the acceptability rule inconsistency, the agents exchange the following arguments:

1. Mitigator: $\{(a_2, Steak\ ad\ too\ unhealthy)\}\ (b_2)$
2. Persuader: $\{(b_2, Steak\ ad\ shows\ quality\ meat)\}\ (c_2)$
3. Mitigator: $\{(c_2, User\ prefers\ sweets)\}\ (d_2)$
4. Persuader: $\{(d_2, This\ preference\ does\ not\ matter)\}\ (e_2)$
5. Mitigator: $\{(e_2, d_2)\}$
6. Persuader: $\{(c_2, b_2)\}$

From the scenario specification, the following argumentation framework can be constructed:

$$AF_2 = \langle\{a_2, b_2, c_2, d_2, e_2\}, \{(b_2, a_2)(b_2, d_2), (c_2, b_2), (d_2, c_2), (e_2, d_2)\}\rangle$$

Applying preferred, complete, grounded, and maximal ideal semantics to solve the framework yields the following results:

- Preferred: $\{a_2, c_2, e_2\}$
- Complete: $\{a_2, c_2, e_2\}, \{\}$
- Grounded: $\{\}$
- Maximal ideal: $\{a_2, c_2, e_2\}$

As can be seen, applying grounded semantics removes acceptability rule a_2, although this rule is not successfully attacked according to the definition stipulated in our requirements. In contrast, applying maximal ideal semantics does not remove a_2. Figure 4 contains a graphical overview of how the different argumentation semantics solve the argumentation framework of the second example.

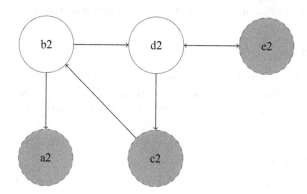

Fig. 4. Example 2: argumentation semantics visualization: preferred (grey) and maximal ideal (dashed border). The grounded extension is $\{\}$. The preferred/maximal ideal argument set is also complete. $\{\}$ is an additional complete set of arguments.

6   Discussion

6.1   Argumentation and Jason

Argumentation in agent-oriented programming languages has been covered by previous research. In particular, both Berariu [4] and Panisson et al. [14] describe the implementation of argumentation capabilities with Jason. In contrast to these works, we present a service-oriented approach for argumentation support that provides one simple function to solve arguments and thus offers a higher level of abstraction for application developers. For future work, we suggest considering advanced abstract argumentation approaches like value-based argumentation [3] to account for agent preferences and possibilistic argumentation [2] to account for uncertainty. Generally, we consider the approach of writing service interface extensions for Jason as promising to integrate with existing relevant technologies, for example to provide technology bridges to tools and frameworks developed by the machine learning community.

6.2   Alternative Negotiation Approaches

We consider the empathic agent approach as fundamentally different from other negotiation approaches (see for an overview: Fatima and Rahwan [11]) in that the empathic agents only optimize their own utility if acceptability rules explicitly permit doing so and strongly consider the utility of other agents in their environment. I.e., empathic agent application scenarios can be considered at the intersection of mixed-motive and fully cooperative games. Comprehensive comparisons, in particular with other argumentation-based negotiation approaches, as for example developed by Black and Atkinson [5], are certainly relevant future work. For example, simulations that elaborate on the benefits and limitations of specific empathic agent implementations could be developed and executed in scenarios similar to the one presented in the examples of this paper.

6.3   Limitations

This paper explains how to implement basic empathic agents with Jason and how to extend Jason to handle scenarios with initially inconsistent beliefs by applying argumentation theory. However, the implemented agents are merely running examples to illustrate the *empathic agent* concept and do not solve any practical problem. The following limitations need to be addressed to facilitate real-world applicability:

- Although we extended Jason to better support the use case in focus, the current implementation does not provide sufficient abstractions to conveniently implement complex empathic agents. Moreover, our empathic agent implementation does not make use of the full range of BDI-related concepts Jason offers.
- While the agents can find consensus in case of simple belief inconsistencies, they do not use scalable methods for handling complex scenarios with large state and action spaces, and are required to explicitly exchange their preferences (utility mappings) with each other, which is not possible in many application scenarios, in which the utility mappings need to be *learned*.
- So far our empathic agent only interacts with other empathic agents. This simplification helps explain the key concepts with basic running examples

and is permissible in some possible real-world scenarios, in particular when designing empathic agents that interact with agents of the same type in closed environments (physically or virtually) that are not accessible to others. To interact with agents of other designs, the empathic agent concept needs to be further extended, for example, to be capable of dealing with malevolent agents.

- The analysis of the argumentation approach is so far limited to the admissible set-based semantics as introduced in the initial abstract argumentation paper by Dung [9] that considers arguments as propositional atoms. The structure of the arguments is not addressed.

6.4 Towards a Generic Empathic Agent

For future research, we suggest further extending Jason–or to work with alternative programming languages and frameworks–to support more powerful abstractions for implementing empathic agents, for example:

- Develop engineering-oriented abstractions for empathic agents, possibly following a previously introduced architecture proposal [13]. While further extending Jason can facilitate scientific applications and re-use, providing a minimally invasive library or framework for a general-purpose programming language could help introduce the concept to the software engineering mainstream. User interfaces could be developed that facilitate the explainability of the empathic agents' decision processes.
- Leverage existing research to better address partial observability and subjectivity. The aforementioned survey by Albrecht and Stone shows that a body of reinforcement learning research from which can be drawn upon exists [1]. Yet, the general problem of scaling intelligent agents in complex and uncertain environments remains an open challenge. It would be interesting to investigate how this problem affects the applicability of our empathic agent.
- Design and implement empathic agents that can meaningfully interact with agents of other architectures and ultimately with humans. As a starting point, one could develop an empathic Jason agent that interacts with a *deceptive* Jason agent as introduced by Panisson et al. [15].
- Extend the analysis of a suitable argumentation semantics for the empathic agent by considering advanced approaches, for example value-based argumentation as introduced by Bench-Capon [3], and by identifying a feasible structure the arguments can use internally.

7 Conclusion

In this paper, we show how to implement proof-of-concept prototypes of empathic agents with an existing multi-agent development framework. Also, we devise an extension of the previously established empathic agent concept that employs argumentation theory to account for inconsistent beliefs between

agents. In our analysis of different admissible set-based argumentation semantics, we have determined that maximal ideal semantics are the most feasible for
addressing the problem in focus, as they provide both an unambiguous solution
of a given argumentation framework and are not overly sceptical (in contrast to
grounded semantics) as to which arguments are acceptable.

However, the provided empathic agent implementations do not yet scale to
solve real-world problems. It is important to identify or devise more powerful methods to handle subjectivity and partial observability, and to ultimately
enable empathic agents to meaningfully interact with agents of other architecture
types. On the technology side, an abstraction specifically for empathic agents
is needed that forms a powerful framework for implementing empathic agents.
We suggest creating a Jason extension that can facilitate the implementation of
empathic agents with a framework that is familiar to members of the academic
multi-agent systems community, but to also consider providing abstractions for
programming frameworks that are popular among industry software engineers
and are not necessarily BDI-oriented or even agent-oriented per se.

Acknowledgements. We thank the anonymous reviewers for their constructive feedback. This work was partially supported by the Wallenberg AI, Autonomous Systems
and Software Program (WASP) funded by the Knut and Alice Wallenberg Foundation.

References

1. Albrecht, S.V., Stone, P.: Autonomous agents modelling other agents: a comprehensive survey and open problems. Artif. Intell. **258**, 66–95 (2018)
2. Alsinet, T., Chesnevar, C.I., Godo, L., Simari, G.R.: A logic programming framework for possibilistic argumentation: formalization and logical properties. Fuzzy
Sets Syst. **159**(10), 1208–1228 (2008)
3. Bench-Capon, T.J.: Persuasion in practical argument using value-based argumentation frameworks. J. Log. Comput. **13**(3), 429–448 (2003)
4. Berariu, T.: An argumentation framework for BDI agents. In: Zavoral, F., Jung,
J., Badica, C. (eds.) Intelligent Distributed Computing VII. SCI, vol. 511, pp.
343–354. Springer, Cham (2014). https://doi.org/10.1007/978-3-319-01571-2_40
5. Black, E., Atkinson, K.: Choosing persuasive arguments for action. In: The 10th
International Conference on Autonomous Agents and Multiagent Systems, vol. 3,
pp. 905–912. International Foundation for Autonomous Agents and Multiagent
Systems (2011)
6. Bordini, R.H., Hübner, J.F.: BDI agent programming in agentspeak using *Jason*.
In: Toni, F., Torroni, P. (eds.) CLIMA 2005. LNCS (LNAI), vol. 3900, pp. 143–164.
Springer, Heidelberg (2006). https://doi.org/10.1007/11750734_9
7. Christman, J.: Autonomy in moral and political philosophy. In: Zalta, E.N. (ed.)
The Stanford Encyclopedia of Philosophy. Metaphysics Research Lab, Stanford
University, Spring 2018 edn. (2018)
8. Coplan, A.: Will the real empathy please stand up? A case for a narrow conceptu-
alization. South. J. Philos. **49**(s1), 40–65 (2011)
9. Dung, P.M.: On the acceptability of arguments and its fundamental role in non-
monotonic reasoning, logic programming and n-person games. Artif. Intell. **77**(2),
321–357 (1995)

10. Dung, P.M., Mancarella, P., Toni, F.: Computing ideal sceptical argumentation. Artif. Intell. **171**(10–15), 642–674 (2007)
11. Fatima, S., Rahwan, I.: Negotiation and bargaining. In: Weiss, G. (ed.) Multiagent Systems, 2nd edn, pp. 143–176. MIT Press, Cambridge (2013). Chap. 4
12. Kampik, T., Nieves, J.C., Lindgren, H.: Coercion and deception in persuasive technologies. In: 20th International TRUST Workshop (2018)
13. Kampik, T., Nieves, J.C., Lindgren, H.: Towards empathic autonomous agents. In: 6th International Workshop on Engineering Multi-Agent Systems (EMAS 2018), Stockholm, July 2018
14. Panisson, A.R., Meneguzzi, F., Vieira, R., Bordini, R.H.: An approach for argumentation-based reasoning using defeasible logic in multi-agent programming languages. In: 11th International Workshop on Argumentation in Multiagent Systems (2014)
15. Panisson, A.R., Sarkadi, S., McBurney, P., Parson, S., Bordini, R.H.: Lies, bullshit, and deception in agent-oriented programming languages. In: 20th International TRUST Workshop, Stockholm (2018)
16. Rahwan, I.: Argumentation among agents. In: Weiss, G. (ed.) Multiagent Systems, 2nd edn, pp. 177–210. MIT Press, Cambridge (2013). Chap. 5
17. Sen, S., Crawford, C., Rahaman, Z., Osman, Y.: Agents for social (media) change. In: Proceedings of the 17th International Conference on Autonomous Agents and Multiagent Systems (AAMAS 2018), Stockholm (2018)
18. Thimm, M.: Tweety - a comprehensive collection of java libraries for logical aspects of artificial intelligence and knowledge representation. In: Proceedings of the 14th International Conference on Principles of Knowledge Representation and Reasoning (KR 2014), Vienna, Austria (2014)

Towards Fully Probabilistic Cooperative Decision Making

Miroslav Kárný[(✉)] and Zohreh Alizadeh

The Czech Academy of Sciences, Institute of Information Theory and Automation,
POB 18, 182 08 Prague 8, Czech Republic
{school,za}@utia.cas.cz
http://www.utia.cz/AS

Abstract. Modern prescriptive decision theories try to support the dynamic decision making (DM) in incompletely-known, stochastic, and complex environments. Distributed solutions single out as the only universal and scalable way to cope with DM complexity and with limited DM resources. They require a solid cooperation scheme, which harmonises disparate aims and abilities of involved agents (human decision makers, DM realising devices and their mixed groups). The paper outlines a distributed fully probabilistic DM. Its flat structuring enables a fully-scalable cooperative DM of adaptive and wise selfish agents. The paper elaborates the cooperation based on sharing and processing agents' aims in the way, which negligibly increases agents' deliberation effort, while preserving advantages of distributed DM. Simulation results indicate the strength of the approach and confirm the possibility of using an agent-specific feedback for controlling its cooperation.

Keywords: Decision making · Cooperation ·
Fully probabilistic design · Bayesian learning

1 Introduction

A decision making theory supports agents to select actions, which aim to influence the closed decision loop, which couples the agent with its environment. DM is a complex process requiring a selection of relevant variables, adequate technical and theoretical tools, knowledge, aim elicitation, etc. Repeatedly-applicable DM procedures rely on a computer support, which needs a quantification of DM elements and, primarily, algorithmic solutions of all DM steps. Any of permanently-evolving DM theories designs a strategy (policy [25], decision function [38]), which maps the agent's knowledge and aims on actions. The optimal design selects the strategy, which meets agent's aims in the best way under the faced circumstances. An excessive DM complexity is tackled here.

Supported by GAČR, grant 16-09848S. This work was strongly influenced by the feedback provided to us by Dr. Tatiana V. Guy. We appreciate her insight and help.

M. Slavkovik (Ed.): EUMAS 2018, LNAI 11450, pp. 156–171, 2019.
https://doi.org/10.1007/978-3-030-14174-5_11

Evolution singled out *distributed* DM as the only universal and fully scalable way of coping with DM complexity and limited resources of individual agents [4]. *Cooperation* of involved agents decides on the success or failure in reaching individual or collective aims [1,22,27,34,35,40]. It must not recur the problems, which make distributed DM inevitable. This limits applicability of theory of Bayesian games [10] and excludes presence of a cooperation-controlling mediator, who has to deal with a quite aggregated knowledge and a small number of actions. Thus, a personalised machine support of selfish (aim oriented) agents, dynamically acting in the changing environment containing other agents, is needed. To our best knowledge, no general support of this type exists. This paper contributes to its creation by inspecting cooperation within the discussed scenario. It relies on theory of fully probabilistic design of decision strategies (FPD, [9,15,36]). FPD is a proper extension [19] of prevailing Bayesian DM [7,29,39]. The paper deals with the cooperation, which assumes that the involved agents use FPD and are wise enough to cooperate to the degree required for achieving their selfish aims.

1.1 Paper Layout

The outline of the considered flat multi-agents system in Sect. 2 provides backbone of the subsequent text. Section 3 recalls a single adaptive agent that uses Bayesian learning and the feasible certainty-equivalent version of FPD. Section 4 describes the employed cooperation concept while commenting on its position with respect to its direct predecessors. The experimental part, Sect. 5, indicates soundness of the adopted concept. Remarks in Sect. 6 primarily outline the further anticipated research.

1.2 Notions and Notation

A simple DM task is considered in order to focus on the central cooperation problem. It is close to Markov decision processes [25] dealing with finite numbers of actions and of fully observable states. Throughout:

- The set of ys with $|\boldsymbol{y}| < \infty$ values y_j is denote $\boldsymbol{y} = \{y_j\}_{j=1}^{|\boldsymbol{y}|}$.
- The same symbol marks a random variable, its realisation, and its possible value. San serif fonts mark mappings. Mnemonic symbols are preferred.
- *Probability mass functions* (pmf) are implicitly conditioned on the known initial state $x_0 \in \boldsymbol{x}$.
- The *observable state* x_t of the modelled stochastic environment evolves in *discrete time* $t \in \boldsymbol{t}$. The evolution is influenced by optional *actions* $a \in \boldsymbol{a}$. A value $a_t \in \boldsymbol{a}$ of the action a is selected by the agent at time $t \in \boldsymbol{t}$.
- The closed decision loop, formed by the agent and its environment, operates on the *behaviour* $b = (x_t, a_t)_{t \in \boldsymbol{t}} \in \boldsymbol{b}$, i.e. on the collection of states and actions up to the decision *horizon* $|\boldsymbol{t}| < \infty$.

- The random behaviour $b \in \boldsymbol{b}$ is described by a joint pmf

$$\mathsf{c_s}(b) \equiv \mathsf{c_s}(b|x_0) = \mathsf{c_s}(x_{|t|}, a_{|t|}, x_{|t|-1}, a_{|t|-1}, \ldots, x_1, a_1|x_0).$$

It is the complete $closed-loop\,model$. Chain rule for pmfs [28] factorises it

$$\mathsf{c_s}(b) = \prod_{t \in t} \mathsf{c_s}(x_t|a_t, x_{t-1}, a_{t-1}, \ldots, a_1, x_0)\mathsf{c_s}(a_t|x_{t-1}, a_{t-1}, \ldots, x_1, a_1, x_0)$$

$$= \prod_{t \in t} \mathsf{m}(x_t|a_t, x_{t-1})\mathsf{r}_t(a_t|x_{t-1}), \quad b = (x_t, a_t)_{t \in t}. \tag{1}$$

The mnemonically renamed factors after the second equality in (1): (a) exemplify the adopted assumption that the Markov environment and agent are considered; (b) focus us on time-invariant environments; (c) recognize that the first generic factor is the *environment* model, describing the probability of transiting to the state x_t from the state x_{t-1} when the action a_t is applied; (d) interpret the second generic factor as the *decision* rule, which assigns the probability of selecting the action a_t, when knowing the state x_{t-1}.

- The optional, generally randomised, decision strategy is the collection of decision rules $\mathsf{s} = (\mathsf{r}_t)_{t \in t}$. The optimising DM selects the optimal strategy $\mathsf{s_o}$. The optimality is defined with respect to agent's decision preferences, which are here quantified by the ideal closed-loop model

$$\mathsf{c_i}(b) = \prod_{t \in t} \mathsf{m_i}(x_t|a_t, x_{t-1})\mathsf{r_i}(a_t|x_{t-1}), \quad b = (x_t, a_t)_{t \in t}. \tag{2}$$

It is the product of ideal environment models $\mathsf{m_i}$ and ideal decision rules $\mathsf{r_i}$. Both are time invariant for simplicity. The ideal closed-loop model assigns high values to preferred behaviours and low values to unwanted ones. The use of this ideal pmf is in Sect. 3.

2   Flat Multi-agents Systems

This section outlines the adopted concept of agents' interactions. The wish to support selfish imperfect agents motivates it. The adjective selfish implies that the agent follows its "personal" aims while the adjective "imperfect" labels the agent's limited knowledge, limited observation, evaluation, and acting abilities. Such an agent acts within the environment containing other imperfect selfish agents, which directly or indirectly influence the agent's degree of success or failure in reaching its personal aim. The considered wise but still imperfect agent takes it into account and makes public a part of information it deals with. This allows other agents to modify their strategy so that mutual inevitable clash is diminished and consequently, the considered interacting imperfect agents get chance to reach their individual unchanged aims in a better way.

The assumed common universal strategy-design methodology (FPD) and the common language (probabilistic descriptions of both environment and aims) allows to process the shared information without a special mediating or even facilitating agent, which would become bottleneck as it is always imperfect in the discussed sense.

The imperfection of each agent implies that it can reach only information provided by a small number of other agents, its recognisable neighbours. This makes the intended processing of the shared information feasible and the whole considered multi-agent interaction fully scalable.

It is worthy to add the following comments to the above outline.

- Aims of individual agents can be close and even identical. Thus, the explicit cooperation is well possible in this scheme.
- Individual agents can be created by a group of cooperating agents with its mediator, facilitator or leader up to the point where the joint resources suffice. Thus, the inspected multi-agent scenario may support all traditional multi-agent systems.
- The considered scheme imitates how complex societies act.
- Naturally, individual agents may publish misleading information and locally exploiting it may deteriorate quality of the strategy, which respects them. The agent that uses such an information can recognise this effect in a longer run and assign small weight (trust) to such an adversary agent.

The subsequent text considers such a flat scheme and describes firstly the considered type of agent and then the exploitation of the limited shared information, namely, shared description of neighbours' aims. Fixed trust to a neighbour is assumed at this research stage.

3   Single Agent Using FPD

This section focuses on single agent. It provides the FPD-optimal strategy along with its certainty-equivalent, receding-horizon approximation [23].

3.1   Formulation and Solution of Fully Probabilistic Design

An agent influences the closed-loop behaviour $b \in \boldsymbol{b}$ by selecting its randomised strategy s. Its choice shapes the closed-loop model (1). Ex post, Bayesian DM [29] evaluates the behaviour desirability (as seen by the agent) via a real-valued *loss function* $\mathsf{L}(b)$, $b \in \boldsymbol{b}$. A priori, the quality of the strategy is evaluated via the expected loss, which is the generalised moment of the closed-loop model

$$\mathsf{E}_s[\mathsf{L}] = \sum_{b \in \boldsymbol{b}} \mathsf{c}_s(b)\mathsf{L}(b). \tag{3}$$

The Bayesian optimal strategy minimises the expected loss (3). FPD generalises this set up and uses the *ideal closed-loop model* c_i (2) instead of the loss function

L. FPD selects the optimal strategy, which makes the closed-loop model closest to the given ideal closed-loop model. FPD axiomatisation [19] implies that Kullback-Leibler divergence (KLD, [21])

$$D(c_s \| c_i) = E_s \left[\ln \left(\frac{c_s}{c_i} \right) \right] = \sum_{b \in b} c_s(b) \ln \left(\frac{c_s(b)}{c_i(b)} \right) \qquad (4)$$

is the adequate proximity measure[1].

The universal loss (4) depends on the optimised strategy, unlike L (3). Thus, FPD defines *FPD-optimal strategy* s_o as

$$s_o \in \text{Arg} \min_{s \in s} D(c_s \| c_i). \qquad (5)$$

Algorithm 1 explicitly provides the FPD-optimal strategy s_o (5). It consists of the *optimal decision rules* $(r_{o;t})_{t \in t}$ and exploits the environment model (1) and the factorised ideal closed-loop model (2). The proof of its optimality is, e.g., in [37]. Its presented form prepares the receding-horizon approximation of FPD.

Algorithm 1. Design of FPD-Optimal Decision Strategy

Inputs: Dimensions $|a|$, $|x|$; initial $\underline{\tau}$ and terminal $\bar{\tau}$ time moments of the design
        Environment model m   % belief description
        Factorised ideal closed model $c_i = m_i r_i$ % preference description
Evaluations:
Initialise $g(x) = 1$, $\forall x \in x$,
for $\tau = \bar{\tau}$ **to** $\underline{\tau}$ **do**
  for $x \in x$ **do**
    for $a \in a$ **do**
      $d(a,x) = \sum_{\tilde{x} \in x} m(\tilde{x}|a,x) \ln \left(\frac{m(\tilde{x}|a,x)}{m_i(\tilde{x}|a,x)g(\tilde{x})} \right)$
    end for
    $g(x) = \sum_{\tilde{a} \in a} r_i(\tilde{a}|x) \exp(-d(\tilde{a},x))$ % $-\ln(g(x))$ is the *value function*

    for $a \in a$ **do**
      $r_{o;\tau}(a|x) = \frac{r_i(a|x) \exp(-d(a,x))}{g(x)}$
    end for
  end for
end for
Outputs: FPD-optimal strategy $s_o = (r_{o;\tau})_{\tau=\underline{\tau}}^{|\tau|}$,   % s_o is optimal iff $\underline{\tau} = 1$, $\bar{\tau} = |t|$

3.2 Certainty-Equivalent Receding-Horizon FPD

The cooperation concept, inspected in Sects. 4 and 5, assumes that the environment model is obtained by learning it. The model candidates are parameterised

---

[1] The axiomatisation [19] also shows that any Bayesian DM formulation can be approximated by an FPD formulation to an arbitrary precision.

by time-invariant pmf values $\mathsf{m}(\tilde{x}|a, x, \theta) = \theta(\tilde{x}|a, x)$. The finite-dimensional parameter $\theta = (\theta(\tilde{x}|a, x))_{\tilde{x}, x \in x, a \in a}$ is unknown to the applied strategy. Thus, the strategy meets natural conditions of control [28] and the Bayesian learning can be used in the closed decision loop. The considered parametric environment model belongs to exponential family [3]. As such, it possesses self-reproducing Dirichlet's prior. Its finite-dimensional sufficient statistic is the occurrence array $\mathsf{v} = (\mathsf{v}(\tilde{x}|a, x))_{\tilde{x}, x \in x, a \in a}$, $\mathsf{v} > 0$, [14]. The environment model corresponding to the knowledge accessible by the agent is the predictor

$$\mathsf{m}(\tilde{x}|a, x, \mathsf{v}) = \frac{\mathsf{v}(\tilde{x}|a, x)}{\sum_{\tilde{x} \in x} \mathsf{v}(\tilde{x}|a, x)} = \hat{\theta}_{\mathsf{v}}(\tilde{x}|a, x), \quad \tilde{x}, x \in x, a \in a.$$

The pair $\mathcal{X} = (x, \mathsf{v})$ becomes the observable hyper-state so that Algorithm 1 is optimal with $\mathcal{X}$ replacing x. Practically, it is mostly infeasible as the time-varying value functions in Algorithm 1 depend on the hyper-state $\mathcal{X}$ of a huge dimension. At time $t \in t$, feasibility is recovered by the standard certainty-equivalent approximation with receding horizon $h \le |t| - t$, see e.g. [23,24], and extensive references there. It replaces the unknown parameter θ in $\mathsf{m}(\tilde{x}|a, x, \theta)$ by the current point estimate $\hat{\theta}_{\mathsf{v}_{t-1}}$. $\mathsf{m}(\tilde{x}|a, x, \hat{\theta}_{\mathsf{v}_{t-1}})$ serves as the environment model during the strategy design, which runs backward for $\tau = \bar{\tau} = t + h - 1, \dots, \underline{\tau} = t$ till $t + h - 1$, $h \le |t| - t + 1$. After applying the action a_t and observing the state x_t, the occurrence array v_{t-1} is updated, $\mathsf{v}_t(x_t|a_t, x_{t-1}) = \mathsf{v}_{t-1}(x_t|a_t, x_{t-1}) + 1$, and the procedure repeats, see Algorithm 2.

Algorithm 2. On-Line Certainty-Equivalent Receding-Horizon FPD

Inputs: Dimensions $|a|$, $|x|$, receding horizon h

            Initial state x_0, occurrence array $\mathsf{v}_0 > 0$ % prior belief description

            Factorised ideal closed model $\mathsf{c}_i = \mathsf{m}_i \mathsf{r}_i$ % preference description

Evaluations:

for real time $t = 1$ **to** $|t|$ **do**

    Get environment-model estimate $\mathsf{m}(\tilde{x}|a, x) = \frac{\mathsf{v}_{t-1}(\tilde{x}|a, x)}{\sum_{\tilde{x} \in x} \mathsf{v}_{t-1}(\tilde{x}|a, x)}$, $\forall \tilde{x}, x \in x, a \in a$

    Get $(\mathsf{r}_{0;\tau})_{\tau=t}^{t+h-1}$=**Algorithm 1**$(|a|, |x|, \underline{\tau} = t, \bar{\tau} = \min(t + h - 1, |t|), \mathsf{m}, \mathsf{m}_i, \mathsf{r}_i)$

    Sample action $a_t \sim \mathsf{r}_{0;t}(a|x_{t-1})$

Closed-loop outputs: Applied action a_t, state x_t observed on the environment

    Learn by updating the occurrence array $\mathsf{v}(x_t|a_t, x_{t-1}) = \mathsf{v}(x_t|a_t, x_{t-1}) + 1$

end for

Remarks

- Both algorithms have versions for continuous state and action spaces [15]. They are feasible for linear Gaussian models and their finite mixtures, [14,28].
- Algorithm 2 is presented in its rudimentary version. For instance, its computational complexity can be significantly decreased by iterations-spread-in-time strategy [14]. Its design omits the reset of g in Algorithm 1 and allows the use of the receding horizon close to one.

- The (approximately) optimal randomised FPD strategy is explorative. It is adaptive when employing forgetting [20], ideally, data-dependent as in [12].
- Undiscussed automated knowledge [6] and preference [5] elicitation would make Algorithm 2 (relatively) universal for single-agent DM. For a range of DM tasks, it is implementable into cheap portable devices. This makes the cooperation discussed in Sect. 4 realistically applicable.

4   Multiple Agents Sharing Ideal Closed-Loop Models

An agent mostly acts in the environment populated by other active agents. The agent should model them and respect their influence [10]. Such Bayesian games soon reach scalability limits as the learning and the strategy design become infeasible due to the quickly growing complexity of the handled DM elements.

The agent may ignore other agents and take them as non-modelled part of its environment. This feasible way may often lead to unfavourable results and calls for feasible countermeasures. Conceptually, the agent is to share an information with its *neighbours*. These are agents with whom its behaviour overlaps. Such sharing enables automated cooperation, negotiation [40] and conditions a conflict resolution. Quest for scalability admits only the information-sharing schemes working without a mediating center, i.e. a *flat cooperation structure*.

Agents exploiting FPD use the joint probabilistic ontology, which describes both their beliefs about environment and their DM preferences. This both enables generic flat cooperation schemes [16] and decreases the information sharing to a combination of probabilistic distributions, a classical pooling problem [8]. Among various possibilities, supra-Bayesian pooling fits the FPD framework. Its lack of a complete algorithmic solution is counteracted in [2, 30–32]. These solutions are *impartial* with respect to the involved agents. They have led to a tuning-knob-free solution [13], which may serve as a "universal" impartial pooling solver.

Tests of impartial solutions were relatively successful. However, except the specialised case [18], they focused on static DM tasks. Also, the universality is not for free. The proposed solutions do not differentiate importance, strength and other specific properties of interacting agents. The decoupling of the processing of the shared information from the ultimate DM aim is the price paid for universality.

This criticism motivates the research whose basic steps are presented here.

4.1   Cooperation Circumstances

Opening the way towards filling the differentiation-gap left by predecessors [30–32] is the main paper aim. To focus on it, a simple, but well-generalisable, flat cooperation is treated. It concerns FPD-using agents in a common environment. Superscript ^{k}, $k \in \boldsymbol{k}$, marks *DM elements of the kth agent*: the behaviour $b^k \in \boldsymbol{b}^k$, the environment model m^k, the ideal closed-loop model c_i^k and its factors, i.e. the ideal environment model m_i^k, and the ideal decision rule r_i^k.

Inspected agents are neighbours of an agent. It means that its behaviour overlaps with behaviours of neighbours and the agent is aware of existence of common variables in them. In the considered case, the environment state $x \in \boldsymbol{x}$ is the commonly accessible behaviour part. The kth agent generates its optimised actions $a^k \in \boldsymbol{a}^k$. Others may at most observe it. This enhances the DM quality but it is unconsidered as it makes no conceptual difference. We focus on yet-untested sharing of information about ideal models, i.e. about neighbours' preferences.

The inspected cooperation concerns *wise agents* who are willing to broadcast parts of pmfs (here ideal pfms) they use. Each agent utilises the information broadcasted by neighbours for the modification of its closed-loop ideal model employed in the strategy design. The agents remain *selfish*, and do not change their ideal closed-loop models according which they evaluate improvements achieved due to the information sharing.

4.2   Question Related to Pooling for FPD

The main questions encountered in pooling pmfs for FPD are:

1. How to pool the shared pmfs?
2. How to cope with the fact that behaviour sets of neighbours differ?
3. How to present the results to agents and how they should use them?
4. How to tune optional pooling parameters in order to support individual agents in approaching their disparate DM aims?

Questions 1, 2, 3 are mostly answered by predecessors, see below. Sect. 5 reflects the search for an insight indispensable for answering the open question 4.

Answer to question 1 follows [30–32]. The *pooled* ideal closed-loop model offered to kth agent $\tilde{\mathsf{c}}_i^k$ is to be a convex combination of the processed pmfs

$$\tilde{\mathsf{c}}_i^k = \sum_{j \in \boldsymbol{k}} \lambda_j^k \mathsf{c}_i^j, \quad \lambda_j^k \geq 0, \quad \sum_{j \in \boldsymbol{k}} \lambda_j^k = 1, \quad \forall k \in \boldsymbol{k}. \tag{6}$$

This excludes, for instance, the popular geometric pooling [8]. The referred papers select the weights $(\lambda_j^k)_{j \in \boldsymbol{k}}$ uniquely using the involved prior pmf and the impartiality requirement. When relaxing the latter, the weights become optional and allow the reflection of the ultimate pooling aim: the support of agent's DM.

Answer to question 2: The combination (6) is meaningful iff all agents operate on the same behaviour $\boldsymbol{b} = \boldsymbol{b}^k$, $k \in \boldsymbol{k}$, i.e. iff the involved agents know and model all variables entering the neighbours' behaviours. This is definitely unrealistic. In the inspected case, this would imply to know and model the behaviour on the super-set $\boldsymbol{b}$ of behaviours treated by all neighbours

$$\boldsymbol{b} = \left(x_t, (a_t^j)_{j \in \boldsymbol{k}} \right)_{t \in \boldsymbol{t}}. \tag{7}$$

Thus, the pooling (6) can be applied iff the shared pmfs are *extended on* $\boldsymbol{b}$.

The original neighbours' pmfs could be interpreted as marginal pmfs of the constructed extensions. The extensions are, however, not unique as proved in connection with the copula theory [26]. Even more importantly, the combined pmfs are generically incompatible. Then no extension having them as marginal pmfs exists. It is well seen on the considered pooling of the ideal closed-loop models. In this case, the cooperation is to counteract the fact that selfish agents have different preferences with respect to the common environment state. This reflects that an agent wants to reach its specific closed-loop behaviour by assigning the highest probability to it by the personally-chosen ideal closed-loop model.

It implies that the extension is to be approached as a search for a compromise. The search is a supporting decision task with the extended pmfs being the optional actions, cf. [13,17]. In the considered case, the solution reduces to application of the maximum entropy principle [33] to time-invariant factors of the ideal closed-loop models. When requiring the preservation of the kth closed-loop model and individual agents' strategies, the gained *extension* e_i^k, $k \in \boldsymbol{k}$, ignores unknown influence of neighbours' actions on the kth state as well as their unknown dependence. The resulting kth extension reads

$$\mathsf{e}_i^k(\tilde{x}, (a^j)_{j \in k}|x) = \mathsf{m}_i^k(\tilde{x}|a^k, x) \prod_{j \in k} \mathsf{r}_i^j(a^j|x), \quad k \in \boldsymbol{k}. \tag{8}$$

Answer to question 3: The use of (6) to extensions (8) gives the pooled closed-loop ideal model on the super-set $\boldsymbol{b}$ (7) of the behaviour sets $\boldsymbol{b}^k$. The kth agent is uninterested and even unaware of actions a^j, $j \in \boldsymbol{k} \setminus \{k\}$, complementing $\boldsymbol{b}^k$ to $\boldsymbol{b}$. Thus, it makes sense to present this agent only the relevant marginal pmf $\tilde{\mathsf{c}}_i^k$ of the pooled closed-loop ideal model. The result offered to kth agent is

$$\tilde{\mathsf{c}}_i^k(\tilde{x}, a^k|x) = \left[\lambda_k^k \mathsf{m}_i^k(\tilde{x}|a^k, x) + \sum_{j \in k \setminus \{k\}} \lambda_j^k \mathsf{f}_i^j(\tilde{x}|x) \right] \mathsf{r}_i^k(a^k|x), \text{ where}$$

$$\mathsf{f}_i^j(\tilde{x}|x) = \sum_{a^j \in a^j} \mathsf{m}_i^j(\tilde{x}|a^j, x)\mathsf{r}_i^j(a^j|x), \quad \tilde{x}, x \in \boldsymbol{x}, \quad a^k \in \boldsymbol{a}^k. \tag{9}$$

The *wise* agent k should use the ideal pmf $\tilde{\mathsf{c}}_i^k$ (9) for *designing* its strategy.

Towards answering question 4: The algorithmic choice of the weight λ_j^k, which kth agent assigns to jth neighbour, is yet unsolved. The solution direction is, however, obvious. As said, the kth agent uses the pooled ideal closed-loop model when designing its approximation of the FPD-optimal strategy. It has at disposal its original ideal. Thus, it can evaluate the action quality, after using the designed action and after observing the realised environment state. This enables to relate the weights $(\lambda_j^k)_{j \in k}$ to the reached DM quality and to design an additional feedback generating better weights for the subsequent design round.

A systematic design of the mentioned feedback is an important auxiliary DM task. Its solution needs a model relating the optional weights $(\lambda_j^k)_{j \in k}$ to the observable DM quality, which is quantified by the reached value of the original closed-loop ideal model c_i^k. The extensive experiments, whose samples are in

Sect. 5, primarily serve to the accumulation of experience needed for a feasible modelling of the relation of the weights to the truly reached DM quality.

Remarks

- The limited resources of an agent are helpful and make the solution scalable as the real agent has a small number $|k|$ of recognised neighbours.
- The weights λ_j^k, $j \in k$, are private for and specifically selected by kth agent.
- Pmf f_i^j (9) is an action-independent, ideal f *orecaster* offered by jth agent.
- Adaptive learning is inevitable as the agent uses, almost by definition, a simplified model of its environment containing other active agents, [11].
- The pooled ideal closed-loop model should be modified at each real time moment. This allows an adaptation to the varying set of neighbours, their changing ideal forecasters, as well as (foreseen) data-dependent changes of the λ-weights, driven by the selfish preferences and built in personal c_i^j.
- Agents' selfishness implies that equilibrium, if reachable, will be of Nash's type. An analysis will be possible after operationally resolving question 4.
- The cooperation via sharing environment models is algorithmically identical and desirable [18]. It should be used jointly with the discussed one.
- Omissions of the mentioned ways to improve "cheaply" agent's performance is driven by the wish to preserve the presentation simplicity.

4.3 Algorithmic Summary

This part summarises the proposed fully-scalable cooperation of wise, selfish, FPD-using agents. It shows that the computational costs paid by an agent for this cooperation are small. Broadcasting the information about shared closed-loop ideals is the probably most demanding operation. It only needs to broadcast the ideal forecasters $(f^j)_{j \in k}$ (9), possibly less often than the agents act.

Each agent $k \in k$ acts according to Algorithm 3, which modifies its ideal closed-loop model c_i^k to the pooled ideal $\tilde{c}_i^k$ (9). Otherwise, it coincides with Algorithm 2. The boxed $\boxed{\text{text}}$ in Algorithm 3 stresses the made changes.

Algorithm 3. FPD by Wise Selfish Cooperating Agent

Inputs: Agent's identifier $\underline{k} \in k$, dimensions $|a^k|$, $|x|$, receding horizon h^k

Initial state x_0, occurrence array $v_0^k > 0$ % prior belief description

Factorised ideal closed model $c_i^k = m_i^k r_i^k$ % preference description

The neighbours' ideal forecasters $\boxed{(f_i^j)_{j \in k \setminus \{k\}}}$ % of the state evolution

The cooperation weights $\boxed{(\lambda_j^k)_{j \in k}}$ % $\sum_{j \in k} \lambda_j^k = 1$, $\lambda_j^k \geq 0$

Evaluations:

Get the pooled ideal $\boxed{\tilde{c}_i^k = \lambda_k^k m_i^k r_i^k + \sum_{j \in k \setminus \{k\}} \lambda_j^k f_i^j}$ % cooperation

Outputs: $(a_{t^k}^k, x_{t^k})_{t^k \in t^k} = $**Algorithm 2**$(|a^k|, |x|, h^k, x_0, v_0^k, \tilde{c}_i^k)$

Remarks

- The agent identifier ^{k} delimits, which non-marginalised closed-loop ideal is used. Importantly, it stresses that all DM elements are fully under the agent's control, except of the environment state and external forecasters.
- The agent may work in a fully asynchronous mode and use a "personal" real time $t^k \in t^k$. This makes the advocated cooperation way quite flexible.

5   Experimental Part

The adopted concept is demonstrated on a simple well-understandable example. It exhibits all features of the general case and illustrates all notions used.

5.1   Simulation Set Up

The considered pair of agents, $k \in k = \{1, 2\}$ is interpreted as independent heaters influencing the common room temperature x. The quantised temperature is the observable state $x \in x = \{1, \ldots, 20\}$. Agents' actions are $a^k \in a^k = a = \{1, 2\} \equiv \{\text{off}, \text{on}\}$. The closed-loop behaviours are $b^k = (x_t, a_t^k)_{t \in t}$ up to $|t| = 500$.

The room is the common simulated environment modeled by the transition probabilities $\pi(x_t | a_t^1, a_t^2, x_{t-1})$, $x_t, x_{t-1} \in x$, $a_t^1, a_t^2 \in a$, $t \in t$. They are obtained via quantisation of the linear Gaussian model with the conditional moments

$$x_t\text{-mean} = 0.65(a_t^1 + a_t^2 - 2) + 0.96x_{t-1} - 0.02, \ x_t\text{-variance} = 0.25, \ x_0 = 10. \ (10)$$

The constants in (10) are chosen to: (a) imitate a slow response of the heated room; (b) make the influence of both actions the same; (c) make the highest temperature $x = 20$ reachable when one agent is heating only; (d) let the temperature fall to the lowest temperature $x = 1$ if both heaters are permanently off; and (e) let random effect be visible but not excessive.

A cooperation is vital as the *agents differ in ideal (desired) room temperatures*

$$x_i^1 = 12, \quad x_i^2 = 15.$$

The agents model their wishes by the ideal environment model, for $a^k = \text{"on"}$,

$$m_i^k(\tilde{x} | a^k = 2, x) = \begin{cases} 0.9 & \text{if } \tilde{x} = x_i^k \\ 0.05 & \text{if } \tilde{x} = x_i^k - 1 \\ 0.025 & \text{if } \tilde{x} = x_i^k + 1 \\ \text{uniform} & \text{otherwise} \end{cases} \quad k \in k = \{1, 2\}.$$

For the actions $a^k =$ "off" probabilities of $\tilde{x} = x_i^k \pm 1$ are swapped.

The ideal decision rules try to spare energy and prefer the action "off"

$$r_i^k(a^k = 1 = \text{"off"}|x^k) = \begin{cases} 0.9 \text{ if } x^k \geq x_i^k \\ 0.1 \text{ if } x^k < x_i^k \end{cases}, \quad k \in \boldsymbol{k} = \{1,2\}.$$

The agents recursively learn Markov models starting from the occurrence arrays

$$v_0^k = (\text{the model } (10) \text{ with the gain of the other action set to zero}) \times v^k.$$

The optional degrees of freedom $v^k > 0$ determine precision of the prior Dirichlet's distribution. The presented results correspond to the choice $v^k = 1$, $k \in \boldsymbol{k}$.

For $|\boldsymbol{k}| = 2$, each agent selects single cooperation weight $\lambda^k = \lambda_k^k$. The weights $\lambda_{j\neq k}^k = 1 - \lambda^k$. Simulations run for all pairs (λ^1, λ^2) on the grid $\lambda^k \in \boldsymbol{\lambda} = \{0, 0.1, \ldots, 0.9, 1.0\}$. The option $\lambda^k = 1$ means no cooperation. The kth agent accepts the ideal pmf of its neighbour as its own if $\lambda^k = 0$.

Each agent applies FPD, Algorithm 3, with the receding horizon $h = 5$ and the pooled ideal closed-loop model (6), $\tilde{x}, x \in \boldsymbol{x}$, $a^k \in \boldsymbol{a}^k$, for $\lambda^k \in \boldsymbol{\lambda}$, $k \in \boldsymbol{k} = \{1,2\}$,

$$\tilde{c}_i^1(\tilde{x}, a^1|x) = \left[\lambda^1 m_i^1(\tilde{x}|a^1, x) + (1 - \lambda^1) \sum_{a^2 \in a^2} m_i^2(\tilde{x}|a^2, x) r_i^2(a^2|x)\right] r_i^1(a^1|x)$$

$$\tilde{c}_i^2(\tilde{x}, a^2|x) = \left[\lambda^2 m_i^2(\tilde{x}|a^2, x) + (1 - \lambda^2) \sum_{a^1 \in a^1} m_i^1(\tilde{x}|a^1, x) r_i^1(a^1|x)\right] r_i^2(a^2|x).$$

The reached quality is judged via logarithm of the agent's ideal closed-loop model evaluated in the realised behaviour b^k. To make the result comparable with the value of the neighbour, this value is shifted by the absolute maximum

$$q^k = \ln(c_i^k(b^k)) - \max_{b^k \in b^k} \ln(c_i^k(b^k)), \quad k \in \boldsymbol{k}. \tag{11}$$

5.2   Commented Results

The results are influenced by the inherent asymmetry of the problem: the agent may contribute to the temperature increase but the decrease depends only on the environment dynamics and on the realisation of random influences.

Figures 1, 2, and 3 present time courses of the room temperature. Figure 1 corresponds to the cooperation coefficients $(\lambda^1, \lambda^2) = (0.1, 0.1)$ for which q^1 reaches its highest value. Figure 2 corresponds to the cooperation coefficients $(\lambda^1, \lambda^2) = (0.8, 0.2)$ for which q^2 reaches its highest value. Figure 3 corresponds to the combination of cooperation coefficients $(\lambda^1, \lambda^2) = (0.2, 0.3)$ for which the impartially judged joint quality $q^1 + q^2$ reaches its maximum.

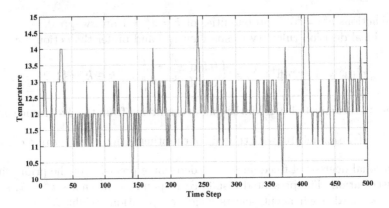

Fig. 1. The best temperature trajectory, maximising q^1 (11), for the agent $k = 1$ wishing the temperature $x_i^1 = 12$. It is reached for the weights $(\lambda^1, \lambda^2) = (0.1, 0.1)$.

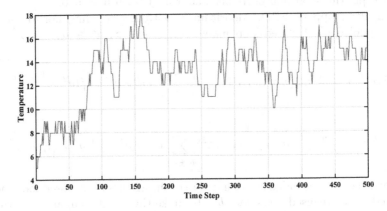

Fig. 2. The best temperature trajectory, maximising q^2 (11), for the agent $k = 2$ wishing the temperature $x_i^2 = 15$. It is reached for the weights $(\lambda^1, \lambda^2) = (0.8, 0.2)$.

The results primarily show that the value of the unchanged ideal closed-loop model in the measured data (11) is indeed a good indicator of the closed-loop quality from the agent's view point. This confirms the chance for a successful data-dependent choice of λ^k. Also, the example: (a) illustrates the theory; (b) confirms that the information sharing influences the achieved closed-loop behaviour; (c) shows asymmetry of the chosen environment; (d) indicates that possible Nash's equilibria could be searched around the maximum of $\sum_{k \in k} q^k$.

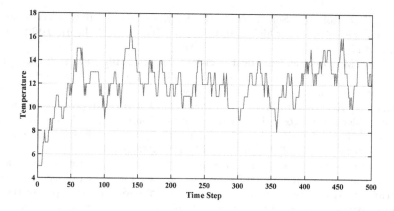

Fig. 3. The temperature trajectory, which is expected to be the best reachable compromise, maximising $q^1 + q^2$ (11), between wishes $x_i^1 = 12$ d $x_i^2 = 15$. It is reached for the weights $(\lambda^1, \lambda^2) = (0.2, 0.3)$.

6   Concluding Remarks

This paper inspects a cooperation methodology for a flatly-interacting multiple agents. It is based on sharing of ideal closed-loop models for FPD. It confirms the chance to adapt the cooperation-controlling weights in closed loop according to selfish aims of respective agents. At the same time, the accumulated experience demonstrates that sharing solely some DM elements does not guarantee a high decision quality. Other DM elements, as the learnt environment model, have to be shared. FPD makes it an identical task. Also, other unused possibilities as prior knowledge elicitation, forgetting, exploitation/exploration balance have to be exploited to get a robust practical tool. These measures will be addressed in near future. Good news is that this way is promising and feasible. All induced tasks are solvable at single-agent level with a negligible deliberation overhead caused by acting in multi-participant environment. The paper exemplifies this.

References

1. Aknine, S., Caillou, P., Pinson, S.: Searching Pareto optimal solutions for the problem of forming and restructuring coalitions in multi-agent systems. Group Decis. Negot. **19**, 7–37 (2010)
2. Azizi, S., Quinn, A.: Hierarchical fully probabilistic design for deliberator-based merging in multiple participant systems. IEEE Trans. Syst. Man Cybern. PP(99), 1–9 (2016). https://doi.org/10.1109/TSMC.2016.2608662
3. Barndorff-Nielsen, O.: Information and Exponential Families in Statistical Theory. Wiley, Hoboken (1978)
4. Bond, A., Gasser, L.: Readings in Distributed Artificial Intelligence. Morgan Kaufmann, Burlington (2014)
5. Chen, L., Pu, P.: Survey of preference elicitation methods. Technical report IC/2004/67, HCI Group Ecole Politechnique Federale de Lausanne, Switzerland (2004)

6. Daee, P., Peltola, T., Soare, M., Kaski, S.: Knowledge elicitation via sequential probabilistic inference for high-dimensional prediction. Mach. Learn. **106**(9), 1599–1620 (2017)
7. DeGroot, M.: Optimal Statistical Decisions. McGraw-Hill, New York City (1970)
8. Genest, C., Zidek, J.: Combining probability distributions: a critique and annotated bibliography. Stat. Sci. **1**(1), 114–148 (1986)
9. Guan, P., Raginsky, M., Willett, R.: Online Markov decision processes with Kullback Leibler control cost. IEEE Trans. Autom. Control **59**(6), 1423–1438 (2014)
10. Harsanyi, J.: Games with incomplete information played by Bayesian players. I-III. Management Science **50**(Suppl. 12), 1763–1893 (2004)
11. Kárný, M.: Adaptive systems: local approximators? In: Workshop on Adaptive Systems in Control and Signal Processing, pp. 129–134. IFAC, Glasgow (1998)
12. Kárný, M.: Recursive estimation of high-order Markov chains: approximation by finite mixtures. Inf. Sci. **326**, 188–201 (2016)
13. Kárný, M.: Implementable prescriptive decision making. In: Guy, T., Kárný, M., D., D.R.I., Wolpert, D. (eds.) Proceedings of the NIPS 2016 Workshop on Imperfect Decision Makers, vol. 58, pp. 19–30. JMLR (2017)
14. Kárný, M., et al.: Optimized Bayesian Dynamic Advising: Theory and Algorithms. Springer, London (2006)
15. Kárný, M., Guy, T.V.: Fully probabilistic control design. Syst. Control Lett. **55**(4), 259–265 (2006)
16. Kárný, M., Guy, T.V., Bodini, A., Ruggeri, F.: Cooperation via sharing of probabilistic information. Int. J. Comput. Intell. Stud. **1**, 139–162 (2009)
17. Kárný, M., Guy, T.: On support of imperfect Bayesian participants. In: Guy, T., et al. (eds.) Decision Making with Imperfect Decision Makers. Intelligent Systems Reference Library, vol. 28, pp. 29–56. Springer, Berlin (2012). https://doi.org/10.1007/978-3-642-24647-0_2
18. Kárný, M., Herzallah, R.: Scalable harmonization of complex networks with local adaptive controllers. IEEE Trans. SMC: Syst. **47**(3), 394–404 (2017)
19. Kárný, M., Kroupa, T.: Axiomatisation of fully probabilistic design. Inf. Sci. **186**(1), 105–113 (2012)
20. Kulhavý, R., Zarrop, M.B.: On a general concept of forgetting. Int. J. Control **58**(4), 905–924 (1993)
21. Kullback, S., Leibler, R.: On information and sufficiency. Ann. Math. Stat. **22**, 79–87 (1951)
22. Lewicki, R., Weiss, S., Lewin, D.: Models of conflict, negotiation and 3rd party intervention - a review and synthesis. J. Organ. Behav. **13**(3), 209–252 (1992)
23. Mattingley, J., Wang, Y., Boyd, S.: Receding horizon control. IEEE Control Syst. Mag. **31**(3), 52–65 (2011)
24. Mayne, D.: Model predictive control: recent developments and future promise. Automatica **50**, 2967–2986 (2014)
25. Mine, H., Osaki, S.: Markovian Decision Processes. Elsevier, New York (1970)
26. Nelsen, R.: An Introduction to Copulas. Springer, Heidelberg (1999). https://doi.org/10.1007/978-1-4757-3076-0
27. Nurmi, H.: Resolving group choice paradoxes using probabilistic and fuzzy concepts. Group Decis. Negot. **10**, 177–198 (2001)
28. Peterka, V.: Bayesian system identification. In: Eykhoff, P. (ed.) Trends and Progress in System Identification, pp. 239–304. Pergamon Press, Oxford (1981)
29. Savage, L.: Foundations of Statistics. Wiley, Hoboken (1954)

30. Sečkárová, V.: Cross-entropy based combination of discrete probability distributions for distributed decision making. Ph.D. thesis, Charles University in Prague, Faculty of Mathematics and Physics, Department of Probability and Mathematical Statistics., Prague (2015). Submitted in May 2015. Successfully defended on 14 September 2015
31. Sečkárová, V.: Weighted probabilistic opinion pooling based on cross-entropy. In: Arik, S., Huang, T., Lai, W.K., Liu, Q. (eds.) ICONIP 2015. LNCS, vol. 9490, pp. 623–629. Springer, Cham (2015). https://doi.org/10.1007/978-3-319-26535-3_71
32. Sečkárová, V.: On supra-Bayesian weighted combination of available data determined by Kerridge inaccuracy and entropy. Pliska Stud. Math. Bulgar. **22**, 159–168 (2013)
33. Shore, J., Johnson, R.: Axiomatic derivation of the principle of maximum entropy and the principle of minimum cross-entropy. IEEE Trans. Inf. Theory **26**(1), 26–37 (1980)
34. Simpson, E.: Combined decision making with multiple agents. Ph.D. thesis, Hertford College, Department of Engineering Science, University of Oxford (2014)
35. Simpson, E., Roberts, S., Psorakis, I., Smith, A.: Dynamic Bayesian combination of multiple imperfect classifiers. In: Guy, T., Kárný, M., Wolpert, D. (eds.) Decision Making and Imperfection. Studies in Computation Intelligence, vol. 474, pp. 1–35. Springer, Berlin (2013). https://doi.org/10.1007/978-3-642-36406-8_1
36. Todorov, E.: Linearly-solvable Markov decision problems. In: Schölkopf, B., et al. (eds.) Advances in Neural Information Processing, pp. 1369–1376. MIT Press, NY (2006)
37. Šindelář, J., Vajda, I., Kárný, M.: Stochastic control optimal in the Kullback sense. Kybernetika **44**(1), 53–60 (2008)
38. Wald, A.: Statistical Decision Functions. Wiley, London (1950)
39. Wallenius, J., Dyer, J., Fishburn, P., Steuer, R., Zionts, S., Deb, K.: Multiple criteria decision making, multiattribute utility theory: recent accomplishments and what lies ahead. Manag. Sci. **54**(7), 1336–1349 (2008)
40. Zlotkin, G., Rosenschein, J.: Mechanism design for automated negotiation and its applicatin to task oriented domains. Artif. Intell. **86**, 195–244 (1996)

Endorsement in Referral Networks

Ashiqur R. KhudaBukhsh[✉] and Jaime G. Carbonell

Carnegie Mellon University, Pittsburgh, USA
{akhudabu,jgc}@cs.cmu.edu

Abstract. Referral networks is an emerging research area in the intersection of Active Learning and Multi-Agent Systems where experts—humans or automated agents—can redirect difficult instances (tasks or queries) to appropriate colleagues. *Learning-to-refer* involves estimating topic-conditioned skills of colleagues connected through a referral network for effective referrals. *Proactive skill posting* is a learning setting where experts are allowed a one-time local network advertisement of a subset of their top skills. The learning challenge is exploiting partially available (potentially noisy) self-skill estimates, including adversarial strategic lying to attract unwarranted referrals. In this paper, we introduce the notion of endorsement typically found in professional networks where one colleague endorses another on particular topic(s). We first augment proactive skill posting with endorsements and propose modifications to existing algorithms to take advantage of such endorsements, penalizing subsequent referrals to agents with bogus skill reporting. Our results indicate that truthful endorsements improve performance as they act as an additional cushion to early failures of strong experts. When combined with truthful endorsements, extensive empirical evaluations indicate performance improvement in proactive-DIEL and ϵ-Greedy in both market-aware and market-agnostic skill posting setting while retaining desirable properties like tolerance to noisy self-skill estimates and strategic lying.

Keywords: Active Learning · Referral networks ·
Proactive skill posting

1 Introduction

Referral networks is an emerging research area in the intersection of Active Learning and Multi-Agent Systems [1] where experts—humans or automated agents—can redirect difficult instances (tasks or queries) to appropriate colleagues. The *learning-to-refer* challenge involves estimating the topic-conditioned skills of colleagues for effective referrals. Inspired by the real-world phenomenon that experts often tell their colleagues about their areas of strength, *proactive skill posting* [2] is a learning setting where experts are allowed a one-time local-network advertisement (only restricted to directly connected colleagues) of a subset of their skills.

© Springer Nature Switzerland AG 2019
M. Slavkovik (Ed.): EUMAS 2018, LNAI 11450, pp. 172–187, 2019.
https://doi.org/10.1007/978-3-030-14174-5_12

From a distributed Machine Learning perspective, proactive skill posting is effectively learning in presence of partially available (potentially noisy) priors, and the key challenges are (1) suitable algorithm initialization to use such priors when available or bound priors when unavailable (2) devising penalty mechanism to discourage wilful misreporting and (3) tolerance to noisy self-skill estimates. Previous work in referral networks has proposed modifications to existing action selection algorithms and demonstrated superior cold-start performance while ensuring incentive compatibility in simpler settings without endorsements.

In this paper, we significantly generalize the notion of side information beyond self-advertised noisy priors to include endorsements from other colleague experts. Such endorsements are common in real professional networks, such as LinkedIn. In addition to self-advertisement of skills, in professional networks, we often see endorsements coming from colleagues: *Joe is strong at Java Programming which is endorsed by eleven people including Sally who is an established expert in the field.* In the proposed augmented setting, a colleague can endorse another colleague on a topic. Endorsements can help in two ways: Endorsements can nudge referral decisions towards a candidate with sparse observations (few trials or samples). Endorsements can also offset bad experiences, e.g., a few early observed failures from an otherwise strong expert may relegate that expert to never again receive referrals on her topic of expertise, unless offset by colleague endorsements.

Our contributions are the following: (1) we augment proactive skill posting with endorsements and (2) propose algorithmic modifications to take endorsements into account. (3) Extensive empirical evaluation demonstrates our proposed algorithms outperform proactive-DIEL, the state-of-the-art, both across *market-aware* and *market-agnostic* settings, while retaining desirable properties like tolerance to noisy self-skill estimates and robustness to strategic lying. (4) Our algorithmic modification translates into proactive algorithms of other mean-based method such as ϵ-Greedy [3].

2   Related Work

The referral framework draws inspiration from *referral chaining*, first proposed in [4] and subsequently extended in [5–8]. Previous research in referral networks has primarily focused on three major directions (1) evaluating viability of the referral networks under uninformed prior settings [1] (2) proactive skill posting setting where the main focus is to design truthful algorithms to deal with (potentially noisy) partially available prior [2,9] and (3) robustness criteria such time-varying expertise [10], evolving networks [11] and capacity constraints [1]. To the best of our knowledge, our main contribution in this paper, augmenting skill posting with endorsements, has not been previously considered in the referral networks literature.

Proactive skill posting is also related to the bandit literature with *side-information* [12,13] in the sense that algorithms do not start from scratch. However, a key difference is that, instead of requiring observed trials [13] or knowing the shape of the reward distribution [14], the *side-information* in our case is obtained through advertisement of skills by the experts themselves (who may in fact willfully misreport to attract more business) and subsequent endorsements from colleagues. This ties our work broadly to adversarial machine learning [15] and truthful mechanism design [16–19]. Our work is different from the bandit literature and other mechanism design approach (e.g., [17]) in the following key ways: (1) there is a clear scale distinction, as unlike traditional multi-armed bandits (MAB), referral networks can be viewed as several multi-expert multi-topic parallel MAB threads (2) proactive skill posting deals with partially available (potentially noisy) priors.

Endorsement connects our work to the vast literature of computational trust and reputation based models [20–23]. Our proposed approach compares and contrasts with the existing literature on computational trust and reputation based systems in the following key ways: Unlike [24–26], in our conditional sampling commitment approach, we consider both *direct trust evidence* (directly observed performance) and *indirect trust evidence* (endorsements). Similar to [27], our trust aggregation method depends on the number of observed samples as we only consider endorsements seriously when direct experience is sparse.

3 Background

3.1 Motivation

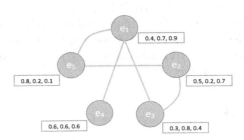

Fig. 1. A referral network with five experts.

Before presenting the technical background for this work, we first illustrate the effectiveness of appropriate referrals with a small simplified example of a referral network with five experts shown in Fig. 1 (this example is taken from [1]). The nodes of the graph represent the experts, and the edges indicate a potential referral link, i.e., the experts 'know' each other and can send or receive referrals and communicate results. Note that, even for a small network of five experts, we chose a graph that is not fully connected reflecting that in real-world, it is infeasible to assume every expert knows every other expert in the network. Consider three different topics – call them t_1, t_2, and t_3 – and the figures in brackets indicate an expert's topical expertise (probability of solving a given task) in each of these.

In the example, with a query belonging to t_2, without any referral, the client may consult first e_2 and then possibly e_5. This would lead to a solution probability of $0.2 + (1 - 0.2) \times 0.2 = 0.36$ (the probability that e_2 fails to solve a task belonging to t_2 is $1 - 0.2 = 0.8$). With referrals, an expert handles a problem she knows how to answer, and otherwise if she had knowledge of all the other connected colleagues' expertise, e_2 could refer to e_3 for the best skill in t_2, leading to a solution probability of $0.2 + (1 - 0.2) \times 0.8 = 0.84$.

As shown in the above example, appropriate referrals can substantially improve performance, however, the key challenge for *learning-to-refer* is to estimate the topic-conditioned skills. Prior research has demonstrated with suitable algorithm initialization and a penalty mechanism to discourage strategic lying, *proactive skill posting*, a setting where experts report a subset of (potentially noisy) priors to their colleagues, can effectively address the cold-start problem. In this work, we move one step further where experts can endorse certain advertised skills of their colleagues. We divide the necessary technical background for this work into three parts. First, we present basic notation, definitions and assumptions for the general referral framework. Next, we provide technical details for proactive skill posting. We consider (a) market-agnostic and [2,28] (b) market-aware proactive skill posting [9]. Finally, we present the mechanism for endorsement, and describe algorithmic modifications to take advantage of endorsements.

3.2   Referral Network Preliminaries

We summarize our basic notation, definitions, and assumptions, mostly from [1, 2], where further details regarding expertise, network parameters, proactive skill posting mechanism and simulation details can be found.

Referral network: Represented by a graph (V, E) of size k in which each vertex v_i corresponds to an expert e_i $(1 \leq k)$ and each bidirectional edge $\langle v_i, v_j \rangle$ indicates a *referral link* which implies e_i and e_j can co-refer problem instances.

Subnetwork of an expert e_i: The set of experts linked to an expert e_i by a referral link.

Scenario: Set of m instances $(q_1, \ldots, q_m)$ belonging to n topics $(t_1, \ldots, t_n)$ addressed by the k experts $(e_1, \ldots, e_k)$.

Expertise: Expertise of an expert/question pair $\langle e_i, q_l \rangle$ is the probability with which e_i can solve q_l.

Referral mechanism: For a query budget Q (following [1,2], we kept fixed to $Q = 2$ across all our current experiments), this consists of the following steps.

1. A user issues an *initial query* q_l to a randomly chosen *initial expert* e_i.
2. The initial expert e_i examines the instance and solves it if possible. This depends on the *expertise* of e_i w.r.t. q_l.

3. If not, a *referral query* is issued by e_i to a *referred expert* e_j within her subnetwork, with a query budget of $Q - 1$. *Learning-to-refer* involves improving the estimate of who is most likely to solve the problem.
4. If the referred expert succeeds, she sends the solution to the initial expert, who sends it to the user.

Note that, while privacy is not a key focus in this work, there is no direct communication between the user and the *referred expert*, hence user and the *referred expert*'s identities are protected from each other.

Parallel to Multi-armed Bandits: In a multi-armed bandit setting, there are K arms with unknown reward distributions $D_1, \ldots, D_K$; each pull of arm i reveals a reward drawn from D_i. The goal is to maximize the sum of expected rewards. It is intuitive to spot the parallel between the classic multi-armed bandit problem and referral decision since, from the point of view of a single expert on a given topic, the referral decision is indeed an action selection problem. Similar to the unknown rewards distributions of the arms, in this case, the challenge is to estimate the unknown expertise distributions. Consequently, the *learning-to-refer* challenge in an uninformed prior setting has been suitably addressed by several well-known multi-armed bandit algorithms such as DIEL [29], Thompson Sampling [30], and ϵ-Greedy [3]. In fact, several of these algorithms form the basic building blocks for *proactive skill posting* and proactive skill posting with endorsements.

In a distributed setting, each expert maintains an action selection thread for each topic in parallel. In order to describe an action selection thread, we first fix topic to T and expert to e. Let $q_1, \ldots, q_N$ be the first N referred queries belonging to topic T issued by expert e to any of her K colleagues denoted by $e_1, \ldots, e_K$. For each colleague e_i, e maintains a reward vector $\mathbf{r}_{j,n_{e_j}}$ where $\mathbf{r}_{j,n_{e_j}} = (r_{j,1}, \ldots, r_{j,n_{e_j}})$, i.e., the sequence of rewards observed from expert e_j on issued n_{e_j} referred queries. Understandably, $N = \sum_{j=1}^{K} n_{e_j}$. Let $m(e_j)$ and $s(e_j)$ denote the sample mean and sample standard deviation of these reward vectors. Some of the algorithms we consider require initializing these reward vectors; we will explicitly mention any such initialization. In addition to the reward vectors, for each colleague e_j, e maintains S_{e_j} and F_{e_j} where S_{e_j} denotes the number of observed successes (reward $= 1$) and F_{e_j} denotes the number of observed failures (reward $= 0$). Clearly, without any initialization of the reward vectors, $\forall (S_{e_j} + F_{e_j}) > 0$, $m(e_j) = \frac{S_{e_j}}{S_{e_j} + F_{e_j}}$ (i.e., empirical mean is the ratio of total number of observed successes and total number of observations).

Like any other action selection problem, *learning-to-refer* also poses the classic exploration-exploitation trade-off: on one hand, we would like to refer to an expert who has performed well in the past on this topic (exploitation), while ensuring enough exploration to make sure we are not missing out on stronger experts. At a high level, each of the algorithms computes a score for every expert e_j (denoted by $score(e_j)$) and selects the expert with the highest combined score breaking any remaining ties randomly. We first provide the preliminaries for proactive skill posting and then a description of proactive-DIEL[1] for the more general *market-agnostic* setting.

3.3 Proactive Skill Posting Preliminaries

Advertising Unit: A tuple $\langle e_i, e_j, t_l, \mu_{e_j,t_l}^{adv} \rangle$, where e_i is the *target expert*, e_j is the *advertising expert*, t_l is the topic and μ_{e_j,t_l}^{adv} is e_j's (advertised) topical expertise. When e_j is truthful and can accurately estimate her own skills, $\mu_{e_j,t_l}^{adv} = \mu_{e_j,t_l}$, the true topical expertise.

Algorithm 1: proactive-DIEL(e, T)

Initialization:

$\forall e_j \in subnetwork(e), n_{e_j} \leftarrow 2$

foreach $e_j \in subnetwork(e)$ **do**

    if T is *explicit bid* **then**

        | $\mathbf{r}_{j,n_{e_j}} \leftarrow (\mu_{e_j,T}^{adv}, \mu_{e_j,T}^{adv})$

    else

        | $\mathbf{r}_{j,n_{e_j}} \leftarrow (\mu_{e_j,t_{secondBest}}^{adv}, \mu_{e_j,t_{secondBest}}^{adv})$

    end

end

Main loop:

foreach *referred query* **do**

    foreach $e_j \in subnetwork(e)$ **do**

        | $score(e_j) = m(e_j) + \dfrac{s_{e_j}}{\sqrt{n_{e_j}}}$

    end

    $best = \arg\max_j score(e_j)$

    Observe reward r and compute penalty p after referring to e_{best}

    Update $\mathbf{r}_{best,n_{e_{best}}}$ with $r - p$, $n_{e_{best}} \leftarrow n_{e_{best}} + 1$

end

Advertising Budget: The number of advertising units available to an expert, following [28], set to twice the size of that expert's subnetwork; each expert reports her top two skills to her subnetwork. Regardless of e_j's strategic behavior or accuracy in self-skill estimates, we further assume that e_j's advertisement on a given topic is identical across all colleagues.

---

[1] In the original paper [2], proactive-DIEL is referred to as proactive-DIEL$_t$ because it uses a trust-based mechanism for incentive compatibility. Since both proactive-DIEL$_t$ and proactive-DIEL$_\Delta$ use the same penalty mechanism, we drop the t subscript from *market-agnostic* algorithms and use the Δ subscript to indicate *market-awareness*.

Advertising Protocol: A one-time advertisement that happens at the beginning of the simulation or when an expert joins the network. The advertising expert e_j reports to each target expert e_i in her subnetwork the two tuples $\langle e_i, e_j, t_{best}, \mu^{adv}_{e_j, t_{best}} \rangle$ and $\langle e_i, e_j, t_{secondBest}, \mu^{adv}_{e_j, t_{secondBest}} \rangle$, i.e., the top two topics in terms of the advertising expert's topic means.

In our experiments, we consider both traditional (*market-agnostic*) [2] and *market-aware* [9] skill posting. In *market-agnostic* setting, at the beginning of the simulation, an expert has knowledge about only her own skills, hence, for a given expert, $\mu_{t_{best}}$ is simply her maximum topical expertise. In *market-aware* proactive skill posting, every expert has access to an estimate of $\overline{\mu_{t_l}}$ (average network skill on each topic t_l) and reports the skills with her largest relative advantage μ_Δ (where for a given expert/topic pair $\langle e_j, t_l \rangle$, $\mu_{\Delta_{e_j, t_l}} = \mu_{e_j, t_l} - \overline{\mu_{t_l}}$).

Explicit Bid: A topic advertised in the above protocol.

Implicit Bid: A topic that is not advertised, for which an upper skill bound is estimated. For *market-agnostic* skill posting, it is simply the second highest topical expertise, $\mu^{adv}_{e_j, t_{secondBest}}$. For *market-aware* skill posting, a similar bound is computed (further details can be found in [9]).

proactive-DIEL: We now describe the state-of-the-art algorithm in *market-agnostic* skill posting setting. Proactive-DIEL is based on Distributed Interval Estimation Learning (DIEL), the known state-of-the-art referral learning algorithm [1] in uninformed prior setting. First proposed in [29], Interval Estimation Learning (IEL) has been extensively used in stochastic optimization [31] and action selection problems [32,33]. Action selection using DIEL works in the following way. At each step, DIEL selects the expert e_j with highest $m(e_j) + \frac{s(e_j)}{\sqrt{n_{e_j}}}$. In the uninformed prior setting, every action is initialized with two rewards of 0 and 1, allowing us to initialize the mean and variance.

The intuition behind selecting an expert with a high expected reward ($m(e_j)$) and/or a large amount of uncertainty with respect to topical expertise in the reward ($s(e_j)$) is the following: A large variance implies greater uncertainty, indicating that the expert has not been sampled with sufficient frequency to obtain reliable skill estimates. Selecting such an expert is an *exploration step* which will increase the confidence of e in her estimate. Also, such steps have the potential of identifying a highly skilled expert, whose earlier skill estimate may have been too low. Selecting an expert with a high $m(e_j)$ amounts to exploitation. Initially, the initial choices made by e tend to be explorative since the intervals are large due to the uncertainty of the reward estimates. With an increased number of samples, the intervals shrink and the referrals become more exploitative.

As shown in Algorithm 1, proactive-DIEL differs from DIEL in two key ways. First, unlike the uninformed DIEL where actions are initialized with two rewards of 0 and 1, proactive-DIEL favors an informed exploration. In case of an *explicit bid*, it initializes with two rewards of the advertised prior. For an *implicit bid*, it uses the second-best skill as an upper-bound. Second, once an expert is chosen and the reward is observed, a penalty is computed using a *penalty on distrust* mechanism. The penalty estimates how much the advertised skills are off from the sample means (for further details, see [2]).

4 Endorsement

Endorsement Unit: We assume that an endorsement unit is a tuple $\langle e_k, e_j, t_l \rangle$, where e_k is the *endorsing expert*, e_j is the *advertising expert*, and t_l is the topic. An endorsement implies that e_k has observed at least M samples from e_j on topic t_l (i.e., e_k has referred M or more tasks/queries belonging to t_l to e_j) and the sample mean is higher or equal to the advertised prior. In this work, we consider truthful endorsement. Let $S(e_k, e_j, t_l)$ and $F(e_k, e_j, t_l)$ denote the number of successes and failures of e_j on topic t_l observed by e_k, respectively. The endorsement condition implies: $\frac{S(e_k,e_j,t_l)}{S(e_k,e_j,t_l)+F(e_k,e_j,t_l)} \geq \mu_{e_k,t_l}^{adv}$, and $S(e_k, e_j, t_l) + F(e_k, e_j, t_l) \geq M$. For all our experiments, we set M to 40.

Endorsement Budget: The maximum number of endorsements an expert can perform. It is unlikely that an expert will exhibit similar level of eagerness to endorse others as she would in advertising her own set of skills. Hence, we assume that the maximum number of endorsements an expert can perform is half the advertisement budget, i.e., equal to the size of that expert's subnetwork.

Endorsement Visibility: We assume *friend-of-friend* visibility, i.e., $\langle e_k, e_j, t_l \rangle$ is visible to everyone who is either a colleague of e_j or has a shared colleague with e_j, or both. We denote the friend-of-friend of e_j as *friend-of-friend*(e_j).

Endorsement Protocol and Proactive-DIEL^E: We are now ready to describe our main algorithmic contribution in this paper, proactive-DIEL^E. As shown in Algorithm 2, endorsement is used as a conditional sampling commitment. During any point in the horizon, an *endorsing expert* can endorse a particular colleague expert on a given topic only once and only if all the conditions are met, i.e., sufficient number of observed samples, and performance is equal or better than the advertised prior. Whenever an expert e_j is endorsed by a new colleague on topic t_l, e_j receives additional sampling commitment of w referrals on t_l from all other experts belonging to *friend-of-friend*(e_j) (excluding the endorsing expert e_k). For all our experiments, we set w to 5.

Proactive-DIEL^E uses the same initialization and *penalty on distrust* mechanism as Proactive-DIEL. Few key points to note: As shown in Algorithm 2, endorsement is applicable only when the colleague in question has not been sampled enough ($\frac{M}{2}$ samples or less); endorsements do not affect colleagues with sufficient observation. Second, since endorsements are additive, multiple endorsements from different colleagues on the same topic will result in higher sampling commitments. However, at the maximum, endorsements can fetch MK referrals ($\frac{M}{2}$ for each of the top two topics, K being the subnetwork size of the concerned expert). Hence, endorsement alone cannot help an expert beyond a constant number of referrals, in order to get a continued stream of referrals, eventually, the expert has to deliver. Third, and the most crucial: endorsement does not

---

Algorithm 2: proactive-DIEL$^E(e, T)$

Initialization:
$\forall e_j \in subnetwork(e), n_{e_j} \leftarrow 2$
foreach $e_j \in subnetwork(e)$ **do**
    if T is *explicit bid* **then**
        $\mid$ $\mathbf{r}_{j,n_{e_j}} \leftarrow (\mu^{adv}_{e_j,T}, \mu^{adv}_{e_j,T})$
    else
        $\mid$ $\mathbf{r}_{j,n_{e_j}} \leftarrow (\mu^{adv}_{e_j,t_{secondBest}}, \mu^{adv}_{e_j,t_{secondBest}})$
    end
end

Main loop:
foreach *referred query* **do**
    foreach $e_j \in subnetwork(e)$ **do**
        if $commitment(e_j, T) > 0$ && $n_{e_j} - 2 \leq \frac{M}{2}$ **then**
        $\mid$ $score(e_j) \leftarrow max((m(e_j) + \frac{s_{e_j}}{\sqrt{n_{e_j}}}), \mu^{adv}_{e_j,T})$
        else
        $\mid$ $score(e_j) \leftarrow m(e_j) + \frac{s_{e_j}}{\sqrt{n_{e_j}}}$
        end
    end
    $best = \arg\max_j score(e_j)$
    $secondBest = \arg\max_{j \neq best} score(e_j)$
    Observe reward r and penalty p after referring to e_{best}
    Update $\mathbf{r}_{best,n_{e_{best}}}$ with $r - p$, $n_{e_{best}} \leftarrow n_{e_{best}} + 1$
    if $score(e_{secondBest}) > m(e_{best}) + \frac{s_{e_{best}}}{\sqrt{n_{e_{best}}}}$ **then**
    $\mid$ $commitment(e_{best,T}) \leftarrow commitment(e_{best,T}) - 1$
    end
    if $n_{e_{best}} - 2 \geq M$&&$\frac{S(e,e_{best},T)}{S(e,e_{best},T)+F(e,e_{best},T)} \geq \mu^{adv}_{e_{best},T}$ **then**
        if $hasEndorsed(e, e_{best}, T) = 0$ **then**
            $endorse(e, e_{best}, T)$
            $hasEndorsed(e, e_{best}, T) \leftarrow 1$
        end
    end
end

---

necessarily imply guaranteed referral; it only acts as an additional cushion such that algorithms do not get blindsided by early failures of strong experts.

The primary difference between proactive-DIEL$^E_\Delta$ and Proactive-DIEL^E is in their initialization scheme; Proactive-DIEL$^E_\Delta$ uses a different prior-bounding technique proposed in [9].

5   Experimental Setup

Data Set: as our synthetic data set[2], we used the same 1000 scenarios used in [2, 28]. Each scenario consists of 100 experts connected through a referral network (average connection density: 16 ± 5) and 10 topics. Further details regarding the data set generator's parameter configuration can be found in [28].

Baselines: For a given proactive skill posting setting (*market-aware* or *market-agnostic*), our baseline is the previously known top-performing algorithm. For *market-aware* setting, we considered proactive-DIEL$_\Delta$ [9] and for the more general, *market-agnostic* setting, we considered proactive-DIEL [9]. Our upper bound is the performance of a network where every expert has access to an oracle that knows the true topical expertise of every expert-topic pair.

Algorithm Configuration: We used the same parameter configurations for proactive-DIEL and proactive-ϵ-Greedy as reported in [2]. M is set to 40 and w is set to 5. The two new parameters are configured using a smaller background data set with similar distributional properties.

Performance Measure: Following [1,2], we use the same performance measure, overall task accuracy of our multi-expert system, as in previous work in referral networks. So if a network receives n tasks of which m tasks is solved (either by the *initial expert* or the *referred expert*), the overall task accuracy is $\frac{m}{n}$. Q, the per-instance query budget, is set to 2. Each algorithm is run on the data set of 1000 referral networks and the average over such 1000 simulations is reported in our results section. In order to facilitate comparability, for a given simulation across all algorithms, we chose the same sequence of initial expert and topic pairs. In order to empirically evaluate Bayesian-Nash incentive compatibility, we followed the same experimental protocol followed in [2].

Computational Environment: Experiments on synthetic data were carried out on Matlab R2016 running Windows 10.

6   Results

6.1   Robustness to Early Failures

Much of DIEL's practical success stems from an early identification of a strong expert and subsequent exploitation. The initialization phase of proactive-DIEL allows informed exploration guided by the advertised priors with stronger experts (based on advertised priors) getting precedence over weaker ones. However, when a strong expert fails at the first few occasions, lowering both the reward mean and variance, and with moderately strong experts performing well in the early rounds, the strongest expert may get sidelined forever not receiving any future referrals to correct her estimated expertise.

---

[2] The data sets along with detailed instructions, algorithm implementations, and parameter configurations are publicly available at https://www.cs.cmu.edu/~akhudabu/referral-networks.html.

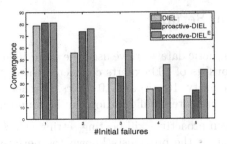

Fig. 2. Robustness to early failure.

In Fig. 2, we analyzed robustness to early failures in the following way. For each *subnetwork*, we only focus on the strongest expert that has an initial streak of m failures. We found that DIEL did not have difficulty to converge to the strongest expert with a single early failure. However, with $m \geq 2$, there was a significant drop in convergence. Proactive-DIEL handled two failures better than DIEL and performed comparably with Proactive-DIEL^E when $m \leq 2$. However, as m grows larger, the benefit of endorsements becomes more prominent.

6.2   Performance Gain

Figure 2 demonstrates that endorsements indeed allow an additional cushion to strong experts for string of early failures. The ability to resurrect strong experts from a string of early failures translates into a steady-state performance improvement. As shown in Fig. 3, proactive-DIEL with endorsement outperformed the state-of-the-art across both *market-aware* and *market-agnostic* settings. A paired t-test revealed that at the steady-state (1000 samples or more per subnetwork), proactive-DIEL with endorsement was better than both the corresponding baselines with p-value less than 0.0001.

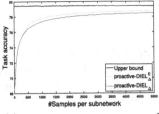
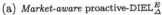

(a) *Market-aware* proactive-DIEL$^E_\triangle$

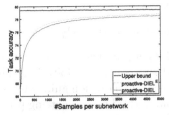

(b) *Market-agnostic* proactive-DIEL^E

Fig. 3. Proactive skill posting with endorsement.

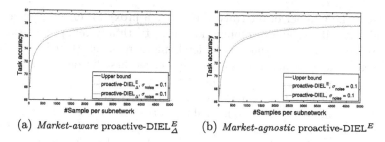

(a) *Market-aware* proactive-DIEL$_\Delta^E$ (b) *Market-agnostic* proactive-DIEL^E

Fig. 4. Proactive skill posting with noisy self-skill estimates.

6.3 Tolerance to Noisy Self-skill Estimates

Even when experts post their skills truthfully, their self-estimates may not be precise (see, e.g., [34]). Imprecise skill estimation in proactive skill posting was first explored in [2, 28]. Note that, since a noisy bid can be interpreted as deliberate misreporting and vice-versa, robustness to noisy self-skill estimates and robustness to strategic lying are two major goals and there lies an inherent trade-off between them.

We assume Gaussian noise on the estimates in the form of $\hat{\mu} = \mu + \mathcal{N}(0, \sigma_{noise})$, where $\hat{\mu}$ is an expert's own estimate of her true topic-mean μ, and σ_{noise} is a small constant (0.05 or 0.1 in our experiments). Figure 4 compares the performance of the proactive referral algorithms with noisy self-skill estimates. We found that even with a higher value of σ_{noise}, proactive skill posting algorithms with endorsements retained a small advantage over their non-endorsing counterparts, thus absolutely accurate estimates of own skill, which is an impractical assumption, is not particularly necessary for the algorithms to succeed. Similar to our earlier experiments, we also conducted a paired t-test which revealed that at the steady-state (1000 samples or more per subnetwork), proactive-DIEL with endorsement was better than both the corresponding baselines with p-value less than 0.0001.

6.4 Incentive Compatibility

Mimicking real-life endorsements, in our setting, endorsements are one-sided; the *endorsing expert* only informs when observed performance is better than or equal to the advertised prior. However, even when we consider truthful endorsements, strategic behavior from the side of the *advertising expert* during skill posting may affect incentive compatibility. Intuitively, over-reporting of skills is unlikely to fetch more referrals since in expectation, over-reporting will fetch fewer endorsements than honest advertisement. However, underbidding could fetch more endorsements and hence could be a viable strategy to attract more business. In order to empirically evaluate Bayesian-Nash incentive compatibility, a weaker form of truthfulness, where being honest is weakly better than lying, we follow the same experimental design as in [2].

We treat the number of referrals received as a proxy for expert benefit, and we consider specific strategy combination (e.g., truthfully report best-skill but overbid second-best skill) that could fetch more referrals (listed in Table 1). For a given strategy s and scenario $scenario_i$, we first fix one expert, say e_l^i. Let $truthfulReferrals(e_l^i)$ denote the number of referrals received by e_l^i beyond a steady-state threshold (i.e., a referral gets counted if the initial expert has referred 1000 or more instances to her subnetwork) when e_l^i and all other experts report truthfully. Similarly, let $strategicReferrals(e_l^i)$ denote the number of referrals received by e_l^i beyond a steady-state threshold when e_l^i misreports while everyone else advertises truthfully. We then compute the following Incentive Compatibility factor ($ICFactor$) as:

$$ICFactor = \frac{\sum_{i=1}^{1000} truthfulReferrals(e_l^i)}{\sum_{i=1}^{1000} strategicReferrals(e_l^i)}.$$

A value greater than 1 implies truthfulness in expectation, i.e., truthful reporting fetched more referrals than strategic lying.

Table 1. Comparative study on empirical evaluation of Bayesian-Nash incentive-compatibility. Strategies where being truthful is no worse than being dishonest are highlighted in bold.

$\mu_{t_{best}}$	$\mu_{t_{secondBest}}$	Proactive DIEL [2]	Proactive DIEL^E	Proactive DIEL$_\triangle$ [9]	Proactive DIEL$_\triangle^E$
Truthful	Overbid	**1.02**	**1.02**	**1.02**	**1.03**
Overbid	Truthful	1.19	1.23	1.19	1.23
Overbid	Overbid	1.25	1.34	1.22	1.33
Truthful	Underbid	**1.15**	**1.07**	**1.10**	**1.08**
Underbid	Truthful	1.16	1.14	1.13	1.16
Underbid	Underbid	1.32	1.30	1.37	1.21
Underbid	Overbid	1.15	1.13	1.12	1.15
Overbid	Underbid	1.50	1.30	1.28	1.29

As shown in Table 1, both proactive-DIEL^E and proactive-DIEL$_\triangle^E$ are robust to strategic lying. As a general pattern, overbidding got more penalized in presence of endorsements. When compared against their algorithmic counterparts without endorsement, we found that in some of the cases underbidding fetched more referrals due to endorsements. However, for any bidding strategy, it was never the case that truthful reporting fetched fewer referrals than strategic lying.

6.5   Generalizability of the Endorsement Framework

In order to test the generalizability of our approach, we performed experiments on an additional referral-learning algorithm, proactive-ϵ-Greedy. Unlike DIEL, ϵ-

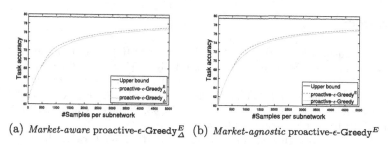

(a) *Market-aware* proactive-ϵ-Greedy$_\Delta^E$ (b) *Market-agnostic* proactive-ϵ-Greedy^E

Fig. 5. Generalization to other mean-based methods.

Greedy only considers the mean observed reward to determine the most promising action [3]. It explores via an explicit probabilistic diversification step – randomly selecting a connected expert for referral (for further algorithm details, see [2]). As shown in Fig. 5, we obtained similar improvements with endorsements.

7   Conclusions and Future Work

In this paper, we augment endorsement to proactive skill posting setting and propose algorithmic modifications for performance gain. Our extensive empirical evaluations demonstrated that (1) endorsements improved performance of proactive-DIEL across both (2) *market-aware* and *market-agnostic* skill posting setting. The endorsement mechanism is (3) tolerant to noise in self-skill estimates and (4) strategic lying and (5) generalizes to other mean-based methods like ϵ-Greedy.

Our work can be extended in the following ways.

– **Truthfulness:** In practice, endorsements may not necessarily be truthful. Devising mechanisms to discourage strategic lying such as early endorsement (endorsing without observing M or more samples giving unfair advantage to endorsed colleague) or false endorsement could be an interesting future research direction.
– **Collusion detection:** Many such endorsements can have mutual benefits. We propose to investigate the extent a collusion-clique of k experts mutually endorsing each other to attract business can affect the overall network performance and methods to detect and remedy such collusion.
– **Incentivising endorsement:** In our current scheme, endorsement is an altruistic action. Similar to *penalty on distrust* mechanism, if we add mechanisms to discourage strategic endorsement, experts would tend to err on the safer side and avoid endorsing others. Hence, another relevant question is how should we incentivise endorsement? One possible direction can be a one-time reward for a truthful endorsement after sufficiently vetting an endorsed expert.

References

1. KhudaBukhsh, A.R., Carbonell, J.G., Jansen, P.J.: Robust learning in expert networks: a comparative analysis. J. Intell. Inf. Syst. **51**(2), 207–234 (2018)
2. KhudaBukhsh, A.R., Carbonell, J.G., Jansen, P.J.: Incentive compatible proactive skill posting in referral networks. In: Belardinelli, F., Argente, E. (eds.) EUMAS/AT -2017. LNCS, vol. 10767, pp. 29–43. Springer, Cham (2018). https://doi.org/10.1007/978-3-030-01713-2_3
3. Auer, P., Cesa-Bianchi, N., Fischer, P.: Finite-time analysis of the multiarmed bandit problem. Mach. Learn. **47**(2–3), 235–256 (2002)
4. Kautz, H., Selman, B., Milewski, A.: Agent amplified communication, pp. 3–9 (1996)
5. Yolum, P., Singh, M.P.: Dynamic communities in referral networks. Web Intell. Agent Syst. **1**(2), 105–116 (2003)
6. Yu, B.: Emergence and evolution of agent-based referral networks. Ph.D. thesis. North Carolina State University (2002)
7. Yu, B., Venkatraman, M., Singh, M.P.: An adaptive social network for information access: theoretical and experimental results. Appl. Artif. Intell. **17**, 21–38 (2003)
8. Yolum, P., Singh, M.P.: Emergent properties of referral systems. In: Proceedings of the Second International Joint Conference on Autonomous Agents and Multiagent Systems, pp. 592–599. ACM (2003)
9. KhudaBukhsh, A.R., Hong, J.W., Carbonell, J.G.: Market-aware proactive skill posting. In: Ceci, M., Japkowicz, N., Liu, J., Papadopoulos, G.A., Raś, Z.W. (eds.) ISMIS 2018. LNCS, vol. 11177, pp. 323–332. Springer, Cham (2018). https://doi.org/10.1007/978-3-030-01851-1_31
10. KhudaBukhsh, A.R., Carbonell, J.G.: Expertise drift in referral networks. In: Proceedings of the 17th International Conference on Autonomous Agents and MultiAgent Systems, pp. 425–433. International Foundation for Autonomous Agents and Multiagent Systems (2018)
11. KhudaBukhsh, A.R., Carbonell, J.G., Jansen, P.J.: Proactive-DIEL in evolving referral networks. In: Criado Pacheco, N., Carrascosa, C., Osman, N., Julián Inglada, V. (eds.) EUMAS/AT -2016. LNCS, vol. 10207, pp. 148–156. Springer, Cham (2017). https://doi.org/10.1007/978-3-319-59294-7_13
12. Langford, J., Strehl, A., Wortman, J.: Exploration scavenging. In: Proceedings of the 25th International Conference on Machine Learning, pp. 528–535. ACM (2008)
13. Shivaswamy, P., Joachims, T.: Multi-armed bandit problems with history. In: Artificial Intelligence and Statistics, pp. 1046–1054 (2012)
14. Bouneffouf, D., Feraud, R.: Multi-armed bandit problem with known trend. Neurocomputing **205**, 16–21 (2016)
15. Huang, L., Joseph, A.D., Nelson, B., Rubinstein, B.I., Tygar, J.: Adversarial machine learning. In: Proceedings of the 4th ACM Workshop on Security and Artificial Intelligence, pp. 43–58. ACM (2011)
16. Babaioff, M., Sharma, Y., Slivkins, A.: Characterizing truthful multi-armed bandit mechanisms. In: Proceedings of the 10th ACM Conference on Electronic Commerce, pp. 79–88. ACM (2009)
17. Biswas, A., Jain, S., Mandal, D., Narahari, Y.: A truthful budget feasible multi-armed bandit mechanism for crowdsourcing time critical tasks. In: Proceedings of the 2015 International Conference on Autonomous Agents and Multiagent Systems, pp. 1101–1109 (2015)

18. Tran-Thanh, L., Stein, S., Rogers, A., Jennings, N.R.: Efficient crowdsourcing of unknown experts using multi-armed bandits. In: European Conference on Artificial Intelligence, pp. 768–773 (2012)
19. Tran-Thanh, L., Chapman, A.C., Rogers, A., Jennings, N.R.: Knapsack based optimal policies for budget-limited multi-armed bandits. In: Proceedings of the Twenty-Sixth AAAI Conference on Artificial Intelligence (2012)
20. Yolum, P., Singh, M.P.: Engineering self-organizing referral networks for trustworthy service selection. IEEE Trans. Syst. Man Cybern.-Part A: Syst. Hum. 35(3), 396–407 (2005)
21. Sabater, J., Sierra, C.: Review on computational trust and reputation models. Artif. Intell. Rev. 24(1), 33–60 (2005)
22. Yu, H., Shen, Z., Leung, C., Miao, C., Lesser, V.R.: A survey of multi-agent trust management systems. IEEE Access 1, 35–50 (2013)
23. Wang, Y., Singh, M.P.: Trust representation and aggregation in a distributed agent system. In: AAAI, vol. 6, pp. 1425–1430 (2006)
24. Jonker, C.M., Treur, J.: Formal analysis of models for the dynamics of trust based on experiences. In: Garijo, F.J., Boman, M. (eds.) MAAMAW 1999. LNCS, vol. 1647, pp. 221–231. Springer, Heidelberg (1999). https://doi.org/10.1007/3-540-48437-X_18
25. Schillo, M., Funk, P., Rovatsos, M.: Using trust for detecting deceitful agents in artificial societies. Appl. Artif. Intell. 14(8), 825–848 (2000)
26. Shi, J., Bochmann, G.V., Adams, C.: Dealing with recommendations in a statistical trust model. In: Proceedings of AAMAS Workshop on Trust in Agent Societies, pp. 144–155 (2005)
27. Mui, L., Mohtashemi, M., Halberstadt, A.: A computational model of trust and reputation. In: 2002 Proceedings of the 35th Annual Hawaii International Conference on System Sciences, HICSS, pp. 2431–2439. IEEE (2002)
28. KhudaBukhsh, A.R., Carbonell, J.G., Jansen, P.J.: Proactive skill posting in referral networks. In: Kang, B.H., Bai, Q. (eds.) AI 2016. LNCS, vol. 9992, pp. 585–596. Springer, Cham (2016). https://doi.org/10.1007/978-3-319-50127-7_52
29. Kaelbling, L.P.: Learning in Embedded Systems. MIT Press, Cambridge (1993)
30. Thompson, W.R.: On the likelihood that one unknown probability exceeds another in view of the evidence of two samples. Biometrika 25(3/4), 285–294 (1933)
31. Donmez, P., Carbonell, J.G., Schneider, J.: Efficiently learning the accuracy of labeling sources for selective sampling. In: Proceedings of KDD 2009, p. 259 (2009)
32. Wiering, M., Schmidhuber, J.: Efficient model-based exploration. In: Proceedings of the Fifth International Conference on Simulation of Adaptive Behavior, SAB 1998, pp. 223–228 (1998)
33. Berry, D.A., Fristedt, B.: Bandit Problems: Sequential Allocation of Experiments (Monographs on Statistics and Applied Probability), vol. 12. Springer, Heidelberg (1985). https://doi.org/10.1007/978-94-015-3711-7
34. MacKay, T.L., Bard, N., Bowling, M., Hodgins, D.C.: Do pokers players know how good they are? Accuracy of poker skill estimation in online and offline players. Comput. Hum. Behav. 31, 419–424 (2014)

Markov Chain Monte Carlo for Effective Personalized Recommendations

Michail-Angelos Papilaris and Georgios Chalkiadakis$^{(\boxtimes)}$

School of Electrical and Computer Engineering, Technical University of Crete,
Chania, Greece
mixpapilaris@gmail.com, gehalk@intelligence.tuc.gr

Abstract. This paper adopts a Bayesian approach for finding top recommendations. The approach is entirely personalized, and consists of learning a utility function over user preferences via employing a sampling-based, non-intrusive preference elicitation framework. We explicitly model the uncertainty over the utility function and learn it through passive user feedback, provided in the form of clicks on previously recommended items. The utility function is a linear combination of weighted features, and beliefs are maintained using a Markov Chain Monte Carlo algorithm. Our approach overcomes the problem of having conflicting user constraints by identifying a convex region within a user's preferences model. Additionally, it handles situations where not enough data about the user is available, by exploiting the information from clusters of (feature) weight vectors created by observing other users' behavior. We evaluate our system's performance by applying it in the online hotel booking recommendations domain using a real-world dataset, with very encouraging results.

Keywords: Adaptation and learning · Recommender systems

1 Introduction

Personalized recommender systems aim to help users access and retrieve relevant information or items from large collections, by automatically finding and suggesting products or services of potential interest [20]. User preferences could be self-contradicting; are often next to impossible for the user to specify; and are notoriously difficult to infer, while doing so often requires a tedious elicitation process relying on evidence of others' behavior [15,20].

In this work we propose a complete model for learning the user preferences and use it to make targeted recommendations. Our system: *(a)* does not rely on user-specified hard constraints; and *(b)* does not require an explicit user preferences elicitation process: rather, it learns a user model through passive observation of her actions. In particular, we follow a Bayesian approach that operates as follows. We model the user utility over items as a linear function of the item features, governed by a weight vector $\boldsymbol{w}$. Our goal is to learn this

© Springer Nature Switzerland AG 2019
M. Slavkovik (Ed.): EUMAS 2018, LNAI 11450, pp. 188–204, 2019.
https://doi.org/10.1007/978-3-030-14174-5_13

vector in order to provide personalized recommendations to our user. We capture the uncertainty of the weight vector w, which parameterize the utility function, through a distribution P_w over the space of possible weight vectors.

Every time a user enters our system, we propose a number of items that comply with feedback we had received from her in previous interactions. The feedback is in the form of clicks, and it can be translated to preferences. This feedback could in principle be used to update the prior weight distribution via Bayes rule. However, even for the simplest case of a uniform prior, the posterior could become very complex [19]. Thus, we adopt a different approach: instead of trying to refit the prior through an algorithm (such as *Expectation-Maximization (EM)* [4]), we keep the constraints infered from previous user interactions, and employ a *Markov Chain Monte Carlo (MCMC)* algorithm (specifically, Metropolis-Hastings [14]), in order to condition the prior and derive efficiently samples from the posterior [2,8,25].

In our work, we show how to find effectively and efficiently a "good" starting point for the MCMC algorithm, a challenging task in a multidimensional space like ours. Moreover, we take into consideration the problem of having conflicting constraints. This situation could be encountered when the user changes radically her preferences, or if we have gathered too many constraints and we could not find a weight vector that could satisfy all of them at the same time. We overcome this by using a linear programming algorithm that finds, effectively and efficiently, a convex region of all the possible weight vectors that satisfy all the user constraints. In addition, we show that the problem of having insufficient information about users can be alleviated by performing clustering over other users' preferred items, and exploiting the clusters to recommend items to the "high uncertainty" users.

As explained above, we have no need to set questions to the user, nor use textual information regarding an item in order to elicit user preferences. Moreover, we do not rely on any user ratings of the recommended items. We tested our system using a real world dataset that consist of 5000 hotels, each of which is being characterized by five main features; and exploited our knowledge of the preferences of synthetic users we constructed, in order to verify the effectiveness of our algorithm. Our simulations indeed confirm that the approach is able to quickly focus on a users' true preferences, and produce top personalized recommendations when operating in a realistically large recommendations space.

2 Background and Related Work

MCMC techniques estimate by simulation the expectation of a statistic in a complex model. Successive random selections form a Markov chain, the stationary distribution of which is the target distribution. Thus, MCMC is particularly useful for the estimation of posterior distributions in complex Bayesian models like ours [1].

Our work was to some extent inspired by work on *inverse reinforcement learning (IRL)*. The usual *reinforcement learning (RL)* problem is concerned with how agents ought to take actions in an environment so as to maximize some notion of cumulative reward, while the goal of IRL is to learn the model's reward function (which guides optimal behaviour). For instance, [18] views agent decisions as a set of linear constraints on the space of possible reward functions. Bayesian IRL [19], on the other hand, assumes that a distribution over possible reword functions exists and has to be inferred. In our case, the goal is also to learn the utility each item has for a user. We model this function as being linearly additive, governed by a weight vector. Conceptually, therefore, our approach is similar to IRL; however, we do not use an MDP to model our problem, as we do not explicitly view it as a sequential decision making one.

2.1 Related Work

In a recent work focused on the movies' recommendations domain, Tripolitakis and Chalkiadakis [24] proposed the use of *probabilistic topic modeling* algorithms to learn user preferences. That work modeled items *and* users as mixtures of latent topics, and was to an extent able to capture changes in user tastes or mood shifts, via the incorporation of the "Win Or Learn Fast" principle [6] found in the reinforcement learning literature. However, their work relies on "crowd-sourced" or otherwise gathered textual information about the items (movies in their case); and it was able to exhibit only marginal improvement in recommendations' performance, when tested against other algorithms used in the movie recommendations literature.[1] By contrast, our work here does not attempt to build preference models relying on crowdsourced information about the items, and uses a rather simple exploration technique. Still, via the combination of MCMC and linear programming techniques, it is able to achieve an outstanding recommendations performance, as demonstrated in our experiments.

Also recently, Nasery *et al.* [17] present a recommender that introduces a novel prediction model that generates item recommendations using a combination of feature-based and item based preferences. More specifically, they propose a novel matrix factorization method (called FPMF), which incorporates user feature preferences to predict user likes, and evaluate it in the movie recommendations domain.

There is naturally much previous work on hotel or travel booking recommenders. However, these typically involve lengthy preference elicitation processes; require the use of textual information; or are "collaborative filtering" in nature, relying on choices made by other users in the past. For instance, Dong and Smyth [12] propose a personalized hotel booking recommendation system. In their work, however, they are attempting to mine features from the user reviews in order to create a user profile and exploit it to recommend items that comply with that information. Then, [22] considers an interactive way of ranking travel packages. However, for each iteration, the user is asked to rank a set of items,

---

[1] That was also the case for the work of [3], which was modeling items *and* users as multivariate normals; like [24], and in contrast to our work here, [3] required users to actually *rate* a (small) number of items.

and the system requires several iterations of explicit preference elicitation before providing recommendations to the user.

One interesting work is that of [8], which considers the broad task of predicting the future decisions of an agent based on her past decisions. To account for the uncertainty regarding the agent's utility function, the authors consider the utility to be a random quantity that is governed by a prior probability distribution. When a new agent is encountered, this estimate serves as a prior distribution over her utility function. The constraints implied by the agent's observed choices are then used to condition the prior, obtaining a posterior distribution through an MCMC algorithm, as we do here.

The work whose model we adopt to a large extent in this paper, however, is that of [25], which generates top-k *packages* for users by capturing personalized preferences over packages using a linear utility function, which the system learns through feedback provided by the user. In their work, the authors of [25] propose several sampling-based methods to capture the updated utility function. They develop an efficient algorithm for generating top-k packages using the learned utility function, where the rank ordering respects any of a variety of ranking semantics. Nevertheless, they do not tackle conflicts in the user constraints, do not employ clustering to reduce uncertainty for new users, and do not deal with the problem of appropriately initializing their MCMC algorithm.

3   A Personalized Recommender System

An overview of our system is the following (Fig. 1). When a user enters, we retrieve the constraints derived from her previous interactions, and check if there are any conflicting constraints in order to handle them. Next we update beliefs by conditioning the prior distribution with the aforementioned constrains. Subsequently, we rank our items according to the posterior samples, and present the top k items to the user. We record which of those items are being clicked on by the user, and assume that clicked items are more appealing to her than unclicked ones. We store the new constraints derived from the feedback received, and progress to the next interaction. Finally, we form clusters with the posterior samples from the already registered users on our system, and use these with new users so as to reduce our initial uncertainty about them.

Model Settings. We assume that we are given a set of n items S_I. Each item is being described by m features $(f_1, f_2, ..., f_m)$. For example, a feature of an item could be the price, rating etc. Every item $i \in I$ is described by an m-dimensional vector $(v_1^i, v_2^i, ..., v_m^i)$, where v_j^i corresponds to the value of the respective feature f_j. For every user we have a *constraint set* based on the feedback that we have received from her. Every constraint has the following form: $i_k \succ i_t$ where i_k, i_t are items in S_I.

Our proposed model needs all of our items $i \in S_I$ in our database to be normalized. To achieve that, we found the maximum and the minimum values of each item feature, and then for each item feature value v_j^i we normalize its

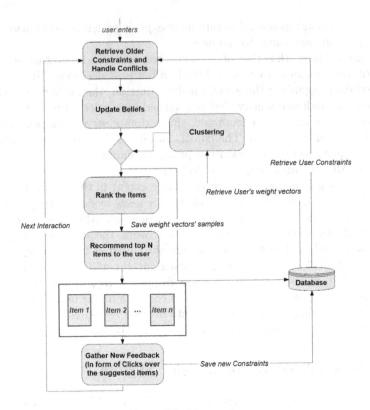

Fig. 1. Model overview.

value: $v_j^i \leftarrow \frac{v_j^i - min(v_j)}{max(v_j) - min(v_j)}$, where $min(v_j)$ and $max(v_j)$ are the min and max values of the respective j feature in S_I. After normalization, all items' feature values lie in $[0, 1]$.

Utility Function. The user-specific utility function U over items $i \in S_I$, that we wish to learn, directly depends on its feature vector. The space of all mappings between possible combinations of the feature values and utility values is uncountable, making this task challenging. Since we need to express the structure of an item concisely, we assume that the utility function is linearly additive and governed by a weight vector. This is a structure for the utility function commonly assumed in practice [8,15]. For each item, the utility function is computed as follows:

$$U(i) = \sum_{j=1}^{m} w_j v_j^i \tag{1}$$

where v_j^i is the value of the respective feature, and w_j is the weight value associated with this feature. The value for m depends on the number of features that each item has. Each w_j takes values in $[-1, 1]$. A positive w_j weight means that larger rather than smaller values are preferred for the corresponding feature, a negative w_j weight means that smaller rather than larger values are preferred for the corresponding feature, while a zero weight means that the user is indifferent about the specific feature.[2] Our final goal is to learn the weight vector and suggest items to the user that have the maximum utility taking into consideration this weight vector.

Users often only have a rough idea of what they want in a desirable item, and also find such preference very difficult to quantify. For this reason, users are not able to specify or even know the exact values of the w_j weights that drive utility function U. We model this uncertainty in a Bayesian manner, assuming that the vector of weights, $\boldsymbol{w}$, is not known in advance, but it can be described by a probability distribution P_w. We model this distribution as a *multivariate normal*. Such a distribution is often used to describe, at least approximately, any set of (possibly) correlated real-valued random variables, each of which clusters around a mean value (see also the work of [3]). The P_w can be learned by the feedback received.

3.1   Constraints

We now discuss how to generate constraints to reduce our uncertainty over weight vectors. To begin, our system suggests to the user a number of items, and we record which of them are being selected (clicked on) by the user. We make the assumption that clicked items are more appealing to the user than the unclicked ones.[3] Thus, the feedback in form of new clicks, produces a set of pairwise preferences as follows. Assume that we present three items, (i_1, i_2, i_3), to a user, each of which has (the same) three features (f_1, f_2, f_3). If a user selects the second item, we can derive the following pairwise preferences: $i_2 \succ i_1$ and $i_2 \succ i_3$ (to put this otherwise, the user's selection is translated to $U(i_2) > U(i_1)$ and $U(i_2) > U(i_1)$). We can use, e.g., $i_2 \succ i_1$ to get:

$$U(i_2) > U(i_1) \tag{2}$$

$$\Rightarrow \sum_{j=1}^{m} w_j * v_j^{i_2} > \sum_{j=1}^{m} w_j * v_j^{i_1} \tag{3}$$

$$\Rightarrow w_1 * v_1^{i_2} + w_2 * v_2^{i_2} + w_3 * v_3^{i_2} >$$
$$w_1 * v_1^{i_1} + w_2 * v_2^{i_1} + w_3 * v_3^{i_1} \tag{4}$$

The feature values of each item are known in advance, but we do not know the weights. We can focus on possible weight vectors by employing the following lemmas:

---

[2] To illustrate, assume a "hotel price" is "low", say 0.1. If the user prefers really cheap hotels, she might have a weight of -0.9 for "hotel price", thus deriving a higher utility for this hotel compared to that derived for an expensive hotel of price, say, 0.9 (since $-0.09 > -0.81$).

[3] Moreover, all clicked items are originally equally appealing. However, as interactions with the system increase, beliefs about the desirability of the items get updated.

Lemma 1 *[25]: Given feedback $i_1 \succ i_2$, if $i_1 \succ i_2$ does not hold under the weight vector $\boldsymbol{w}$, $P_w(\boldsymbol{w}|i_1 \succ i_2) = 0$.*

That is, every feedback received in terms of clicks, rules out all the weight vectors that do not satisfy the generated constraint. Additionally,

Lemma 2 *[25]: The set of valid weight vectors which satisfy a set of preferences forms a convex set. So valid weight vectors form a continuous and convex region.*

Handling Conflicting Constraints. A major challenge encountered is that, after a few interactions with our system, some of the generated constraints may have conflicts with newer ones *(a)* when the user changed radically his preferences; or *(b)* if we have gathered too many constraints, and cannot find a weight vector that satisfies all of them at the same time. To deal with this problem we used a well-known linear programming algorithm, Simplex [10]. We employ Simplex in order to find whether a feasible weight vector is available. In the case Simplex returns that there is no feasible weight vector which satisfies all constraints, we remove the oldest constraint we have gathered and rerun the Simplex algorithm. We do this as many times required to get a valid weight vector; notice that this process always leaves the set of constraints with at least one member, thus Simplex never faces the degenerate case of dealing with an empty constraints set. In this way, the set of constraints we have stored is always free of conflicts.

3.2 Beliefs Updating

Even if the prior distribution is a "nice" one with a compact closed form representation, the posterior distribution can be quite complex [8,19]. Refitting the weight distribution P_w after each feedback from the user would be a very inefficient and time-consuming task for a real-time recommender. We avoid this by employing the sampling-based framework of [25]: items preferences resulting from user feedback can be readily translated into constraints on the samples drawn from P_w. In our approach we follow a technique that is also used in the work of [8] and [25]: we condition our weights prior distribution with the constraints derived from each user in order to acquire samples from the posterior via employing MCMC, as we detail below. Note that if the amount of the feedback received is small, simple sampling techniques like the popular rejection sampling [13] could be effective. But as the feedback increases, those sampling methods prove to be inefficient, because the valid convex region drastically shrinks. Sampling techniques like MCMC are more suitable for cases like this, because they are 'aware' of the feedback received, and can handle cases with higher dimensionality [25].

A major problem encountered in the sampling process was that in order for our MCMC algorithm to start collecting samples from the posterior distribution, it needs to start from a point that is located inside the valid convex region. Otherwise, the Markov chain remains stuck to the initial point. To overcome

---

Algorithm 1: Our Metropolis Hastings Algorithm

Initialize x(0) with the sample from Simplex Algorithm;
$i = 1$;
repeat
    With $p = 0.5$ accept the current sample x_i; continue;
    Propose: $x_i \sim$ Stochastic walk (described in text);
    Acceptance Prob: $\alpha(x_i|x_{i-1}) = \min(1, \frac{\pi(x_i)}{\pi(x_{i-1})})$;
    $p \sim$ Uniform$(0, 1)$;
    if x_i *satisfies all constraints and* $p < \alpha$ **then**
        accept x_i,
    else
        accept x_{i-1},
    end
until $i >= number of samples$;

---

this problem we used Simplex algorithm. More specifically, for each constraint $i_j > i_k$, we introduce a variable ϵ_t so that we have:

$$U(i_j) > U(i_k) + \epsilon_t \tag{5}$$

and the objective function that we want to maximize is

$$max \sum_{t=1}^{T} \epsilon_t \tag{6}$$

where T is the number of the constraints. This results in finding a weight vector that is closer to the center of the convex region. Otherwise, if we start from a 'bad' location in the distribution (the limits of the convex region), the sampler spends the first iterations to slowly move towards the main body of the distribution. With Simplex we were able to find a starting point that does not violate our constraints, which was necessary in order to iterate and start collecting posterior samples. After initiating the algorithm with a Simplex sample, we are ready to carry out the main sequence of its MCMC steps.

Metropolis Hastings Algorithm. We first construct a regular grid in the m-dimensional hypercube, where m is the number of features that each item has. Each w_i takes values in $[-1, 1]$ and our grid interval is set to 0.02. The main loop of our Metropolis-Hastings[4] is known Algorithm 1 consists of three components:

1. Generate a proposal (or a candidate) sample x_{cand}.

---

[4] We note that ours is essentially a standard version of the Metropolis-Hastings algorithm, which is known to be almost always convergent [21].

2. Compute an "acceptance probability" α for the candidate sample, via a function based on the candidates' distribution (usually termed *the proposal distribution*), and the full joint density. In fact, we use an estimate of the probability density function that corresponds to a clustering of the prior samples we have gathered from previous interactions.[5]
3. Accept the candidate sample with probability α, the acceptance probability, or reject it with probability $1 - \alpha$.

Specifically, we choose a candidate as follows. With probability 0.5 we keep the old sample x_{i-1}, and with the remaining probability, x_{cand} (or x_i as in Algorithm 1) is chosen with a stochastic walk from among x_{i-1}'s $2m$ neighbours as follows. After choosing (uniformly) the neighbour that we will visit first, we move forward with 0.5 probability; and move backwards with 0.5 probability also. After we have chosen the direction, we choose the number of steps that we are going to move in the grid. With 0.9 probability we move one step in the grid. With probability 0.1, we choose to move, with equal probability, either 2*step, 3*step or 4*step. We observed empirically that making, with small probability, relatively bigger "jumps" in the grid, helps in preventing the Markov chain from getting stuck in a particular part of the distribution. All probability values above were chosen empirically.

The next step after choosing x_{cand}, is to check if it satisfies all generated constraints. If this sample violates one or more constraints, then we reject it and we keep the old one; otherwise we decide whether to keep it based on the acceptance function. After we start collecting samples from the posterior, to avoid collecting samples that are highly correlated, it is common to pick a subset of them. So we introduce a "lag" parameter, set to 100 iterations in our implementation.

Now, there are mainly two kinds of "proposal distributions", symmetric and asymmetric. A proposal distribution is a symmetric distribution if: $q(x_i|x_{i-1}) = q(x_{i-1}|x_i)$ Standard choices of symmetric proposals include Gaussians or uniform distributions centered at the current state of the chain. Here we work with a symmetric (uniform) proposal distribution, as it makes the algorithm more straightforward, both conceptually and computationally. In cases with symmetric distributions like ours, the latter simplifies to [9]:

$$\alpha(x_i|x_{i-1}) = \min(1, \frac{\pi(x_i)}{\pi(x_{i-1})}) \tag{7}$$

---

[5] In some detail, we divide *each* of the m dimensions into a fixed number of "segments"–ten (10) in our implementation—and use this segmentation to generate "buckets" to place our samples into. In this way, we create 10^m buckets in total: for instance, if we had only two dimensions, e.g. "price" and "distance to city center", we would be creating 100 buckets. Then, each prior sample is allocated to its corresponding bucket, based on Euclidean distance. When Algorithm 1 picks a sample, it checks which bucket it belongs to, and uses the number of samples in the buckets to estimate the $\pi(\cdot)$ density in Eq. 7. Thus, this method uses the prior samples to estimate the posterior joint density.

where $\pi(\cdot)$ is the full joint density. Simply put, when the proposal distribution is symmetric, the probability α becomes proportional to how likely each of the current state x_{i-1} and the proposed state x_i are under the full joint density.

3.3   Ranking the Items

The ultimate goal of a recommender system is to suggest a short ranked list of items, namely the top-k recommendations that are ideally the most appealing for the end user. A framework based on utility function, like ours, essentially defines a total order over all items. Moreover, a recommender naturally faces the dilemma of recommending items that best match its current beliefs about the user, or items that could improve user satisfaction and help form more accurate beliefs. This corresponds to the typical exploration vs exploitation problem in learning environments [23]. Although several ranking methods exist, there is no universally accepted ranking semantics given the uncertainty in the utility function. In our work, we use a ranking method based on expectation, used widely in the AI literature (see, e.g., [5,7,23]).

Ranking Based on the Expectation Algorithm. The expectation algorithm *(Exp)* is defined as follows [25]. Given an item space I and probability distribution P_w over weight vectors w, find the set of top-k items with respect to their expected utility value. More specifically, for each item and every (sample) weight vector, we calculate its utility $U_{w_l}(i)$ under each sample w_l vector, using Eq. 1. Subsequently, the expected utility value for each item i can be calculated given $U_{w_l}(i)$, and the probability of the corresponding w_l weight vector:

$$EU(i) = \sum_{l=1}^{L} U_{w_l}(i) * Prob(w_l) \tag{8}$$

where L is the number of sample weight vectors. Finally, we sort the dataset in ascending order according to the items expected utility, and present the top k ranked items to the user, after adding an explicit exploration component as follows.

Suggesting Items. It is very important to introduce an exploration component in the system. User preferences are in many cases very complex and difficult to map completely. We confront this challenge by presenting some[6] random items along with those that the *Exp* algorithm returns. Those items serve the purpose of correcting the bias introduced from the initial distribution of P_w and combating mistakes and noise from user feedback. Moreover, the uncertainty inherent in the utility function aids exploration, in a true Bayesian manner.

---

[6] Specifically, 2 out of 7 items presented to the user are chosen randomly; see Sect. 4.2 below.

Now, a recommender makes suggestions to the user based on knowledge acquired through user's previous interactions with the system. A major problem occurs when we have gathered little or no data on what the user prefers. In those situations, the suggestions made, could be almost "blind". Such a process would converge slowly, and especially the initial beliefs could be far from the user's true preferences. In what follows, we propose a solution to address this problem.

3.4 Clustering

It is often the case in recommender systems that we have no knowledge of the preferences of a user, or that we have not gathered enough constraints in order to limit the uncertainty that the weight distribution P_w has. This is the notorious "cold start" problem. The most common approach to tackle this problem is to randomly suggest items to the user in order to receive feedback and update the distribution. However, this has the disadvantage that suggested items in the initial interactions, might be far from the users' true preferences and this could result in producing suboptimal recommendations and therefore slow convergence of the weight distribution. Thus, instead of recommending random items in the initials interactions, we make use of a novel method that employs "clustered" weight vectors from other, existing users (essentially recommending items matching the preferences of corresponding formed user categories).

Employing the Clusters. Every time we run *Metropolis-Hastings* to derive posterior samples for a user, we save those samples to our database. After having collected several weight vector samples from the posteriors of existing users, we cluster them into k groups (k being the number of recommendations made to a user in every interaction) using the popular *k-means* algorithm [16]. As mentioned, we use them to address scenarios of limited or no information regarding a new user. Instead of considering all weight vectors as candidates, we use the clusters' centroids, and suggest to the new user the most "appropriate" items given the preferences of each group. Specifically, the k cluster centroids are used in the ranking algorithm, to rank and present to new users the items that are best fits for the clusters' centroids. By suggesting items based on the cluster centroids, and receiving feedback from the user, we drastically decrease the uncertainty that we have about their preferences, as demonstrated in our simulations.[7]

4   System Evaluation

In our work we used a real-world (anonymized) dataset consisting of 5000 hotels in the city of Paris, retrieved from a popular international hotel booking website

---

[7] The idea of employing clustering to address the "cold start" problem has also appeared in [26]. However, that work uses averaging over *user ratings* to produce recommendations that are appropriate for each cluster. In our work, we make no use of user ratings over items, and make recommendations based on the clusters' centroids rather than employing some averaging-over-cluster-contents process.

in JSON format. In order to compose a user's preferences, we focus on the following hotel features: the price per day, the number of hotel 'stars', the rating it has received from clients[8], the distance from the city center, and the amenities that it has (e.g., breakfast, pool, spa etc.). The features we focused on were chosen or made quantifiable. In our experiments, we assume that the user has chosen her destination, and we try to learn her preferences in order to make personalized recommendations, so that she should not need to search manually through the thousands of hotels available. Our users are synthetic: this allows us to measure our model's effectiveness more accurately, since we can compare recommendations to the "true" user preferences (which we have perfect knowledge of).

4.1 Building the Synthetic Users

We generated 100 simulated users that was used in order to evaluate our system's performance. Their preferences values were sampled from distributions. It was important that the simulated users represented as best as possible the preferences of the real users, in order to have a realistic assessment of the model performance. By creating the user preferences by sampling distributions, we introduce an uncertainty necessary to account even for users with unconventional preferences. The preferences of each user consist of five values.

After sampling the distributions we normalize all preferences values to lie in the $[0, 1]$ interval. In order to determine the price per day that a user is willing to pay for a hotel, we were inspired by the income distribution of the UK citizens for the year 2011.[9] Thus, for extracting the price preference, we used a Burr type XII distribution ($a = 25007, c = 2.0, k = 2.0$). To derive the hotel's stars preference, we assume that if the 'can pay' price is high, then it is expected that a hotel with a high number of stars (4–5) is preferred, so we sample a Beta distribution with the following parameters a = 8, b = 2. Respectively, we use a Beta with parameters $a = 2, b = 8$ if the price value is low. Otherwise, we take sample from a Beta distribution with $a = 8, b = 8$. Here, we choose a Beta distribution, as we want the number of stars to have mean values between zero and one. Additionally, such distributions are more appropriate for modelling uncertainties in the real world, as they can concentrate probability in a desired range [11]. Subsequently we choose to sample a Beta distribution with parameters $a = 1, b = 3$ in order to derive the proximity to the city center preference. Using this distribution we want to simulate most people's tendency to prefer a hotel near the city center rather than in the countryside. A sampled value closer to zero corresponds to a preference for a hotel close to the center, while value closer to 1 means the opposite. In order to specify the value of 'ratings by previous guests' value a user demands, we used a Beta with parameters $a = 6$ and $b = 2$. We choose this distribution in order to simulate the tendency that most users prefer hotels with

---

[8] Note that "clients' rating" is just an item's (a hotel's) feature. We stress that we do not ask our system's users to rate the items, and it is not the system's aim to produce recommendations based on such ratings.

[9] https://www.gov.uk/government/statistics.

high ratings. Finally, the last feature that characterizes the user preferences was the required hotel amenities. To derive the "amenities" preference, we used a uniform distribution in $[0, 1]$, with a higher number meaning that a hotel with more amenities is preferred. All amenities are assumed to have equal impact on a user.

4.2   Experiments and Results

In our experiments we used the synthetic users described above, each of whom had twenty (20) interactions with our system. In each interaction, seven (7) hotels are recommended to the user, and she chooses one. Five (5) of those are the top ranked ones according to our current beliefs regarding user preferences, while two (2) are picked completely randomly from our dataset. The choice made by the user is based on Euclidean distance:

$$d(h, u) = \sqrt{(h_{F1} - u_{F1})^2 + \cdots + (h_{Fn} - u_{Fn})^2} \tag{9}$$

where $h_{F1}, h_{F2}, ..., h_{Fn}$ the values of the corresponding features of the hotel, and $u_{F1}, u_{F2}, ..., u_{Fn}$ are the actual user preferences. Thus the user will always choose the hotel with the minimum Euclidean distance from her preferences.

$$Selection = \min(d(h_1, u), d(h_2, u), ..., d(h_k, u)) \tag{10}$$

where k is the number of suggestions made to the user. In all experiments, our ranking algorithm uses 50 samples randomly picked from 10000 samples returned by the MCMC algorithm. This allows us *to keep response time under 2 s*, on a PC with an i7@4.5 GHz processor and 16 GB of RAM.

Regarding our results, we first report the (average) mean square error (MSE) for each user interaction, calculated given the user's true value function and her "ideal" choices. This helps us assess the accuracy of the recommendation made as the interactions increases. As seen in Fig. 2, the MSE of our method drops quite sharply, after only a few interactions.

The use of MSE does provide us with an insight of the overall method performance. We can provide further testimony to the effectiveness of our approach, however, by using the following "metric". For each simulated user, we found and saved the "top 200" ranked hotels for the 5000 hotels of our dataset, based on the minimum Euclidean distance from the user's true preferences. Then, we were able to compare the suggestions our model made with the top 200 hotels for each user. As shown in Fig. 3, after only five interactions, our system was able to recommend hotels with 38% of them being among the best 200 hotels of each user (38% of the hotels suggested during the last 15 interactions were in the "top-200"). Note that the "top-200" constitutes only 4% of our dataset of 5000 hotels. Similarly, about 20% of the recommended hotels belong in the "top 50" (1% of our dataset), and 8% belong in the "top 10" (0.2% of our dataset).

We also report on the observed performance when using clustering to reduce our initial uncertainty regarding a user. In Fig. 2, we observe a faster reduction of

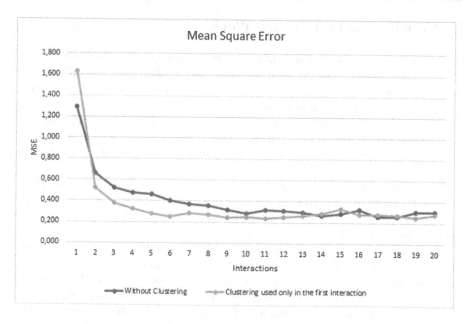

Fig. 2. Mean Squared Error with and without clustering.

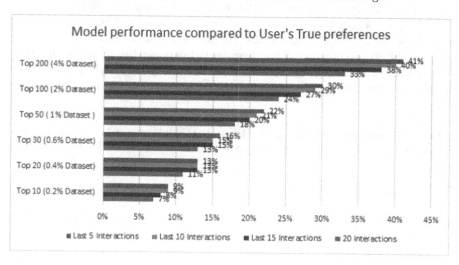

Fig. 3. Performance wrt to a user's true preferences (averages).

the MSE when clusters are used. Naturally, in the long run the MSE converges to the same low value, with and without clustering (the latter loses its importance once enough information is available).

5   Conclusions and Future Work

In this work, we presented a lightweight recommendation system which uses a Bayesian preference elicitation framework, and applied it in the online hotel booking recommendations domain. The algorithm makes personalized recommendations, is simple and fast, but still generic and effective. Our system is non-intrusive, and does not ask the user to rate items. It tackles conflicting user preferences and the problem of having scarse information about a user, while it possesses a simple exploration component. Our simulation experiments confirm the effectiveness of our method.

There are many possible extensions to this work. An obvious future step is to directly compare its effectiveness against other hotel booking recommender systems. This is however a non-trivial task, given that most existing systems are not open-source or fully described in research articles, and thus a direct comparison with the same dataset and user pool is next to impossible to conduct. Another direction we aim to take is experimenting with real users, employing questionnaires in order to evaluate and measure the system's performance. Additionally, it is very common that a user is affected by unexpected factors or unquantifiable features such as the presentation of the item or the available pictures. For instance, in a booking recommendation scenario like ours, the user may not choose based solely on features: on the contrary, she may get influenced by the beautiful photos of a hotel. There exist numerous works in the marketing and advertising domains that describe how the presentation of an item can strongly influence the choices of a user. Thus, in order to take cases like these into account, we aim to incorporate a noise model to simulate unexpected and unquantifiable factors affecting user choices, and thus obtain a more realistic user preferences model.

Acknowledgements. The authors would like to thank Professor Michail Lagoudakis for extremely useful suggestions for improving an earlier version of this work.

References

1. Andrieu, C., de Freitas, N., Doucet, A., Jordan, M.I.: An introduction to MCMC for machine learning. Mach. Learn. **50**(1), 5–43 (2003). https://doi.org/10.1023/A:1020281327116
2. Applegate, D., Kannan, R.: Sampling and integration of near log-concave functions. In: Proceedings of the Twenty-Third Annual ACM Symposium on Theory of Computing, pp. 156–163 (1991)
3. Babas, K., Chalkiadakis, G., Tripolitakis, E.: You are what you consume: a Bayesian method for personalized recommendations. In: Proceedings of the 7th ACM Conference on Recommender Systems (ACM RecSys 2013), Hong Kong, China (2013)
4. Bishop, C.M.: Neural networks for pattern recognition (1995)
5. Boutilier, C.: A POMDP formulation of preference elicitation problems. In: Proceedings of the 18th AAAI Conference, AAAI 2002 (2002)

6. Bowling, M.H., Veloso, M.M.: Multiagent learning using a variable learning rate. Artif. Intell. (AIJ) **136**(2), 215–250 (2002)

7. Chajewska, U., Koller, D., Parr, R.: Making rational decisions using adaptive utility elicitation. In: Proceedings of the 17th AAAI Conference, AAAI 2000, pp. 363–369 (2000)

8. Chajewska, U., Koller, D., Ormoneit, D.: Learning an agent's utility function by observing behavior. In: Proceedings of the International Conference on Machine Learning (ICML) (2001)

9. Chib, S., Greenberg, E.: Understanding the Metropolis-Hastings algorithm, pp. 327–335 (2012)

10. Dantzig, G.B., Orden, A., Wolfe, P.: The generalized simplex method for minimizing a linear form under linear inequality restraints. Pacific J. Math. **5**(2), 183–195 (1955)

11. DeGroot, M.: Probability and Statistics. Addison-Wesley Series in Behavioral Science, Addison-Wesley Publishing Company, Boston (1975). https://books.google.gr/books?id=fxPvAAAAMAAJ

12. Dong, R., Smyth, B.: From more-like-this to better-than-this: hotel recommendations from user generated reviews. In: Proceedings of the 2016 Conference on User Modeling Adaptation and Personalization (UMAP 2016), pp. 309–310 (2016)

13. Gilks, W.R., Wild, P.: Adaptive rejection sampling for Gibbs sampling. J. Roy. Stat. Soc. Ser. C (Appl. Stat.) **41**(2), 337–348 (1992)

14. Hastings, W.K.: Monte Carlo sampling methods using Markov chains and their applications. Biometrika **57**(1), 97–109 (1970). http://www.jstor.org/stable/2334940

15. Keeney, R.L., Raiffa, H.: Decisions with Multiple Objectives: Decisions with Preferences and Value Tradeoffs. Cambridge University Press, Cambridge (1993)

16. MacQueen, J.B.: Some methods for classification and analysis of multivariate observations. In: Proceedings of 5th Berkeley Symposium on Mathematical Statistics and Probability (1967)

17. Nasery, M., Braunhofer, M., Ricci, F.: Recommendations with optimal combination of feature-based and item-based preferences. In: Proceedings of the 2016 Conference on User Modeling Adaptation and Personalization (UMAP 2016), pp. 269–273 (2016)

18. Ng, A.Y., Russell, S.J.: Algorithms for inverse reinforcement learning. In: Proceedings of the Seventeenth International Conference on Machine Learning, pp. 663–670. ICML (2000)

19. Ramachandran, D., Amir, E.: Bayesian inverse reinforcement learning. In: Proceedings of IJCAI-2007, pp. 2586–2591 (2007)

20. Ricci, F., Rokach, L., Shapira, B., Kantor, P.B.: Recommender Systems Handbook, 1st edn. Springer, New York (2010). https://doi.org/10.1007/978-0-387-85820-3

21. Robert, C.P., Casella, G.: Monte Carlo Statistical Methods. Springer, Heidelberg (2004). https://doi.org/10.1007/978-1-4757-4145-2

22. Roy, S.B., Das, G., Amer-Yahia, S., Yu, C.: Interactive itinerary planning. In: Proceedings of the IEEE International Conference on Data Engineering (ICDE) (2011)

23. Russell, S., Norvig, P.: Artificial Intelligence: A Modern Approach, 3rd edn. Prentice Hall, Upper Saddle River (2009)

24. Tripolitakis, E., Chalkiadakis, G.: Probabilistic topic modeling, reinforcement learning, and crowdsourcing for personalized recommendations. In: Criado Pacheco, N., Carrascosa, C., Osman, N., Julián Inglada, V. (eds.) EUMAS/AT -2016. LNCS (LNAI), vol. 10207, pp. 157–171. Springer, Cham (2017). https:// doi.org/10.1007/978-3-319-59294-7_14
25. Xie, M., Lakshmanan, L.V., Wood, P.T.: Generating top-k packages via preference elicitation. Proc. VLDB Endow. **7**(14), 1941–1952 (2014)
26. Yanxiang, L., Deke, G., Fei, C., Honghui, C.: User-based clustering with top-n recommendation on cold-start problem. In: 2013 Third International Conference on Intelligent System Design and Engineering Applications (2013)

Computing Consensus: A Logic for Reasoning About Deliberative Processes Based on Argumentation

Sjur Dyrkolbotn[1] and Truls Pedersen[2(✉)]

[1] Western Norway University of Applied Sciences, Bergen, Norway
sdy@hvl.no
[2] University of Bergen, Bergen, Norway
truls.pedersen@uib.no

Abstract. Argumentation theory can encode an agent's assessment of the state of an exchange of points of view. We present a conservative model of multiple agents potentially disagreeing on the views presented during a process of deliberation. We model this process as iteratively adding points of view (arguments), or aspects of points of view. This gives rise to a modal logic, deliberative dynamic logic, which permits us to reason about the possible developments of the deliberative state. The logic we propose applies to all natural semantics of argumentation theory. Furthermore, under a very weak assumption that the consensus considered by a group of agents is faithful to their individual views, we show that model checking these models is feasible, as long as the argumentation frameworks, which may be infinite, does not have infinite branching.

Keywords: Argumentation theory · Multi agent systems · Modal logic · Dynamic logic · Deliberation · Consensus

1 Introduction

We study agency and argumentation, and propose a framework for modeling and reasoning about *deliberation*. In short, we assume that some agents are given, with their own individual view of the argumentation scenario at hand, and that deliberation is a process by which we attempt to reconcile all these individual views to aggregate a joint, common view on the situation, which we may then analyze further using established techniques from argumentation theory. We develop our framework based on the formal theory of argumentation introduced by Dung [6], which has attracted much interest from the AI community, see [10] for a volume devoted to this theory. In keeping with recent trends, we also

This paper was presented at the 1st International Workshop on Argument for Agreement and Assurance (AAA 2013) and made available online, it has since been cited in [3].

M. Slavkovik (Ed.): EUMAS 2018, LNAI 11450, pp. 205–219, 2019.
https://doi.org/10.1007/978-3-030-14174-5_14

take advantage of *logical* tools, relying both on a truth-functional three-valued view of argumentation [1,7], and on the use of modal logic [8,9].

Unlike much previous work in argumentation theory, including work done on multi-agent argumentative interaction (see, for instance, [10, Chap. 13]), we do not worry about attempting to design procedures for "good" deliberation, but focus instead on a logical analysis of the space of possible outcomes, assuming only a minimal restriction on the nature of the deliberative processes that we consider permissible. The restriction encodes the intuition that the common view aggregated by deliberation should be built in such a way that we only make use of information already present in the view of some agent. This, indeed, seems like a safe requirement, and appears to be one that no reasonable group of agents would ever want to deviate from.

In the multi-agent systems literature it is not uncommon to use the term "deliberation" to describe the process by which agents agree on which action to perform. We prefer to rather refer to such processes as "practical reasoning" and reserve "deliberation" to refer to processes where reasoned interactions explore an issue and which may, or may not, result in a common agreement or consensus.

We remark that our concept of deliberation is also somewhat unusual compared to earlier work in that we focus on processes that aim to reconcile *representations* of the argumentative situation, rather than processes where agents interact based on their different *judgments* about a given structure. While we will not argue extensively for the appropriateness of this shift of attention, we mention that it is prima facie reasonable: if two agents judge a situation differently it seems safe to assume that they must have a *reason* for doing so. Moreover, while this reason could sometimes be due to disagreement about purely logical principles or atomic (i.e., unanalyzable) differences in preference, it seems clear that in real life, disagreement arises just as often, perhaps more often, as a result of a difference in *interpretation*, i.e., from the fact that the agents have different mental representations of the situation at hand.

In fact, in this paper we will refrain from committing to a particular view on judgments, and we will set up our logical framework in such a way that it allows us to use any semantics from formal argumentation as the source of basic judgments about arguments given a framework. Our own focus is solely on the stepwise, iterative development of a common framework, and on the logical analysis of the different ways in which such a process may unfold, by way of a logical treatment of the modalities that arise from quantifying over the space of possible deliberative futures.

The structure of the paper is as follows. In Sect. 2 we give a background on abstract argumentation, concluding by an example that motivates our logic and a definition of *normality*, which allows us to parameterize our constructions by any normal argumentation semantics (all semantics of which we are aware are normal). Then in Sect. 3 we introduce *deliberative dynamic logic* (DDL), a concrete suggestion for a logical framework allowing us to reason about deliberative processes using modal logic. Then in Sect. 4 we show that while our models are generally infinite, model checking is still feasible since we may "shrink" them, by restricting attention only to the relevant parts of the structure. Then, in Sect. 5, we conclude and discuss directions for future work.

2   Background on Abstract Argumentation

Following Dung [6] we represent argumentation structures by directed graphs $F = (S, E)$, such that S is a set of arguments and $E \subseteq S \times S$ encode the attacks between them, i.e., such that if $(x, y) \in E$ then the argument x attacks y. Traditionally, most work in formal argumentation theory has focused on defining and investigating notions of successful sets of arguments, in a setting where the argumentation framework is given and remains fixed. Such notions are typically formalized by an *argumentation semantics*, an operator ε which returns, for any AF F, the set of sets of arguments from $F = (S, E)$ that are regarded as successful combinations, i.e., such that $\varepsilon(F) \subseteq 2^S$. Many proposals exists in the literature, we point to [2] for a survey and formal comparison of different semantics. While some semantics, such as the grounded and ideal semantics, return a unique set of arguments, the "winners" of the argumentation scenario encoded by F, most semantics return more than one possible collection of arguments that *would* be successful if they were held together. For instance, the admissible semantics, upon which many of the other well-known semantics are built, returns, for each AF F, the following sets of arguments:

$$a(F) = \{A \subseteq S \mid E^-(A) \subseteq E^+(A) \subseteq S \setminus A\}$$

Where $E^-(X)$ are the nodes which has an arrow into X, and $E^+(X)$ are the nodes which the nodes in X has an arrow into. That is, the admissible sets are those that can defend themselves against attacks (first inclusion), and do not involve any internal conflicts (second inclusion). A strengthening that is widely considered more appropriate (yet incurs some computational costs) is the *preferred* semantics p, which is defined by taking only those admissible sets that are set-theoretically maximal, i.e., such that they are not contained in any other admissible set. In general, an AF admits many preferred sets, and even more admissible ones. Indeed, notice that the empty set is always admissible by the default (the inappropriateness of which provides partial justification for using preferred semantics instead). As a simple example, consider F below.

$$F : p \overset{\longleftarrow}{\underset{\longrightarrow}{}} q \qquad\qquad a(F) = \{\emptyset, \{p\}, \{q\}\}, \; p(F) = \{\{p\}, \{q\}\}$$

Indeed, it seems hard to say which one of p and q should be regarded as successful in such a scenario. In the absence of any additional information, it seems safest to concede that choosing either one will be a viable option. Alternatively, one may take the view that due to the undetermined nature of the scenario, it should not be permitted to regard either argument as truly successful. This, indeed, is the view taken by unique status semantics, such as the grounded and ideal semantics. However, while such a restrictive view might be appropriate in some circumstance, it seems unsatisfactory for a general theory of argumentation. Surely, in most real-world argumentation situations, it is not tenable for an arbitrator to refrain from making a judgment whenever doing so would involve some degree of discretion on his part.

Since argumentation semantics typically only restrict the choice of successful arguments, without determining it completely, a modal notion of *acceptance* arises, usually referred to as *skeptical* acceptance in argumentation parlance, whereby an argument is said to be skeptically accepted by F under ε if $\forall S \in \varepsilon(\mathsf{F}) : p \in S$. The dual notion is called *credulous* acceptance, and obtains just in case $\exists S \in \varepsilon(\mathsf{F}) : p \in S$. Moreover, since the choice among elements of $\varepsilon(\mathsf{F})$ can itself be a contentious issue, and is not one which can be satisfactorily resolved by single-agent argumentation theory, there has been research devoted to giving an account of multi-agent interaction concerning the choice among members of $\varepsilon(\mathsf{F})$, see [5,11]. While this is interesting, it seems that another aspect of real-world argumentation has an even stronger multi-agent flavor, namely the process by which one arrives at a common AF in the first place. Certainly, two agents, a and b, might disagree about whether to choose p or q in F considered above, but as it stands, such a choice appears arbitrary and, most likely, the two agents would also be willing to admit as much. Arguably, then, the disagreement itself is only superficial. The agents disagree, but they provide no *reason* for their different preferences, and do not provide any content or structure to substantiate them. This leaves an arbitrator in much the same position as he was in before: he might note the different opinions raised, but he has no basis upon which to inquire into their merits, and so his choice must, eventually, still be an exercise in discretion.

In practice, however, it would have to be expected that if the agents a and b were really committed to their stance, they would not simply accept that F correctly encodes the situation and that the choice is in fact arbitrary. Rather, they would produce *arguments* to back up their position. It might be, for instance, that agent a, who favors p, claims that q is inconsistent for some reason, while agent b, who favors q, makes an analogous accusation against the argument p. Then, however, we are no longer justified in seeing this as disagreement about which choice to make from $\varepsilon(\mathsf{F})$. Rather, the disagreement concerns the nature of the argumentation structure itself. The two agents, in particular, put forth different *views* on the situation. For instance, in our toy example, we would have to consider the following two AFs, where V_a, V_b encode the views of a and b respectively.

$$V_a : \quad \bigcirc p \longrightarrow\longleftarrow q \qquad\qquad V_b : \quad p \longrightarrow\longleftarrow q \bigcirc \qquad\qquad (1)$$

Then the question arises: what are we to make of this?

In the following, we address this question, and we approach it from the conceptual starting point that evaluating (higher-order) differences of opinion such as that expressed by V_a, V_b takes place iteratively, through a process of *deliberation*, leading, in a step-by-step fashion, to an aggregated *common* F. Such a process might be instantiated in various ways: it could be the agents debating the matter among themselves and reaching some joint decision, or it could be an arbitrator who considers the different views and reasons about them by emulating such a process. Either way, we are not interested in attempting to provide any guidance towards the "correct" outcome, which is hardly possible in general. Rather, we are interested in investigating the modalities that arise when we consider the space of all possible outcomes (where possible will be defined in due course). Moreover, we are interested in investigating structural questions, asking, for instance, about the importance of the order in which arguments are considered, and the consequences of limiting attention to only a subset of arguments.

We use a dynamic modal logic to facilitate this investigation, and in the next section we define the basic framework and show that model checking is decidable even on infinite AF's, as long as the agent's views remains finitely branching, i.e., as long as no argument is attacked by infinitely many other arguments. We will parameterize our logic by an argumentation semantics, so that it can be applied to any such semantics which satisfies a normality condition. In particular, let $C(\mathsf{F}) = \{C_1^{\mathsf{F}}, \ldots, C_i^{\mathsf{F}}, \ldots\}$ denote the (possibly infinite) set of maximal connected components from F (the set of all maximal subsets of S such that any two arguments in the same set are connected by a sequence of attacks). Then we say that a semantics ε is *normal* if we have, for any $\mathsf{F} = (S, E)$

$$A \in \varepsilon(\mathsf{F}) \Leftrightarrow A = \bigcup_i A_i \text{ for some } A_1, \ldots, A_i, \ldots \text{ s.t. } A_i \in \varepsilon(C_i^{\mathsf{F}}) \text{ for all } i \quad (2)$$

That is, a semantics is normal if the status of an argument depends only on those arguments to which it has some (indirect) relationship through a sequence of attacks. We remark that all argumentation semantics of which we are aware satisfies this requirement, hence we feel justified in dubbing it normality.

3   Deliberative Dynamic Logic

We assume given a finite non-empty set $\mathcal{A}$ of agents and a countably infinite set Π of arguments.[1] The basic building block of dynamic deliberative logic is provided in the following definition.

Definition 1. *A basis for deliberation is an $\mathcal{A}$-indexed collection of digraphs $\mathcal{B} = (V_a)_{(a \in \mathcal{A})}$, such that for each $a \in \mathcal{A}$, $V_a \subseteq \Pi \times \Pi$.*

Given a basis which encodes agents' view of the arguments, we are interested in the possible ways in which agents can deliberate to reach *agreement* on how arguments are related. That is, we are interested in the set of all AFs that can plausibly be seen as resulting from a *consensus* regarding the status of the arguments in Π. What restrictions is it reasonable to place on a consensus? It seems that while many restrictions might arise from pragmatic considerations, and be implemented by specific protocols for "good" deliberation in specific contexts, there are few restrictions that can be regarded as completely general. For instance, while there is often good reason to think that the position held by the majority will be part of a consensus, it is hardly possible to stipulate an axiomatic restriction on the notion of consensus amounting to the principle of

---

[1] Possibly "statements" or "positions", depending on the context of application.

majority rule. Indeed, sometimes deliberation takes place and leads to a single dissenting voice convincing all the others, and often, these deliberative processes are far more interesting than those that transpire along more conventional lines. However, it seems reasonable to assume that whenever *all* agents agree on how an argument p is related to an argument q, then this relationship is part of any consensus. This, indeed, is the only restriction we will place on the notion of a consensus; that when the AF F is a consensus for a *basis*, it must satisfy the following *faithfulness* requirement.

– For all $p, q \in \Pi$, if there is no disagreement about p's relationship to q (attack/not attack), then this relationship is part of F

This leads to the following definition of the set $\Upsilon(\mathcal{B})$, which we will call the set of *complete assents* for $\mathcal{B}$, collecting all AFs that are faithful to $\mathcal{B} = (V_a)_{(a \in \mathcal{A})}$.

$$\Upsilon(\mathcal{B}) = \left\{ \mathsf{F} \subseteq \Pi \times \Pi \ \middle| \ \bigcap_{a \in \mathcal{A}} V_a \subseteq \mathsf{F} \subseteq \bigcup_{a \in \mathcal{A}} V_a \right\} \tag{3}$$

An element of $\Upsilon(\mathcal{B})$ represents a possible consensus among agents in $\mathcal{A}$, but it is an *idealization* of the notion of assent, since it disregards the fact that in practice, assent tends to be *partial*, since it results from a dynamic process, emerging through *deliberation*. Indeed, as long as the number of arguments is not bounded we can *never* hope to arrive at complete assent via deliberation. We can, however, initiate a process by which we reach agreement on more and more arguments, in the hope that this will approximate some complete assent, or maybe even be *robust*, in the sense that there is *no* deliberative future where the results of current partial agreement end up being undermined. Complete assent, however, arises only in the limit.

When and how deliberation might successfully lead to an approximation of complete assent is a question well suited to investigation with the help of dynamic logic. The dynamic element will be encoded using a notion of a deliberative event – centered on an argument – such that the set of ways in which to relate this argument to arguments previously considered gives rise to a space of possible deliberative time-lines, each encoding the continued stepwise construction of a joint point of view. This, in turn, will be encoded as a monotonically growing AF $\mathsf{F} = (S, E)$ where $S \subseteq \Pi, E \subseteq S \times S$ and such that faithfulness is observed by all deliberative events. That is, an event consists of adding to F the agents' combined view of p with respect to the set $S \cup \{p\}$. This leads to the following collection of possible events, given a basis $\mathcal{B}$, a partial consensus[2] $\mathsf{F} = (S, E)$ and an argument $p \in \Pi$:

$$\mathcal{U}_{\mathcal{B}}(\mathsf{F}, p) = \left\{ X \ \middle| \ \bigcap_{a \in \mathcal{A}} V_a|_{S \cup \{p\}} \subseteq X \subseteq \bigcup_{a \in \mathcal{A}} V_a|_{S \cup \{p\}} \right\} \tag{4}$$

---

[2] These "partial consensuses" are sometimes referred to as "contexts" when they are used to describe graphs inductively, as we will do later.

To provide a semantics for a logical approach to deliberation based on such events, we will use Kripke models.

Definition 2 (Deliberative Kripke model). *Given an argumentation semantics ε and a set of views $\mathcal{B}$, the deliberative Kripke models induced by $\mathcal{B}$ and ε is the triple $\mathcal{K}_{(\mathcal{B},\varepsilon)} = (Q_\mathcal{B}, R_\mathcal{B}, \mathbf{e}_\varepsilon)$ such that*

- *$Q_\mathcal{B}$, the set of points, is the set of all pairs of the form $q = (q_S, q_E)$ where $q_S \subseteq \Pi$ and*

$$\bigcap_{a \in \mathcal{A}} Val_{q_S} \subseteq q_E \subseteq \bigcup_{a \in \mathcal{A}} Val_{q_S}$$

The basis $\mathcal{B}$ together with our definition of an event, given in Eq. 4, induces the following function, mapping states to their possible deliberative successors, defined for all $p \in \Pi, q \in Q_\mathcal{B}$ as follows

$$succ(p, q) \ := \ \{(q_S \cup \{p\}, q_E \cup X) \mid X \in \mathcal{U}_\mathcal{B}(q, p)\}$$

We also define a lifting, for all states $q \in Q_\mathcal{B}$:

$$succ(q) \ := \ \{q' \mid \exists p \in \Pi : q' \in succ(q, p)\}$$

- *$R_\mathcal{B} : \Pi \cup \{\exists\} \to 2^{Q_\mathcal{B} \times Q_\mathcal{B}}$ is a map from symbols to relations on $Q_\mathcal{B}$ such that*
  - *$R_\mathcal{B}(p) = \{(q, q') \mid q' \in succ(p, q)\}$ for all $p \in \Pi$ and*
  - *$R_\mathcal{B}(\exists) = \{(q, q') \mid q' \in succ(q)\}$,*
- *$\mathbf{e}_\varepsilon : Q_\mathcal{B} \to 2^{(3^\Pi)}$ maps states to labellings such that for all $q \in Q_\mathcal{B}$ we have $\mathbf{e}_\varepsilon(q) = (\pi_1, \pi_0, \pi_{\frac{1}{2}})$ with*

$$\mathbf{e}_\varepsilon(q) = \{\pi \mid \pi_1 \in \varepsilon(q), \pi_0 = \{p \in q_S \mid \exists q \in \pi_1 : (q, p) \in q_E\}\}$$

Notice that in the last point, we essentially map q to the sets of extensions prescribed by ε when q is viewed as an AF. We encode this extension as a three-valued labeling, however, following [4]. Notice that the default status, attributed to all arguments not in q_S, is $\frac{1}{2}$. The logical language we will use consists in two levels. For the lower level, used to talk about static argumentation, we follow [1,7] in using Łukasiewicz three-valued logic. Then, for the next level, we use a dynamic modal language which allows us to express consequences of updating with a given argument, and also provides us with existential quantification over arguments, allowing us to express claims like "there is an update such that ϕ". This leads to the language $\mathcal{L}_{DDL}$ defined by the following BNF's

$$\phi \ ::= \ \blacklozenge\alpha \mid \neg\phi \mid \phi \wedge \phi \mid \langle p\rangle\phi \mid \Diamond\phi$$

where $p \in \Pi$ and $\alpha \in \mathcal{L}^\blacklozenge$ where $\mathcal{L}^\blacklozenge$ is defined by the following grammar:

$$\alpha ::= p \mid \neg\phi \mid \phi \to \phi$$

for $p \in \Pi$.

We also use standard abbreviations such that $\Box\phi = \neg\Diamond\neg\phi$, $[p]\phi = \neg\langle p\rangle\neg\phi$ and $\blacksquare\alpha = \neg\blacklozenge\neg\alpha$. We also consider that standard boolean connectives abbreviated as usual for connectives not occurring inside a $\blacklozenge$-connective and abbreviations for connectives of Łukasiewicz logic in the scope of $\blacklozenge$-connectives.

Next we define truth of formulas on deliberative Kripke models. We begin by giving the valuation of complex formulas from $\mathcal{L}^{\blacklozenge}$, which is simply three-valued Łukasiewicz logic.

Definition 3 (α-satisfaction). *For any three-partitioning $\pi = (\pi_1, \pi_0, \pi_{\frac{1}{2}})$ of Π, we define*

$$\overline{\pi}(p) = x \; s.t \; p \in \pi_x$$
$$\overline{\pi}(\neg\alpha) = 1 - \overline{\pi}(\alpha)$$
$$\overline{\pi}(\alpha_1 \to \alpha_2) = \min\{1, 1 - (\overline{\pi}(\alpha_1) - \overline{\pi}(\alpha_2))\}$$

Now we can give a semantic interpretation of the full language as follows.

Definition 4 ($\mathcal{L}_{\mathrm{DDL}}$-satisfaction). *Given an argumentation semantics ε and a basis $\mathcal{B}$, truth on $\mathcal{K}_{(\mathcal{B},\varepsilon)}$ is defined inductively as follows, in all points $q \in Q_{\mathcal{B}}$.*

$$\begin{aligned}
\mathcal{K}_{(\mathcal{B},\varepsilon)}, q \vDash \blacklozenge\alpha &\iff & \text{there is } \pi \in \mathbf{e}_\varepsilon(q) \text{ s.t. } \overline{\pi}(\phi) = 1 \\
\mathcal{K}_{(\mathcal{B},\varepsilon)}, q \vDash \neg\phi &\iff & \text{not } \mathcal{K}_{(\mathcal{B},\varepsilon)}, q \vDash \phi \\
\mathcal{K}_{(\mathcal{B},\varepsilon)}, q \vDash \phi \wedge \psi &\iff & \text{both } \mathcal{K}_{(\mathcal{B},\varepsilon)}, q \vDash \phi \text{ and } \mathcal{K}_{(\mathcal{B},\varepsilon)}, q \vDash \psi \\
\mathcal{K}_{(\mathcal{B},\varepsilon)}, q \vDash \langle p\rangle\phi &\iff & \exists (q, q') \in \mathsf{R}_{\mathcal{B}}(p) : \mathcal{K}_{(\mathcal{B},\varepsilon)}, q' \vDash \phi \\
\mathcal{K}_{(\mathcal{B},\varepsilon)}, q \vDash \Diamond\phi &\iff & \exists (q, q') \in \mathsf{R}_{\mathcal{B}}(\exists) : \mathcal{K}_{(\mathcal{B},\varepsilon)}, q' \vDash \phi
\end{aligned}$$

To illustrate the definition, we return to the example depicted in (1). In Fig. 1, we depict this basis together with a fragment of the corresponding Kripke model, in particular the fragment arising from the p-successors of $(\emptyset, \emptyset)$.

Let us assume that $\varepsilon = \mathsf{p}$ is the preferred semantics. Then the following list gives some formulas that are true on $\mathcal{K}_{(\mathcal{B},\varepsilon)}$ at the point $(\emptyset, \emptyset)$, and the reader should easily be able to verify them by consulting the above fragment of $\mathcal{K}_{(\mathcal{B},\varepsilon)}$.

$$\langle p\rangle\blacksquare p, \quad \Diamond\blacksquare p, \quad [p]\Diamond\blacksquare q,$$
$$\neg[p]\Diamond\blacklozenge p, \langle p\rangle\Diamond\blacksquare\neg p, \Diamond\Diamond(\blacklozenge p \wedge \blacklozenge q)$$

We can also record some validities that are easy to verify against Definition 2.

Proposition 1. *The following formulas are all validities of $\mathcal{L}_{\mathrm{DDL}}$, for any $p, q \in \Pi$, $\phi \in \mathcal{L}_{\mathrm{DDL}}$.*

1. $\langle p\rangle\langle q\rangle\phi \leftrightarrow \langle q\rangle\langle p\rangle\phi$
2. $\langle p\rangle[q]\phi \to [q]\langle p\rangle\phi$
3. $\Diamond\Box\phi \to \Box\Diamond\phi$
4. $\langle p\rangle\langle p\rangle\phi \to \langle p\rangle\phi$

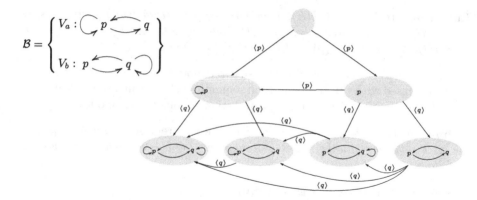

Fig. 1. A fragment of the deliberative Kripke model for $\mathcal{B}$.

We remark that $[q]\langle p\rangle\phi \rightarrow \langle p\rangle[q]\phi$ is *not* valid, as witnessed for instance by the following basis $\mathcal{B}$, for which we have $\mathcal{K}_{(\mathcal{B},p)}, (\emptyset, \emptyset) \models [q]\langle p\rangle\blacksquare p$ but also $\mathcal{K}_{(\mathcal{B},p)}, (\emptyset, \emptyset) \models [p]\langle q\rangle\blacksquare q$ (as the reader may easily verify by considering the corresponding Kripke model).

$$\mathcal{B} = \left\{ \begin{array}{l} V_a : p \longrightarrow q \\ V_b : p \longleftarrow q \end{array} \right\}$$

Finally, let us notice that as Π is generally infinite, we must expect to encounter infinite bases. This means, in particular, that our Kripke models are often infinite. However, in the next section we show that as long as $\mathcal{B}$ is *finitary*, meaning that no agent $a \in \mathcal{A}$ has a view where an argument is attacked by infinitely many other arguments, we can solve the model-checking problem also on infinite models.

4   Model Checking on Finitary Models

Towards this result, we now introduce some notation and a few abstractions to simplify our further arguments. We will work with labeled trees, in particular, where we take a tree over labels X to be some non-empty, prefix-closed subset of X^* (finite sequences of elements of X). Notice that trees thus defined contain no infinite sequences. This is intentional, since we will "shrink" our models (which may contain infinite sequences of related points), by mapping them to trees. To this end we will use the following structures.

Definition 5. *Given a basis $\mathcal{B}$, we define $\mathcal{I}(\mathcal{B})$, a set of sequences over $\Pi \times 2^{\Pi}$ labeled by AFs, defined inductively as follows*

Base case: $\epsilon \in \mathcal{I}(\mathcal{B})$ *and is labeled by the AF* $\mathsf{F}(\epsilon) = (S(\epsilon), E(\epsilon))$ *where* $S(\epsilon) = \emptyset = E(\epsilon)$.

Induction step: *If* $x \in \mathcal{I}(\mathcal{B})$, *then for any* $p \in \Pi$ *and any partial assent* $X = \mathcal{U}_{\mathcal{B}}(x, p)$, *we have* $x; (p, X) \in \mathcal{I}(\mathcal{B})$ *labeled by the AF* $\mathsf{F}(x; (p, X))$ *where* $S(x; (p, X)) = S(x) \cup \{p\}$ *and* $E(x; (p, X)) = E(x) \cup X$.

To adhere to standard naming we use ϵ to denote the empty string. It should not be confused with the argumentation semantics ε. This will also be clear from the context. We next define tree-representations of our Kripke models.

Definition 6. *Let* $\mathcal{K}_{(\mathcal{B}, \varepsilon)}$ *be some model. The* tree representation *of* $\mathcal{K}_{(\mathcal{B}, \varepsilon)}$ *is the set* T, *together with the* representation map $\gamma : Q_{\mathcal{B}} \to 2^T$, *defined inductively as follows*

Base case $\epsilon \in T$ *is the root with* $\gamma((\emptyset, \emptyset)) = \{\epsilon\}$.
Induction step *For any* $x \in T, q \in \mathcal{K}_{(\mathcal{B}, \varepsilon)}$ *with* $x \in \gamma(q)$ *and* $q' \in succ(q)$ *witnessed by* $p \in \Pi$ *and* $X \in \mathcal{U}_{\mathcal{B}}(x, p)$, *we have* $x; (p, X) \in T$ *with* $q' \in \gamma(x; (p, X))$.

Notice that the tree-representation is a tree where each node is an element of $\mathcal{I}(\mathcal{B})$. Some single states in $\mathcal{K}_{(\mathcal{B}, \varepsilon)}$ will have several representations in a tree. That is, $\gamma(q)$ may not be a singleton. On the other hand, it is easy to see that for every state $q \in \mathcal{K}_{(\mathcal{B}, \varepsilon)}$, and every path from $(\emptyset, \emptyset)$ to q, there will be a node $x \in T$ such that $q \in \gamma(x)$.

The main result of our paper is that model checking $\mathcal{L}_{\mathrm{DDL}}$-truth at $(\emptyset, \emptyset)$ is decidable as long as all views are *finitely branching*, i.e., such that for all $a \in \mathcal{A}, p \in \Pi$, p has only finitely many attackers in V_a. Clearly this requires shrinking the models since the modality $\Diamond$ quantifies over an infinite domain whenever Π is infinite. We show, however, that attention can be restricted to arguments from Π that are *relevant* to the formula we are considering. To make the notion of relevance formal, we will need the following measure of complexity of formulas.

Definition 7. *The* white modal depth *of* $\phi \in \mathcal{L}_{\mathrm{DDL}}$ *is* $|\phi|^{\Diamond} \in \mathbb{N}$, *which is defined inductively as follows*

$$
\begin{aligned}
|\alpha|^{\Diamond} &:= 0 &&\text{\textit{no white connectives in these formulas}}\\
|\blacklozenge\alpha|^{\Diamond} &:= 0 \\
|\neg\phi|^{\Diamond} &:= |\phi|^{\Diamond} &&\text{\textit{depth is deepest nesting of}}\\
|\phi \wedge \psi|^{\Diamond} &:= \max\{|\phi|^{\Diamond}, |\psi|^{\Diamond}\} &&\text{\textit{white connectives}}\\
|\Diamond\phi|^{\Diamond} &:= 1 + |\phi|^{\Diamond} \\
|\langle p\rangle\phi|^{\Diamond} &:= 1 + |\phi|^{\Diamond}
\end{aligned}
$$

We let $\Pi|_{\phi}$ denote the set of arguments occurring in ϕ in sub-formulas from $\mathcal{L}^{\blacklozenge}$. Notice that given a state $q \in Q_{\mathcal{B}}$, the satisfaction of a formula of the form $\phi = \blacklozenge\alpha$ at the AF encoded by q is not dependent on the entire digraph $q = (q_S, q_E)$.

Indeed, this is what motivated our definition of normality for an argumentation semantics, leading to the following simple lemma, which is the first step towards shrinking Kripke structures for the purpose of model checking. Given a model $\mathcal{K}_{(\mathcal{B},\varepsilon)}$ and a state $q \in Q_\mathcal{B}$, we let $C(q, \Phi)$ denote the digraph consisting of all connected components from q which contains a symbol from Φ. Then we obtain the following.

Lemma 1. *Given a semantics ε and two bases $\mathcal{B}$ and $\mathcal{B}'$, we have, for any two states $q \in \mathcal{K}_{(\mathcal{B},\varepsilon)}$ and $q' \in \mathcal{K}_{(\mathcal{B}',\varepsilon)}$ and for any formula $\phi \in \mathcal{L}_{\mathrm{DDL}}$ with $|\phi|^{\Diamond} = 0$:*

$$\left(C(q, \Pi|_\phi) = C(q', \Pi|_\phi)\right) \Rightarrow \left(\mathcal{K}_{(\mathcal{B},\varepsilon)}, q \vDash \phi \Leftrightarrow \mathcal{K}_{(\mathcal{B}',\varepsilon)}, q' \vDash \phi\right)$$

In order to complete our argument in this section, we will make use of n-bisimulations modulo a set of symbols.

Definition 8. *Given two models (with possibly different bases, but with common set of symbols Π and semantic ε) $K_\mathcal{B} = \langle Q_\mathcal{B}, R, \varepsilon \rangle$ and $K'_\mathcal{B} = \langle Q_{\mathcal{B}'}, R', \varepsilon \rangle$, states $q \in Q_\mathcal{B}$ and $q' \in Q'_\mathcal{B}$, a natural number n and a set $\Phi \subseteq \Pi$, then we say that q and q' are n-bisimilar modulo Φ (denoted $(K_\mathcal{B}, q) \underleftrightarrow{}_n^\Phi (K_{\mathcal{B}'}, q')$), if, and only if, there are $n + 1$ relations relation $Z_n \subseteq Z_{n-1} \subseteq \cdots \subseteq Z_0 \subseteq Q_\mathcal{B} \times Q'_\mathcal{B}$ such that*

1. $qZ_n q'$,
2. *whenever $(v, v') \in Z_0$, then $C(v, \Phi) = C(v', \Phi)$,*
3. *whenever $(v, v') \in Z_{i+1}$ and vRu, then there is a u' s.t. $v'R'u'$ and $uZ_i u'$,*
4. *whenever $(v, v') \in Z_{i+1}$ and $v'R'u'$, then there is a u s.t. vRu and $uZ_i u'$.*

Let us now also define a particular subset of arguments, the arguments which have at most distance n from some given set of arguments:

Definition 9. *Given a basis $\mathcal{B} = (V_a)_{(a \in \mathcal{A})}$, a subset $\Phi \subseteq \Pi$ and a number n, the n-vicinity of Φ is $D(\mathcal{B}, \Phi, i) \subseteq \Pi$, defined inductively as follows*

$$D(\mathcal{B}, \Phi, 0) = \Phi$$
$$D(\mathcal{B}, \Phi, n + 1) = D(\mathcal{B}, \Phi, n)$$
$$\cup \left\{ p \in \Pi \mid \exists q \in D(\mathcal{B}, \Phi, n) : \{(p, q), (q, p)\} \cap \bigcup_{a \in \mathcal{A}} V_a \neq \emptyset \right\}$$

Notice that as long as Φ is finite and all agents' views have finite branching, then the set D is also finite. Also notice that an equivalent characterization of the set $D((V_a)_{(a \in \mathcal{A})}, \Phi, i)$ can be given in terms of paths as follows: an argument $p \in \Pi$ is in $D(\mathcal{B}, \Phi, i)$ if, and only if, there is a path $p = x_1 x_2 \ldots x_n$ in $\bigcup_{a \in \mathcal{A}} V_a$ such that $x_n \in \Phi$ and $n \leq i$ (we consider an argument p equivalently as an empty path at p).

Definition 10. *Given a formula $\phi \in \mathcal{L}_{\mathrm{DDL}}$. Let $(V_a)_{(a \in \mathcal{A})}$ be a possibly infinite basis, we define $\rho_\phi((V_a)_{(a \in \mathcal{A})})$ such that*

– *for every* $a \in \mathcal{A}$, $\rho_\phi(V_a) := V_a \cap D(V_a, \Pi|_\phi, |\phi|^\Diamond)$

Notice that the Kripke model for $\rho(\mathcal{B})$ will have finite branching as long as the argument symbols in the $|\phi|^\Diamond$-vicinity of the argument symbols in ϕ have finite branching in all agents' views. In the following, we will show that for any finitely branching $\mathcal{B}$ and normal ε, we have $\mathcal{K}_{(\mathcal{B}, \varepsilon)}, (\emptyset, \emptyset) \models \phi$ if, and only if, $\mathcal{K}_{(\rho_\phi(\mathcal{B}), \varepsilon)}, (\emptyset, \emptyset) \models \phi$.

Theorem 1. *Let $\mathcal{B}$ be an arbitrary basis, and $\phi \in \mathcal{L}_{DDL}$.*

$$\left(\mathcal{K}_{(\mathcal{B}, \varepsilon)}, (\emptyset, \emptyset)\right) \quad \underset{|\phi|^\Diamond}{\overset{\Pi|_\phi}{\leftrightarrows}} \quad \left(\mathcal{K}_{(\rho_\phi(\mathcal{B}), \varepsilon)}, (\emptyset, \emptyset)\right)$$

Proof. Let $\mathcal{K}_{(\mathcal{B}, \varepsilon)}$ be an arbitrary model and let T denote its tree representation, while T' denotes the tree representation of $\mathcal{K}_{(\rho_\phi(\mathcal{B}), \varepsilon)}$.

We take $n = |\phi|^\Diamond$ and let Φ be the atoms occurring in ϕ inside the scope of some $\blacklozenge$-operator. Moreover, for brevity, we denote $D = D(\mathcal{B}, \Phi, n)$.

Definition of $(Z_i)_{(0 \leq i \leq |\phi|^\Diamond)}$: We define all the relations Z_i inductively using the tree-representations as follows.

Base case: $(i = 0)$ For all $0 \leq i \leq n$, we let $\epsilon Z_i \epsilon$.

Induction step: $(0 < i \leq n)$ For all $y = x; (v, X) \in T$ and $y' = x'; (v', X') \in T'$, both of length i, with $x(Z_{i+1})x'$. We let, for every $k \leq i$, $y(Z_k)y'$ if, and only if, $v = v'$, and $X \cap (D \times D) = X'$.

Notice that if $x(Z_i)x'$, then $S(x) = S(x')$ and $|S(x)| \leq (n - i)$. Moreover, by consulting Definition 6 it is not hard to see that for all $q \in Q_\mathcal{B}, q' \in Q_{\rho(\mathcal{B})}$ we have, for all $0 \leq i \leq n$ and all $q \in Q_\mathcal{B}, q' \in Q_{\rho(\mathcal{B})}$:

$$\forall x_1, x_2 \in \gamma(q) : \forall x_1', x_2' \in \gamma(q') : x_1(Z_i)x_2 \iff x_1'(Z_i)x_2'$$

This means, in particular, that the following lifting of $(Z_i)_{0 \leq i \leq n}$ to models is well-defined, for all $q \in Q_\mathcal{B}, q' \in Q_{\rho(\mathcal{B})}$ and all $0 \leq i \leq n$:

$$q(Z_i)q' \iff x(Z_i)x'$$

for some $x \in \gamma(x), x' \in \gamma(q')$.

Next we show that $(Z_i)_{0 \leq i \leq n}$ so defined is an n-bisimulation between $\mathcal{K}_{(\mathcal{B}, \varepsilon)}$ and $\mathcal{K}_{(\rho(\mathcal{B}), \varepsilon)}$.

$(Z_i)_{0 \leq i \leq n}$ *witnesses n-bisimulation:* We address all the points of the definition of n-bisimulation modulo Φ in order.

1. Clearly, $(\emptyset, \emptyset) Z_n (\emptyset, \emptyset)$. Hence the first condition of the definition is satisfied.
2. Consider any arbitrary states q, q' and let $x = x_1; x_2; \ldots; x_m$ and $x' = x_1'; x_2'; \ldots; x_m'$ be the corresponding nodes from T, T' that witnesses to $q(Z_0)q'$. By definition of Z_0 we have $S(x) = S(x')$, but it is possible that we have $E(x) \neq E(x')$. However, we must have $C(\mathsf{F}(x), \Phi) = C(\mathsf{F}(x'), \Phi)$, and to see this, it is enough to observe that as $m \leq n$, each of x and x'

contains at most n nodes. Then, since $F(x) = q$ and $F(x') = q'$ are the same on D, and the distance from $\Pi \setminus D$ to Φ is greater than n. That is, any path from an argument in $\Pi \setminus D$ to an argument in $\Phi = \Pi|_\phi$ would be a path consisting of at least $n + 1$ nodes. It follows that no element from Φ can be in a connected components containing elements outside of D.

3. Consider now q, q' corresponding to x and x' such that $x(Z_{i+1})x'$. Notice that $(q, r) \in R_B(\exists)$ if, and only if, there is a (p, X) such that $xR(x; (p, X))$. So all we need to show is that $X \cap (D \times D)$ is in $\mathcal{U}_{\rho_\phi(B)}(x', p)$. Then it will follow that there is a successor to x', namely $(p, X \cap (D \times D))$, with $(x')R'(x'; (p, X \cap (D \times D)))$. This is a straightforward consequence of the Definition 10 of ρ. The argument for the particular sub relations $R_B(p)$ is analogous.

4. Finally consider q, q' corresponding to x and x' such that $x(Z_{i+1})x'$ for (p, X) such that $x'R(x'; (p, X'))$. Again we need to ensure that there is an $X \in \mathcal{U}_B(x, p)$ such that $X' = X \cap (D \times D)$, and again this follows from the Definition 10 of ρ. The argument for the particular sub relations $R_B(p)$ is analogous.

Proposition 2. *Let $\phi \in \mathcal{L}_{DDL}$ and B, B' arbitrary bases. If states $q \in \mathcal{K}_{(B,\varepsilon)}$ and $q' \in \mathcal{K}_{(B',\varepsilon)}$ are $|\phi|^\diamond$-bisimilar modulo $\Pi|_\phi$, then $\mathcal{K}_{(B,\varepsilon)}, q \models \phi \Leftrightarrow \mathcal{K}_{(B',\varepsilon)}, q' \models \phi$. Or, succinctly*

$$\left((\mathcal{K}_{(B,\varepsilon)}, q) \underset{|\phi|^\diamond}{\overset{\Pi|_\phi}{\leftrightarrow}} (\mathcal{K}_{(B',\varepsilon)}, q') \right) \;\Rightarrow\; \left(\mathcal{K}_{(B,\varepsilon)}, q \models \phi \Leftrightarrow \mathcal{K}_{(B',\varepsilon)}, q' \models \phi \right).$$

Proof. The proof is by induction on $|\phi|^\diamond$.

Base case: $(|\phi|^\diamond = 0)$ There are no white connectives, and our states, q and q', are clearly 0-bisimilar modulo Φ. It is also easy to see, consulting Definition 2, that the truth of a formula of modal depth 0 is only dependent on the AF q. Then it follows from the fact that ε is assumed to be normal that the truth of ϕ is in fact only dependent on $C(q, \Phi)$. From $q(Z_0)q'$, we obtain $C(q, \Phi) = C(q', \Phi)$ and the claim follows.

Induction step: $(|\phi|^\diamond > 0)$ We skip the boolean cases as these are trivial, so let $\phi := \Diamond\psi$ (the case of white connectives with an explicit argument is similar). Suppose $|\phi|^\diamond = i+1$ and $q(Z_{i+1})q'$. Suppose further that $\mathcal{K}_{(B,\varepsilon)}, q \models \Diamond\psi$. Then there is a successor of q, $v \in succ(q)$ such that $\mathcal{K}_{(B,\varepsilon)}, v \models \psi$. All successors of q will be i-bisimilar to a successor of q' (point 3. of Definition 8). So we have $(\mathcal{K}_{(B,\varepsilon)}, v) \underset{i}{\overset{\Phi}{\leftrightarrow}} (\mathcal{K}_{(B',\varepsilon)}, v')$. As $|\psi|^\diamond < |\Diamond\psi|^\diamond$ we can apply our induction hypothesis to obtain $\mathcal{K}_{(B',\varepsilon)}, v' \models \psi$, and $\mathcal{K}_{(B,\varepsilon)}, q' \models \Diamond\psi$ as desired.

5   Conclusion and Future Work

We have argued for a logical analysis of deliberative processes by way of modal logic, where we avoid making restrictions that may not be generally applicable, and instead focus on logical analysis of the space of possible outcomes. The deliberative dynamic logic (DDL) was put forth as a concrete proposal, and we showed some results on model checking.

We notice that DDL only allows us to study deliberative processes where every step in the process is explicitly mentioned in the formula. That is, while we quantify over the arguments involved and the way in which updates take place, we do not quantify over the *depth* of the update. For instance, a formula like $\Diamond\Box p$ reads that there is a deliberative update such that no matter what update we perform next, we get ϕ. A natural next step is to consider instead a formula $\Diamond\Box^*\phi$, with the intended reading that there is an update which not only makes ϕ true, but ensures that it remains true for all possible future *sequences* of updates. Introducing such formulas to the logic, allowing the deliberative modalities to be iterated, is an important challenge for future work. Moreover, we would also like to consider even more complex temporal operators, such as those of computational tree logic, or even μ-calculus.

Finding finite representations for the deliberative truths that can be expressed in such languages appears to be much more challenging, but we would like to explore the possibility of doing so.

Also, we would like to explore the question of validity for the resulting logics, and the possibility of obtaining some compactness results. Indeed, it seems that if we introduce temporal operators we will be able to express truths on arbitrary points $q \in Q_B$ by corresponding formulas that are true at $(\emptyset, \emptyset)$, thus capturing the way in which complete assent can be faithfully captured by a finite (albeit unbounded) notion of iterated deliberation.

If the history of the human race is anything to go by, it seems clear that we never run out of arguments or controversy. But it might also be that some patterns or structures are decisive enough that they warrant us to conclude that the *truth* has been settled, even if deliberation may go on indefinitely. A further logical inquiry into this and related questions will be investigated in future work.

References

1. Arieli, O., Caminada, M.W.: A QBF-based formalization of abstract argumentation semantics. J. Appl. Log. **11**(2), 229–252 (2013)
2. Baroni, P., Giacomin, M.: On principle-based evaluation of extension-based argumentation semantics. Artif. Intell. **171**(10–15), 675–700 (2007)
3. Bodanza, G., Tohmé, F., Auday, M.: Collective argumentation: a survey of aggregation issues around argumentation frameworks. Argument Comput. **8**(1), 1–34 (2017)
4. Caminada, M.: On the issue of reinstatement in argumentation. In: Fisher, M., van der Hoek, W., Konev, B., Lisitsa, A. (eds.) JELIA 2006. LNCS (LNAI), vol. 4160, pp. 111–123. Springer, Heidelberg (2006). https://doi.org/10.1007/11853886_11
5. Caminada, M., Pigozzi, G., Podlaszewski, M.: Manipulation in group argument evaluation. In: The 10th International Conference on Autonomous Agents and Multiagent Systems, vol. 3, pp. 1127–1128. International Foundation for Autonomous Agents and Multiagent Systems (2011)

6. Dung, P.M.: On the acceptability of arguments and its fundamental role in non-monotonic reasoning, logic programming and n-person games. Artif. Intell. **77**(2), 321–357 (1995)

7. Dyrkolbotn, S., Walicki, M.: Propositional discourse logic. Synthese **191**(5), 863–899 (2014)

8. Grossi, D.: Argumentation in the view of modal logic. In: McBurney, P., Rahwan, I., Parsons, S. (eds.) ArgMAS 2010. LNCS (LNAI), vol. 6614, pp. 190–208. Springer, Heidelberg (2011). https://doi.org/10.1007/978-3-642-21940-5_12

9. Grossi, D.: On the logic of argumentation theory. In: Proceedings of the 9th International Conference on Autonomous Agents and Multiagent Systems, vol. 1, pp. 409–416. International Foundation for Autonomous Agents and Multiagent Systems (2010)

10. Rahwan, I., Simari, G.R.: Argumentation in Artificial Intelligence, vol. 47. Springer, Heidelberg (2009)

11. Sakama, C., Caminada, M., Herzig, A.: A logical account of lying. In: Janhunen, T., Niemelä, I. (eds.) JELIA 2010. LNCS (LNAI), vol. 6341, pp. 286–299. Springer, Heidelberg (2010). https://doi.org/10.1007/978-3-642-15675-5_25

Decentralized Multiagent Approach
for Hedonic Games

Kshitija Taywade$^{(\boxtimes)}$, Judy Goldsmith, and Brent Harrison

Department of Computer Science, University of Kentucky, Lexington, USA
kshitija.taywade@uky.edu, {goldsmit,harrison}@cs.uky.edu

Abstract. We propose a novel, multi-agent, decentralized approach for hedonic coalition formation games useful for settings with a large number of agents. We also propose three heuristics which, can be coupled with our approach to find sub-coalitions that prefer to "bud off" from an existing coalition. We found that our approach when compared to random partition formation gives better results which further improve when it is coupled with the proposed heuristics. As matching problems are a common type of hedonic games, we have adapted our approach for two matching problems: roommate matching and bipartite matching. Our method does well for additively separable hedonic games, where finding the optimal partition is NP-hard, and gives near optimal results for matching problems.

1 Introduction

A coalition formation game is a cooperative game in which agents must come together to form groups, or *coalitions*. A *hedonic game* is a coalition formation game in which agents have preferences over the coalitions they can be a part of.

Many real world problems can be modeled as hedonic games. Examples include team formation and social group formation. Typically these problems are solved using centralized approaches in which one controlling agent contains perfect knowledge about each agent's preferences and determines how coalitions should be formed. While these techniques have been shown to be effective, it is unlikely that all preferences will be known a priori in most real world situations, limiting their effectiveness. In this paper, we address these issues by introducing a decentralized algorithm for solving hedonic games.

For simplicity, we focus on solving *additively separable* hedonic games (ASHGs). In an additively separable hedonic game, an agent's utility for a coalition is equal to the sum of its utility for every other agent in the coalition. While this may be simple formulation of the problem, Aziz et al. [2] show that finding optimal partitions for ASHGs is NP-hard in the strong sense. Therefore, finding optimal coalition structures requires a heuristic approach. To this end, we model coalition formation as autonomous agents exploring a grid world and forming coalitions with other agents they meet. Unlike a centralized approach, in this formulation each agent is responsible for finding its own coalition and does not initially know preferences of other agents.

© Springer Nature Switzerland AG 2019
M. Slavkovik (Ed.): EUMAS 2018, LNAI 11450, pp. 220–232, 2019.
https://doi.org/10.1007/978-3-030-14174-5_15

We also propose an extension to our decentralized technique inspired by [19] and [8]. This extension, which we refer to as *budding*, is an additional search heuristic in which new coalitions are formed by breaking apart previously formed coalitions in a locally centralized way. We hypothesize that this will lead to, overall, higher utility coalitions for all agents involved. We introduce three budding heuristics in Sect. 4.2.

We explore how our decentralized approach performs on both unconstrained coalition formation and constrained coalition formation hedonic games. In the unconstrained case there are no restrictions on how coalitions can be formed, whereas in constrained coalition formation explicit restrictions, such as limiting the total number of agents allowed in a coalition or limiting the number of total coalitions allowed, are present. Constrained coalition formation hedonic games are often referred to as *matching problems*, and we will be using the two terms interchangeably throughout the remainder of the paper. By exploring how our decentralized approach performs in these two types of environments, we aim to show that our technique is applicable in a variety of potential problems. In experiments we model all types of environments as grid-worlds. The motivation behind using grid-world is that it can better real-world problems where agents must expend some effort to seek out information about other agents or form coalitions with others. For example, consider a situation in which students are asked to form teams for an assignment on the first day of class in school. Lacking any other information on their fellow students, they will physically explore around and inquire about forming teams with other students. This process is better modeled using a grid-world so that the effort spent navigating is taken into account.

In sumary, the primary contributions of this paper are as follows:

- We introduce a novel decentralized approach for hedonic games.
- We introduce the concept of budding in coalition formation and also propose several budding heuristics.
- We show how our approach can be used for hedonic games where there are no constraints and on problems where there are constraints on the size and/or number of coalitions (e.g., matching problems).
- We evaluate the quality of the coalitions found by our proposed decentralized approach in both unconstrained hedonic games and constrained hedonic games.

2   Preliminaries

Definition 1. [5,6] *A* hedonic coalition formation game *(also just "hedonic game") G is defined by a set of agents, and the utilities each agent, a, holds for each coalition C containing a. A* coalition structure π *for G is a partition of the agents into coalitions. The goal of a hedonic game may be to find a utility-maximizing coalition structure, or a stable one.*

There are many notions of *stability* for coalition structures for hedonic games. In this work, we use a myopic notion of stability within a coalition, called *proximal stability* [21]. This is related to "budding," introduced in Sect. 4.2, and is based on the standard hedonic games notion of core stability.

Definition 2. [18] *Given a hedonic game G and a coalition structure π, a blocking coalition $C \notin \pi$ is a set of agents that have higher or equal utility for being in C as for being in their assigned coalition in π, and where at least one agent has strictly higher utility in C.*

We have decided to model problems as additively separable hedonic game which is defined as follows:

Definition 3. [5,6] *In an* Additively Separable Hedonic Game (ASHG), *each agent a_i has a utility, $p_{i,j}$, for each other agent a_j. The total utility $u_i(C)$ of agent a_i from the coalition C is $\Sigma_{j \neq i} p_{i,j}$.*

In hedonic games, preferences can be either *symmetric* or *asymmetric*. When preferences are symmetric, $p_{i,j} = p_{j,i}$ for all pairs i, j. In the asymmetric case, i's utility for j might differ from j's utility for i (so it's possible that $p_{i,j} \neq p_{j,i}$).

We consider two cases: where the size of the coalition is constrained, and where it is not. We consider the following two constrained cases in particular.

Matching problems are best example of constrained case, having restrictions on the number of agents per coalition and the total number of coalitions. We have implemented our approach for two matching problems. The first, roommate matching, has a restriction on the number of agents per coalition i.e., a limit on the number of roommates allowed per room; and a limit on the number of coalitions (the number of rooms available). Our second problem is bipartite matching, where agents of one type are each matched to a single agent of the other type. We define each of those problems below:

Definition 4. *An instance of* Roommate Matching *consists of a set of agents $A = a_1, a_2, \ldots, a_n$ and R, the available number of rooms, and a uniform maximum capacity per room of c. The goal is to form a partition P of agents where each coalition in P has at most C agents, and there are at most R coalitions.*

Given these constraints, agents try to maximize their utility for their own coalition.

Definition 5. *An instance of* Bipartite Matching *consists of two equal size sets of agents $X = x_1, x_2, \ldots, x_n$ and $Y = y_1, y_2, \ldots, y_n$ of distinct types. The goal is to partition $X \times Y$.*

Here, each agent tries to maximize their utility for their match.

3   Related Work

Much of the past work on hedonic games has focused on centralized approaches where a single controller contains knowledge of agent preferences and determines how each agent should act in order to get stable outcome [3, 4, 9] or optimal coalition structure which improves performance of overall system [15–17, 20]. There have been extensions made to these techniques recently, however, in which each agent is responsible for forming its own coalitions [11, 12]. In each of these types of algorithms, a centralized is used to determine which solution is the best overall solution. This is possible because it knows the preferences of all agents. Our technique is truly decentralized since there is no controller that contains this additional information.

In this paper, we develop a decentralized approach for solving hedonic coalition formation games. The idea of using decentralized approaches to solving this problem is not new. One common approach used to solve this problem in a decentralized way is to use reinforcement learning [1, 13]. In these approaches, each agent explores their environment and learns how to form coalitions based on its own reward signal. Researchers have also explored the idea of augmenting reinforcement learning-based approaches by using additional heuristics [14] or through the use of Bayesian reinforcement learning [7]. The primary limitation of these approaches is that one must be able to model the hedonic game as a Markov decision process (MDP) in order to make use of reinforcement learning, which can be difficult to do depending on the nature of the hedonic game. In addition, reinforcement learning-based approaches also rely on reward functions that can be difficult for humans to define. Our decentralized algorithm does not rely on the environment having an MDP structure or on external reward functions and, thus, is more generally applicable than reinforcement learning-based approaches.

In this work, we also propose a novel heuristic for generating new candidate coalitions, which we call budding. This technique is inspired by the work done by Roth and Vate [19] and by the work of Diamantoudi, et al. [8] which involve using sequences of blocking pairs as a means to find stable coalitions in matching problems.

4   Methods

The core element of our decentralized algorithm for hedonic games is that we model the problem as a multi-agent system where agents explore a grid world environment. Each agent is initialized with no prior knowledge except for some basic information about the environment (such as its dimensions). As agents explore the environment, they will encounter other agents by moving into the same grid position as other agents and, in doing so, obtain preference information about them. Coalitions are formed when agents occupy the same grid world position and then choose to enter into a coalition. We go into more detail on this algorithm and the budding extension to this algorithm below.

4.1   Decentralized Algorithm for Hedonic Games

The core of our approach is that we model a hedonic game as a multi-agent system in which each agent navigates a grid-based environment and builds up knowledge of its own preferences. The agents themselves are autonomous, selfish, and myopic. They are selfish in that they are only concerned with their own utility, and they are myopic in that the only part of the environment they can see is their current grid location. In other words, each agent is unable to comprehend the whole environment. After each agent is initialized, it will explore the environment for a given number of time steps. This constitutes one *episode* of learning, and our algorithm will return the set of discovered coalitions after a set number of learning episodes. Both the number of time steps in an episode and the number of learning episodes are tunable parameters that can be customized for a specific hedonic game.

We will now discuss two key elements of our approach: what each agent has the ability to remember and recognize in the environment, and how agents operate in the grid world environment.

What Agents Remember: Agents store limited knowledge of the environment. Specifically, agents in our environment remember the following information:

1. After seeing other agent for the first time, an agent assigns a utility value to that agent and stores it for the future reference. Agents only know their utility for other agents after they interact with them. Note that agents can only see other agents when they are on the same cell location on grid.
2. Agents also remember the location where they last saw their *friends*. In this algorithm, we define an agent's friends to be other agents for whom it has a positive utility. This encourages agents to revisit these locations while exploring as there is an increased likelihood of it being able to form a coalition with a friendly agent in that location.
3. Each agent also contains information about the highest utility values it has received from joining coalitions in the past. In our experiments, we store 10 values, but this is a parameter that can be tuned to the specific problem being solved. This helps the agent evaluate the quality of its current coalition in terms of its own past experience. This further encourages the agent to seek out higher utility coalitions. One important thing to note is that this does not guarantee that an agent will always discover the coalition that results in its overall highest utility. This only encourages the agent to form coalitions that improve upon its past coalitions.

Operating in Grid-World: At the beginning of each episode each agent is placed at the random position on grid. At each time step, every agent will act in the grid world environment at the same time. The most common action for an agent to take is to *explore*, i.e., move to a different location in the grid world. The agent can also take the following actions depending on the specific situation it finds itself in. If an agent occupies the same grid position as other agent, these two can decide to form a coalition. The decision to form a coalition must be mutual, so it will never be the case that a coalition will be formed with only one willing party. The agents will enter into a coalition if the resultant coalition will result in positive utility for both agents. If a coalition is formed, then that coalition will remain in that location on subsequent time steps. After entering into a coalition, agents will remain in that coalition for several time steps waiting for others to join. After a sufficient amount time has passed, the agents in this coalition can consider leaving to form better ones or simply explore the environment.

If an agent moves into a square occupied by a pre-existing coalition, it can lobby to join the coalition. If the agent wishes to join this coalition, then each coalition member casts a vote to either let the new agent in, or reject them. Agents will vote to accept the candidate if the resulting coalition has higher utility than their current one, and agents will vote to reject the candidate if the resulting coalition has lower utility. If more than half of the agents in the coalition vote to let the new agent into the coalition, then the new agent becomes a member of the coalition.

If an agent is already in a coalition, it is possible that changes in coalition membership could cause the agent to want to leave. Since an agent's utility towards its coalition is affected by its preference over coalition members, new members being added or old members leaving will have a large effect on how an individual agent views the coalition. If an agent's utility towards its own coalition becomes sufficiently small, then it may decide to leave the coalition to seek out better ones. This decision is influenced not only by utility values, but also by the amount of time remaining in the episode. If there is still sufficient time remaining in the episode, it may choose to explore more to find higher scoring coalitions. If there is a small amount of time left, however, then agents may choose to exploit knowledge and remain in high scoring coalitions. Note that just after joining the coalition, for few steps agents remain in the coalition (while still having positive utility) and wait to see if changes in coalition caused by leaving or joining of other agents can improve its utility. Agents follow the following mechanism to maximize their utility and settle down in coalitions:

- For the first $(1/3)^{rd}$ of the total steps, an agent compares their current utility from the current coalition with the average of the top 10 highest utilities it has got so far. If its current utility is greater or equal to that average utility then it stays in the coalition.
- For the second $(1/3)^{rd}$ steps, an agent compares their current utility with the average value of the top 10 utilities seen multiplied by 0.75. If its current utility is greater or equal to that value then it stays in the coalition. This means that an agent is more likely to stay in a coalition rather than further explore.
- For the next $(1/6)^{th}$ steps, an agent compares their current utility with the average value of the top 10 utilities seen multiplied by 0.5. If its current utility is greater or equal to that value then it stays in the coalition.

– For the last $(1/6)^{th}$ steps, agents only leave a coalition if they have a negative utility for that coalition.

In this way, agents first set high expectations and then gradually lower their expectations as the clock runs down. The coalition structure at the end of the last episode is considered the solution coalition structure.

4.2 Budding

In the base form of our approach, coalitions can only be formed when multiple agents occupy the same location and all agree to enter into a coalition. To further improve our approach, we have proposed an extension to our baseline approach which introduces a new way to form coalitions. This technique, called budding, allows for large coalitions to break apart into smaller coalitions. This helps avoid the situation in which many agents keep joining a pre-existing coalition when there may be one or more sub-coalitions that could be formed from this one that would result in higher overall utility for the agents involved. Formally, we define budding as follows:

Definition 6. *Consider a set of agents $A = \{a_1, a_2, \ldots, a_n\}$, among which some agents form coalition C. Budding is when a sub-coalition, B, is formed from C. This happens when the number of agents in C reaches a certain threshold T and forming a sub-coalition produces better total utility for agents in B than their total utility from original coalition C.*

We have the following three budding heuristics:

1. Random: When a coalition, C, reaches a sufficient size, a random sub-coalition, B, is formed and the total utility of agents in B which they can get from being part of B is compared with their total utility from being part of C. This process is repeated until a B with higher total utility is found, or until specified number of iterations of trying new sub-coalitions is reached.
2. Greedy: A random agent, a, is chosen from C to "seed" B. While it is possible to increase the total utility of the agents in B by adding an agent from C to B, we do so.
3. Clique Detection: We define a graph on C where vertices are agents in C and edges are between friends. We set B to be the first clique we find where, the total utility of agents in B which they are getting from B is higher than their total utility from being part of C.

We consider these budding strategies to be only locally centralized in that they only occur on a subset of agents that happen to be in a coalition. Since this requires only minimal information sharing between agents that are in a coalition, we still consider this approach to be generally decentralized.

4.3   Adapting the Decentralized Approach for Constrained Coalition Formation

The algorithm as we have described it thus far is suitable for hedonic games where there are no additional restrictions placed on how coalitions are formed or how many coalitions can be formed. Some of the most common types of hedonic games, however, do have these types of restrictions placed on them. In these constrained cases we must, thus, place extra constraints on how our agents form coalitions and on how members are accepted. In our experiments, we consider two specific constrained hedonic games: roommate matching and bipartite matching. Here we will outline the adaptations made to our technique that enable it to be used on these two problems. While we focus on these two problems in this paper, similar alterations to our technique can be made to make it applicable to other matching problems.

Roommate Matching. Recall that roommate matching places the additional constraint on the number of agents that can be in a coalition and the number of coalitions that can be formed. To implement our algorithm for roommate matching we must take these constraints into account when agents form coalitions. To do this, we make the following alteration to our algorithms:

- Because the number of roommates per room is limited to C, if more agents converge on a cell, we choose the set of C that has highest utility, in a centralized manner. All other agents must leave the cell and explore further.
- Since the number of coalitions must not exceed R, our approach has certain pre-designated time steps on which the total number of coalitions on grid is checked. If there are more than R coalitions, then the coalition with the lowest total utility across all its agents will be forced to dissolve and agents in that coalition will search for other coalitions to join.

Here, we make the assumption that agents do not have preferences over specific rooms, only over their roommates.

Bipartite Matching. Recall that bipartite matching involves two sets of agents where agents from one set are trying to form coalitions with agents in the other set. Note that a coalition contains two agents, one from each type. To implement our proposed decentralized approach for bipartite matching we changed the grid-world environment such that it contains two types of agents in equal numbers. When an agent encounters an agent of a different type, it stores information about that agent including its location and its utility for that agent. If two agents from the same type encounter each other, they ignore each other. Each agent also has basic information about the total number of agents of the opposite type present in the environment. Using this, they can calculate how many potential coalitions could be formed and can, from that, reason about how many candidate agents they have not encountered yet. This helps determine if an agent forms a coalition and stays there, or if the agent chooses to explore to form potentially better coalitions.

5 Experiments

To evaluate the effectiveness of our technique, we examine its performance on both constrained coalition formation hedonic games and unconstrained coalition formation hedonic games. In both cases, we initialize the grid world by placing agents in random positions on the grid. We then run the algorithm for 100 learning episodes of 4000 time steps each. Since our algorithm is non-deterministic, we run 10 instances with different preference profiles and report the average total utility of all the agents, averaged over those 10 instances.

Below we will provide more details on the specific test cases we explored.

5.1 Unconstrained Coalition Hedonic Games

For evaluating our approach on unconstrained hedonic games we explored how it would perform in an environment with 10 agents exploring a 10×10 grid. One aspect of the environment that can have a large effect on the outcome is the range of expected utility values. In this test case we experimented with two possible ranges of utility values: the range $(-5, 10)$ and the range $(-10, 10)$. When one agent encounters another agent for the first time, both agents will uniformly sample a preference from these ranges and assign it to the other agent. We evaluate for both asymmetric and symmetric preferences. For this evaluation, we compare the results of our approach with the results of an algorithm that generates random partitions. We do not compare our results against the optimal solution as finding the optimal involves evaluating the utility of every possible partition, which is an NP-hard problem [2].

We also evaluate the effect that our budding technique has on the quality of discovered coalitions. To do this we performed experiments where we compared versions of our decentralized algorithm with budding implemented using each heuristic introduced earlier against a version of our algorithm that did not have budding implemented. We tested these methods in a 10×10 grid world with 10 agents, and also with 20 agents. For this study, we only consider a utility value range of $(-10, 10)$, and we make the assumption that agents have asymmetric preferences. Since we are using budding, we must define a threshold value that determines when the budding process starts. We set this value to 9 when there are 20 agents on the grid (meaning that budding will begin when a coalition has 9 members), and we set this value to 6 when we examined 10 agents on the grid.

5.2 Constrained Coalition Hedonic Games

For our experiments on constrained coalition hedonic games we consider two problems: roommate matching and bipartite matching. For roommate matching we consider a 10×10 grid with 9 agents. In this environment, a "room" can contain 3 agents and there are 3 rooms available. In other words, it is possible for each agent to be a member of a coalition as long as each coalition has 3 members. In these experiments, we compare the results of our decentralized algorithm against the utility associated with the optimal matching. We consider 3 different utility ranges in these experiments: a range of $(1, 10)$, a range of $(-5, 10)$, and a range of $(-10, 10)$. In this problem, we assume that agents have asymmetric preferences. For our evaluation using a bipartite matching problem we use a 5×5 grid containing 20 agents. Here, agents are evenly divided between matching types. Hence, we have 10 agents belonging to one type and the remaining 10 belonging to the other type. In this experiment, we compare the results of our algorithm against the utility value obtained through an optimal matching as well as the average utility of matchings found using the Gale-Shapley algorithm [10]. The Gale-Shapley algorithm is a common technique used to find bipartite matchings and serves as an baseline for this problem.

6   Results and Discussion

In this section we outline the results of the experiments we performed in both unconstrained and constrained coalition formation hedonic games. We also discuss these results in greater detail below.

6.1   Results in Unconstrained Coalition Hedonic Games

The performance of our decentralized algorithm compared to random partitioning can be found in Table 1. As you can see in the table, our decentralized approach was able to significantly outperform our random baseline. It is also notable that this behavior is consistent across all utility ranges and regardless of whether preferences were symmetric or asymmetric. While this baseline is not very sophisticated, this does show that our approach is able to drastically outperform this worst case scenario.

The results of the comparison between versions of our decentralized algorithm with and without budding can be seen in Table 2. The results of this comparison show that the versions of our decentralized algorithm that had the ability to produce new coalitions through budding consistently outperformed the base version of our technique. This shows that our intuition about the potential benefits of budding for coalition formation were well founded. Of the three heuristics, it appears as though the clique-based approach performed better than both random and greedy heuristics. A more rigorous evaluation is required, however, before a more definitive conclusions can be drawn.

Table 1. Comparison of decentralized approach with random partitions for both asymmetric and symmetric cases

Preference type	Utility range	Decentralized approach	Random partitions
Asymmetric	$(-5, 10)$	188.3	57.8
	$(-10, 10)$	67.3	-9.1
Symmetric	$(-5, 10)$	224.2	37.2
	$(-10, 10)$	134.7	19.6

Table 2. Comparison of decentralized approach with and without budding heuristics for the 10-agent environment and the 20-agent environment.

Number of agents	Without budding	Random	Greedy	Clique
10	69.55	72.44	74.55	75.88
20	229.6	235.8	235.2	241.1

6.2   Results in Constrained Coalition Hedonic Games

The results for the experiments we ran on the roommate matching problem can be seen in Table 3. In these experiments we were able to compare against the optimal matching. The first thing to note is that our technique is not able to reproduce the optimal matchings. That being said, we feel that our performance was comparable to this upper bound. We can see from Table 3 that our approach gives results close to the optimal matching for all three utility ranges.

Table 3. Comparison of average total utility over 10 instances obtained by adaption of our approach for roommate matching with the utilitarian optimal matching

Utility range	Decentralized approach	Optimal matching
(1,10)	102.3	129.6
(−5, 10)	65.8	91.3
(−10, 10)	42.9	64.6

Table 4. Comparison of average total utility over 10 instances of 10 couples, obtained by adapting of our approach for bipartite matching, compared with utilities from the Gale-Shapley algorithm and the utilitarian optimal matching

Utility range	Gale-Shapley algorithm	Decentralized approach	Optimal matching
(1,10)	134.8	155.8	166
(−5, 10)	72.6	111.4	124.2
(−10, 10)	23.8	92.2	115

Table 4 contains the results of our experiment on bipartite matching. In this set of experiments we compared our approach against the optimal matching as well as results obtained using the Gale-Shapley algorithm (which only optimizes for one type of agent). While our decentralized approach does not reconstruct the optimal matching, it is interesting to note that our algorithm consistently outperforms the Gale-Shapley algorithm on this problem. Since this was a limited evaluation, this is not enough evidence to claim that our algorithm is superior to the Gale-Shapley algorithm. This does, however, provide some evidence that our algorithm is an effective way to approach the bipartite matching problem. Further evaluations should be done to confirm these results.

7 Conclusion

In this paper we proposed a decentralized approach for solving hedonic games based on modeling the problem as a grid-world exploration problem. We feel that decentralized approaches better simulate how these problems work in a real-world environment and are, thus, more generally applicable than more common, centralized approaches. We also introduce a novel coalition discovery technique called budding, in which large coalitions spawn smaller sub-coalitions if they would increase total utility of agents in the new sub-coalition. Our initial experiments showed promising results as our techniques both with and without budding performed well on a variety of hedonic games. While this work is preliminary, we believe that our experiments provide strong evidence as to the quality of our approach. In the future we want to expand our evaluations to more complex environments and explore additional budding heuristics that can be used to further improve our technique's performance.

References

1. Abdallah, S., Lesser, V.: Organization-based cooperative coalition formation. In: Proceedings of the IEEE/WIC/ACM International Conference on Intelligent Agent Technology, IAT 2004, pp. 162–168. IEEE (2004)
2. Aziz, H., Brandt, F., Seedig, H.G.: Optimal partitions in additively separable hedonic games. In: IJCAI Proceedings-International Joint Conference on Artificial Intelligence, vol. 22, p. 43 (2011)
3. Aziz, H., Brandt, F., Seedig, H.G.: Stable partitions in additively separable hedonic games. In: The 10th International Conference on Autonomous Agents and Multiagent Systems, vol. 1, pp. 183–190. International Foundation for Autonomous Agents and Multiagent Systems (2011)
4. Aziz, H., Brandt, F., Seedig, H.G.: Computing desirable partitions in additively separable hedonic games. Artif. Intell. **195**, 316–334 (2013)
5. Banerjee, S., Konishi, H., Sönmez, T.: Core in a simple coalition formation game. Soc. Choice Welf. **18**(1), 135–153 (2001)
6. Bogomolnaia, A., Jackson, M.O.: The stability of hedonic coalition structures. Games Econ. Behav. **38**(2), 201–230 (2002)
7. Chalkiadakis, G., Boutilier, C.: Bayesian reinforcement learning for coalition formation under uncertainty. In: Proceedings of the Third International Joint Conference on Autonomous Agents and Multiagent Systems, vol. 3, pp. 1090–1097. IEEE Computer Society (2004)
8. Diamantoudi, E., Miyagawa, E., Xue, L.: Random paths to stability in the roommate problem. Games Econ. Behav. **48**(1), 18–28 (2004)

9. Gairing, M., Savani, R.: Computing stable outcomes in hedonic games. In: Kontogiannis, S., Koutsoupias, E., Spirakis, P.G. (eds.) SAGT 2010. LNCS, vol. 6386, pp. 174–185. Springer, Heidelberg (2010). https://doi.org/10.1007/978-3-642-16170-4_16

10. Gale, D., Shapley, L.S.: College admissions and the stability of marriage. Am. Math. Mon. **69**(1), 9–15 (1962)

11. Janovsky, P., DeLoach, S.A.: Increasing coalition stability in large-scale coalition formation with self-interested agents. In: ECAI, pp. 1606–1607 (2016)

12. Janovsky, P., DeLoach, S.A.: Multi-agent simulation framework for large-scale coalition formation. In: 2016 IEEE/WIC/ACM International Conference on Web Intelligence (WI), pp. 343–350. IEEE (2016)

13. Jiang, J.G., Zhao-Pin, S., Mei-Bin, Q., Zhang, G.F.: Multi-task coalition parallel formation strategy based on reinforcement learning. Acta Automatica Sinica **34**(3), 349–352 (2008)

14. Li, X., Soh, L.K.: Investigating reinforcement learning in multiagent coalition formation. In: American Association for Artificial Workshop on Forming and Maintaining Coalitions and Teams in Adaptive Multiagent Systems, Technical report WS-04-06, pp. 22–28 (2004)

15. Michalak, T., Sroka, J., Rahwan, T., Wooldridge, M., McBurney, P., Jennings, N.R.: A distributed algorithm for anytime coalition structure generation. In: Proceedings of the 9th International Conference on Autonomous Agents and Multiagent Systems, vol. 1, pp. 1007–1014. International Foundation for Autonomous Agents and Multiagent Systems (2010)

16. Rahwan, T., Jennings, N.R.: An improved dynamic programming algorithm for coalition structure generation. In: Proceedings of the 7th International Joint Conference on Autonomous agents and Multiagent Systems, vol. 3, pp. 1417–1420. International Foundation for Autonomous Agents and Multiagent Systems (2008)

17. Rahwan, T., Ramchurn, S.D., Jennings, N.R., Giovannucci, A.: An anytime algorithm for optimal coalition structure generation. J. Artif. Intell. Res. **34**, 521–567 (2009)

18. Richardson, M.: On finite projective games. Proc. Am. Math. Soc. **7**(3), 458–465 (1956)

19. Roth, A.E., Vate, J.H.V.: Random paths to stability in two-sided matching. Econometrica: J. Econ. Soc. **58**(6), 1475–1480 (1990)

20. Sandholm, T., Larson, K., Andersson, M., Shehory, O., Tohmé, F.: Anytime coalition structure generation with worst case guarantees. arXiv preprint cs/9810005 (1998)

21. Schlueter, J., Goldsmith, J.: Proximal stability (2018, in progress)

Deep Reinforcement Learning in Strategic Board Game Environments

Konstantia Xenou[1], Georgios Chalkiadakis[1(✉)], and Stergos Afantenos[2]

[1] School of Electrical and Computer Engineering, Technical University of Crete, Chania, Greece
{diaxenou,gehalk}@intelligence.tuc.gr
[2] Institut de recherche en informatique de Toulouse (IRIT), Université Paul Sabatier, Toulouse, France
stergos.afantenos@irit.fr

Abstract. In this paper we propose a novel Deep Reinforcement Learning (DRL) algorithm that uses the concept of "action-dependent state features", and exploits it to approximate the Q-values locally, employing a deep neural network with parallel Long Short Term Memory (LSTM) components, each one responsible for computing an action-related Q-value. As such, all computations occur simultaneously, and there is no need to employ "target" networks and experience replay, which are techniques regularly used in the DRL literature. Moreover, our algorithm does not require previous training experiences, but trains itself online during game play. We tested our approach in the Settlers Of Catan multi-player strategic board game. Our results confirm the effectiveness of our approach, since it outperforms several competitors, including the state-of-the-art *jSettler* heuristic algorithm devised for this particular domain.

Keywords: Deep Reinforcement Learning · Strategic board games

1 Introduction

Deep Reinforcement Learning (or DRL) is widely used in different fields nowadays, such as robotics and natural language processing [23]. Games, in particular, are a very popular testbed for testing DRL algorithms. Methods like Deep Q-Networks were found to be especially successful for video games, where one can learn using video frames and the instant reward.

Research on RL in strategic board games is particularly interesting, because their complexity can be compared to real-life tasks and their testbed allows comparison of many different players as well as AI techniques. The most known example of Deep RL use in this domain is perhaps AlphaGo [24], but other attempts have been made as well in games like chess and backgammon [7,15].

Now, the popular board game "Settlers Of Catan" (SoC), has recently been used as a framework for machine learning and sequential decision making

© Springer Nature Switzerland AG 2019
M. Slavkovik (Ed.): EUMAS 2018, LNAI 11450, pp. 233–248, 2019.
https://doi.org/10.1007/978-3-030-14174-5_16

algorithms [4,28]. Also, it has been used in the field of natural language understanding (parsing discourse used during multi-agent negotiations) [1], but such work has not dealt with strategic decision making.

In this paper we present a novel algorithm and a novel deep network architecture to approximate the Q-function in strategic board game environments. Our algorithm does not directly approximate the whole Q-function, like standard DRL approaches, but evaluates Q-values "locally": in our case, this means that the Q-value for each possible action is computed separately, as if it were the only possible next action. Standard techniques seen in DRL literature so far, like experience replay and target networks, are not used. Instead, we take advantage of the recurrency of the network, as well as the locality of our algorithm, to achieve stable good performance. Our generic Deep Recurrent Reinforcement Learning (DRRL) algorithm was adapted and tested in the SoC domain. Our results show that it outperforms Monte-Carlo-Tree-Search (MCTS) agents, as well as the state-of-the-art algorithm for this domain [28]. In addition, its performance gets close to that of another DRL agent found in the literature [4], though *it does not*—in contrast to that agent—use network pre-training for learning: as we detail later in the paper, our algorithm trains itself "on-line" while playing a game, using fewer than one hundred (100) learning experiences, as opposed to *hundreds of thousands* used by the DRL agent in [4]. Moreover, when we allowed training our network over a series of games, using $\sim 2,000$ learning experiences, our method's performance improves and matches that of the DRL agent in question.

2 Background and Related Work

In this section we provide the necessary background for our work, and a brief review of related literature.

2.1 Deep Reinforcement Learning

The main goal of DRL is to approximate an RL component, such as the Q-function, the value function or the policy. This function approximation is done by generalizing from samples.

The standard framework for RL problems is provided by Markov Decision Processes (MDPs). An MDP is a tuple of the form $(S, A_s, P^a_{ss'}, \gamma, R^a_{ss'})$ where S is a set of the possible states that represent the dynamic environment, A is the set of possible actions available in state s, $P^a_{ss'}$ is the probability to transit from state $s \in S$ to state $s' \in S$ by taking action $a \in A_s$, $\gamma \in [0,1]$ is a discount factor and $R^a_{ss'}$ is the reward function that specifies the immediate reward for transitioning from state $s \in S$ to state $s' \in S$ by taking action $a \in A_s$.

To measure the value (i.e. performance) of a state-to-action mapping, the fundamental Bellman Optimality Equations [3] are usually used. If we consider a policy $\pi(b)$ as the mapping of beliefs to actions, or else the probability of taking action $a \in A_s$, we describe the expected value of the optimal policy:

$$V^*(s) = \max_a \left[R(s,a) + \gamma \sum_{s'} P(s'|s,a)V^*(s') \right] \qquad (1)$$

An optimal policy is then derived as $\pi(s) = argmaxQ^*(s,a)$, where

$$Q^*(s,a) = R(s,a) + \gamma \max_{a'} Q^*(s',a')$$

With the optimal Q-function known, the optimal policy can be easily found, by choosing the action a that maximizes $Q^*(s,a)$ for state s. When the state-action spaces are very large and continuous, the need for function approximation arises [26], in order to compute the underlying functional form of the Q-function, from a finite set of state-action pairs in an environment. In recent years, using deep neural architectures for function approximation became possible.

DRL Algorithms. Recent algorithms combat problems that were inherent in DRL (such as instability and inefficiency during learning). The first such algorithm, which was also able to be highly effective in a wide range of problems without using problem-specific knowledge or fine tunning, was the Deep Q-Network algorithm (DQN) [17].

DQN, implements Q-learning in a deep Convolutional Neural Network (CNN) and manages to master a range of Atari games with only raw pixels and score as input. For stability and better convergence, DQN also uses "experience replay", by storing state transitions. Furthermore the target Q-values are computed in a separate identical target Q network updated in a predefined number of steps.

Many extensions and improvements have been proposed in the literature (e.g [2,9,19]). One of the most significant contributions is AlphaGo [24], the first Go agent to have actually won a human professional player. Alpha Go combines Monte Carlo Tree Search (MCTS) for position evaluation and CNNs for move selection. Research has also focused lately in Deep Reinforcement Learning in continuous action spaces. In this case the DRL problem is approached with policy gradient methods rather than Q-learning [16].

Neural Network Types and Architectures. The most known deep architectures are Deep Feed Forward (DFF), Convolutional (CNN), and Recurrent Neural Networks (RNN). One RNN type that is especially effective in practical applications is Long Short Term Memory (LSTM) [11]. LSTMs make it possible for current information to be processed by keeping in mind previous states' information, since they actually have internal recurrence, more parameters than normal RNNs, and a system of gating units that controls the information flow.

DRL has been successful in working with RNNs and especially LSTMs [18], because of their good performance in sequential data, or data with temporal relationships, and especially for their capability of avoiding the problem of vanishing gradients. Recent work in RNNs [10] also, showed that recurrency in deep networks provides good belief estimates in Partially Observable MDPs.

2.2   Action-Dependent Features and Q-Decomposition

The concept of "action-dependent state features" was introduced in [25], and was later generalized the idea of Q-function decomposition by [22] (and, later, others—see, e.g., [14]). In [25] the RL process is partitioned in multiple "virtual" sub-agents, and rewards realized by each sub-agent are independent given only the local action of each sub-agent.

In more detail, assume a factored representation of the state space, which entails a feature vector (that is, a vector of state variables) representing the state s. In general, when at a state one needs to select the appropriate action, execute it, and then further update the Q-function values given real-world reward gained by this action selection, and our estimates on the long-term effects of the action. Now, if we assume that the feature vector contains action-dependent features, and each specific instantiation of this vector ("feature value") is strongly related to an action a^i, then the long term reward for choosing action a^i depends only on the feature value related to a^i [25]. Since there is only one feature value related to each a^i, an agent can realize rewards independently, by performing only a specific action evaluated locally (i.e., given the current state).

To elaborate further, consider a possible action $a^i \in A$, let I denote $|A|$, and let $e(s, a^i)$ be a fixed function that takes a^i and s as input and outputs a specific instantiation of the feature vector (i.e., its "feature value"). This essentially relates the a^i action to the specific values of the state variables.[1] Intuitively, this means that the effect of using an action at a particular state is that the feature vector takes a specific value (e.g., because specific state variables take on specific values). The feature values can then be associated with the long-term effects of employing a^i at s via a Q-function over action-feature values pairs. That is, $Q(\langle e(s, a^1), ..., e(s, a^I)\rangle, a^i)$ denotes the long-term value of a^i when s gives rise to (or, to use the language of [25], *generalizes to*) the vector $\langle e(s, a^1), ..., e(s, a^I)\rangle$ via a generalization function f. Now, since the feature values are action-dependent, we can assume that the expected value of employing a^i at s depends only on the feature value related to a^i: for a specific f, the value $Q(f(s), a^i)$ entirely depends on $e(s, a^i)$. That is:

$$Q\Big(\langle e(s, a^1), ..., e(s, a^I)\rangle, a^i\Big) = Q\Big(\langle e(s', a^1), ..., e(s', a^I)\rangle, a^i\Big) \qquad (2)$$

whenever $e(s, a^i) = e(s', a^i)$. In other words, $Q(f(s), a^i)$ is entirely independent of (s, a^j) for $j \neq i$.

In our work we apply this idea in deep networks, considering that the association between each possible action and the feature values (networks' input) is based on the networks weights, which are different for each evaluation function.

2.3   The Settlers of Catan (SoC) Domain

The Settlers Of Catan (SoC) is a multi-player board game, where players attempt to build establishments while trading with other players to acquire the needed

---

[1] We remark that no factored state representation was assumed in [25]; rather, each state was linked to a single action-dependent feature (with its set of values).

resources in order to do so. The actual board of the game is an island representation, composed of hexagonal tiles (hexes), each one representing a different land type and resource (Clay, Wool, Wheat, Ore and Sheep). The players establishments can be settlements, cities and roads. Especially roads are used in order for the players to connect their holdings.

In order to be able to build, a player must spend an amount of resources. The available resources are Clay, Wool, Wheat, Ore and Sheep and are represented by *resource cards*. In each turn, depending on the dice roll, it is decided which hexes produce resources, thus the player with a settlement (city) in this hex gains one (two) resource card of the corresponding resource type. To provide an example, if a number corresponding to a clay hex is rolled, the player who owns a city and a settlement adjacent to this hex, will get three clay resource cards.

There are also five kinds of *development cards* (i.e. knight, victory point, monopoly, road building and year of plenty). When a development card is played, it has a positive outcome for the player. For example, the "road building" card allows the player to build two roads with no cost.

Another way for a player to gain resources is by trading with other players or the game bank. Thus in each turn the player rolls the dice and all players gain resources based on their establishments. Then it is up to the player if she wants to propose a trade, if she is in need of some particular resources. Afterwards she can decide whether she wants to build a settlement, road or city, buy or play a development card. When a trade is offered, the other players can accept, reject the offer or make a new counter-offer.

Each time a player expands its territory, she gains victory points. A settlement (resp. city) awards the player 1 (resp. 2) victory point. Furthermore, the player with longest uninterrupted road is awarded 2 victory points. The game ends when a player gets 10 victory points (at least).

Agents for SoC. The Java Settlers game as well as two jSettler agents included in it, was originally created in [28]. The jSettlers use business negotiation strategies for evaluating offers and trading during a negotiation. For other decision making in the game, [28] implemented two separate algorithms for computing the building speed (each corresponds to an agent respectively). The "fast" strategy takes into account the immediate reward, and the "smart" one evaluates actions beyond that. Both switch between three different strategies (road-building, city-building and monopolizing) for deciding their build plan. Then, [8] altered the original jSettler by improving aspects regarding initial placement of builds, favoring specific types of build plans and the purchase of development cards. In [21], model trees and linear regression were used for Q-function approximation.

SoC has also been a popular testbed for MCTS methods. Several implementations have been put forward [20, 27] but without being tested according to the complete game rule set, or without being able to make trade offers. Three different MCTS agents capable of trading, while playing under the full set of rules are introduced in [12]. Specifically, one uses the known bandit family method UCT, the second agent is an extension of UCT using Bayesian inference (BUCT), and the third employs the VPI RL method [5]. All these agents also use parts of the original jSettler, for tasks not supported by the MCTS (e.g. playing development cards). Although the MCTS agents could not outperform the jSettler, the VPI one appeared to be competitive. In [6], an extension of UCT incorporating knowledge mining from a corpus of human game-play was proposed. It had promising results, but was not tested against the jSettler agent.

According to [13], the only approaches so far capable of outperforming the jSettler, are specific heuristic improvements in agent negotiation strategies, and also, interestingly, a DRL approach [4]. That agent trains and tests a fully-connected neural network using DQN against several opponents, focusing on mapping game instances to dialogue actions. The DRL agents trained playing against jSettlers and supervised agents, both outperformed the jSettler. In contrast to [4], our approach does not require a long sequence of training experiences ranging over a series of games, but actually learns to play effectively within a single SoC game instance, in true "on-line" RL style. As such, learning occurs within a realistically small number of rounds.

3   Our Approach

In this section we explain the algorithm and architecture of the novel agent we designed for decision making in strategic board games. This agent observes the environment (i.e. the game state), and returns an appropriate action, one with maximal Q-value. To do this, we built a deep structure. Figure 1 provides a graphical overview of our approach.

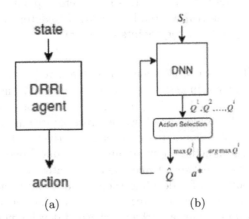

(a)   (b)

Fig. 1. (a) A top-level view of the DRRL agent algorithm, with the state as input and the preferable action as output. (b) The network gets S_t as input at time t and outputs all the Q^is. Then the action with the maximum Q^i for this state is returned as output of the algorithm and the maximum Q^i is used in next network iteration.

We decided to exploit the advantages of recurrent networks and specifically LSTMs, along with the concept of Q-decomposition. This is novel for DRL:

in this way, our network does not approximate the Q-function itself, but the different local Q-values. Considering that the neural network architecture we have implemented uses recurrency, we can employ the Bellman Q-value function as if dealing with an MDP problem [10].

3.1   Local Q-Function Approximation

In order to approximate the Q-function, we will use the Bellman equation as well as the concept of Q-decomposition. We define $s_t = \langle s_j^i \rangle$ as a factored state at time t, with s_j^t being one of N variables (state features) that takes a value in its domain. Then, a_t^i as the action $i = \{1, 2, ...I\}$ selected at t, where $a_i \in \{a^1, a^2, ..., a^I\} \in A$ (and $I = |A|$). S includes all possible states, and A all possible actions. We also consider a reward function $r(s_t, a_t^i)$ which denotes the returned reward when selecting action a_t^i in state s_t at time step t. The real Q-function can be approximated by some $\hat{Q}$ at time t, whose form can be expressed recursively, in terms of the $\hat{Q}$ of the next state, as follows:

$$\hat{Q}(s_t, a_t^i; \theta) = r(s_t, a_t^i) + \gamma \max_a \hat{Q}(s_{t+1}, a; \theta) \tag{3}$$

where θ are some network weight parameters.

Following the concept of Q-decomposition presented in Sect. 2, we assume that the Q function can be decomposed into $i \in [1, I]$ partitions, where $I = |A|$. In this way we can compute the short-term effect of every action a_t^i based on the current environment state, i.e., a local Q-value which is action dependent. Since we try to approximate these Q-values, we will also associate a different set of weights for each one of them, denoted as θ^i, such that:

$$Q^i(s_t, a_t^i; \theta^i) = [\phi_1(s_t) \; \phi_2(s_t) \cdots \phi_N(s_t)] \begin{bmatrix} \theta_1^i \\ \theta_2^i \\ \vdots \\ \theta_N^i \end{bmatrix} = \sum_{j=1}^N \phi_j(s_t) \cdot \theta_j^i \tag{4}$$

where Q^i is the Q-value estimated for action i at timestep t, and the $\phi_j(s_t)$ basis functions are of the form:

$$\phi_j(s_t) = (1 - \sigma(s_t)) \cdot \phi_j(s_{t-1}) + \sigma(s_t) \cdot \tanh(s_t) \tag{5}$$

where σ is the sigmoid function and tanh the hyperbolic tangent function, and each ϕ_j is actually applied to a corresponding state variable s_t^j (the state variables for the SoC domain are listed in Table 1).

Note that with $\phi(s_t)$ being an $1 \times N$ vector, each θ^i, is a $N \times 1$ vector. By multiplying these two vectors, we end up with a unique value for each Q^i.

Now, to come up with better Q^i estimates, we naturally aim to minimize the difference between the "real" Q-value estimate, and Q^i in each time step. This can be framed as an optimization problem that minimizes a loss function with respect to the θ^i parameters, and solved with stochastic gradient descent (SGD) via back-propagation.

Given these calculated Q^is, the action that will now be considered "best" for the state that was observed, is the one with maximal (locally) estimated Q-value:

$$arg \max_i Q^i = arg \max_i Q^i(s_t, a_t^i; \theta^j) = arg \max \begin{cases} Q^1(s_t, a^1; \theta^1) \\ Q^2(s_t, a^2; \theta^2) \\ ... \\ Q^n(s_t, a^I; \theta^I) \end{cases} \quad (6)$$

The Q-value of that a^* action will constitute the new $\hat{Q}$ value estimate for an $s = s_t, a = a^*$ pair: whenever $s = s_t$ is encountered, $a = a^*$ is the *currently assumed* action of choice, and has a value of $\hat{Q}(s, a^*)$; of course, this can be updated in future iterations.

We note again that in our case, there is a separate θ^i for each a^i. Given this, notice that the evaluation for the Q-values at s_t can be computed locally in parallel[2] for all possible actions and corresponding parameters θ.

3.2   Deep Architecture

We implemented a deep network (see Fig. 2) consisting of $I = |A|$ parallel recurrent recursive neural networks (RNN). Each RNN is an LSTM [11] (i.e. LSTM layer) followed by a soft-max activator (soft-max Layer) and outputs a Q-value. The LSTM layer practically summarizes a state input vector retrieved from the environment as a single representation, that contains information about the entire input sequence. The LSTM cells also provide an internal memory to the network, regarding useful information from the previous seen states of the environment, in order for them to be used in the next time step. Those aspects of the LSTM provide us with the necessary mathematical transformations to compute the basis function presented in Sect. 3.1 based on the current state input as well as the previous state's one.

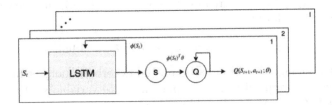

Fig. 2. A visualization of the network for one action. The deep network consists of $I = |A|$ parallel RNNs with LSTMs (LSTM layer). A soft-max activator (soft-max layer) follows each one of them in order to normalize weights θ^i and the Q-value is updated (Q-value layer). Each RNN outputs a Q-value.

---

[2] More accurately, in our implementation in a pseudo-parallel manner: all LSTMs are executed independently and the final action is selected given their outputs.

By iteratively optimizing a selected loss function, the corresponding parameters θ^i for each RNN can be computed. The inner product of the LSTM output with each θ^i vector, is normalized with the soft-max activation function [26], in order to keep values scaled between $[0, 1]$. This information is actually a Q-value estimate. The actual value of each Q^i corresponding to an action is then approximated after the new gradients are computed, (i.e. the θ^i updates from the loss function minimization with SGD) as described earlier. The output of the Q-values at one time step is actually the output of the whole network.

3.3 The DRRL Agent

In this section we present the Deep Recurrent Reinforcement Learning agent, an algorithm to approximate the local Q-values using the deep network above. In each time step t (or a round of a game), the algorithm receives a state of the environment as an input and returns the action corresponding to the maximum Q-value approximated by the network. In response it receives a reward and observes a state transition.

In every t all the Q^is corresponding to actions are updated, as well as the weights θ^i relevant to each one of them. This means, that an optimization step occurs for every Q^i separately, where previous estimated parameters θ^i for a^i are updated. Thus evaluations for all the θ^is are conducted in parallel.

Since all the θ^is are updated, the Q^is are updated too. We preserve each $Q^i(s_t, a_t^i; \theta^i)$ approximation in the Q-values layer as the output of the RNN, and when the back-propagation optimization steps take place, those Q values are updated with the corresponding θ^is. Each Q^i represents the mapping from all states seen so far and parameters to a specific action; thus it is a part of a local Q-function approximator that contains the previous time step Q-value approximation for this action with the corresponding θ^i. The action a_t^i extracted from the relevant Q^i which maximizes it locally, is considered to be the one that maximizes the actual Q-function. So this action is returned by the algorithm. The appropriate $\hat{Q} = Q^i$ for the selected action is fed back in the RNN iteratively as part of the input, representing the Q-value of the previous time-step.

To summarize, the algorithm learns through the network different set of parameters regarding the loss functions, and also saves parameter-dependent function approximators. Thus, there are $I = |A|$ Q-approximators, each one using a different instance of the LSTM layer to update its own parameters in every iteration. A pseudo-code for the algorithm is provided in Algorithm 1.

4 Evaluation

The DRRL agent was evaluated in the Settlers of Catan (SoC) domain, using the jSettlers Java application.[3] In this section we explain how DRRL was instantiated in this domain, and present our evaluation results.

---

[3] http://nand.net/jsettlers/.

Algorithm 1. DRRL Algorithm

1: **procedure** DRRL
2:　　Input $= s_t, \hat{Q}$
3:　　// Returns a "best" action a^* and an estimate of its value at s_t
4:　　Initialize Q^i to zero, $\forall i \in \{1, 2, ..., |A|\}$
5:　　// at t=0 only, all $\hat{Q}s$ and $\theta^i s$ are also initialized to zero
6:
7:　　**for** $\forall i \in \{1, 2, ..., |A|\}$ **do**　　　　▷ All computations and updates in parallel
8:　　　　Set temporary $Q^i(s_t, a_t^i; \theta^i) = \phi(s_t)\theta^i$
9:　　　　Perform SGD with respect to θ^i:
10:　　　$$L(\theta^i) = \nabla_{\theta^i} \left(r(s_t, a_t^i) + \gamma \hat{Q}((s'|a_t, s_t), a') - Q^i(s_t, a_t^i; \theta^i) \right)^2$$
11:　　　　Update θ^i, Q^i
12:　　$a^* = \arg\max_i Q^i(s_t, a_t^i; \theta^i)$
13:　　$\hat{Q}(s_t, a^*) \leftarrow \max_i Q^i(s_t, a_t^i; \theta^i)$
14:　　**return** $\hat{Q}(s_t, a^*), a^*$

4.1 Domain State Representation and Action Set

Our decision making algorithm requires a state of the environment to be fed as an input to the network. The state captures the board of the game, and the information known to the player at time step t. Each state feature (state element) is assigned to a range of integers, and thus the state is an integer vector of 161 elements (see Table 1). In more detail, the first 5 features (hasClay, hasOre, hasSheep, hasWheat, hasWood) represent the available resources of the player whose turn is to move. The board consists of 19 main hexes. Each hex has 6 nodes, thus there are in total 54 nodes. Also there exist 80 edges connecting the nodes to each other. The player can build roads to connect her settlements and cities on those edges. The robber is placed in the desert hex when the game begins, and can be moved according to the game rules.

Since SoC is a board game with numerous possible actions, we focused our learning on a constrained set of actions, in order to reduce computational complexity. Since trading is key to the game, this is the action set we selected. Therefore, we select an action set $A^0 \in A$, containing (i) actions specifying "trade offers", and (ii) "reply" actions towards proposers. For all other actions, the agent simply adopts the choices of the jSettler. Thus, the jSettler part of the agent is responsible for choosing actions like "build a road").[4]

In more detail, the DRRL agent is responsible for "reply" actions to opponent offers (i.e. accept, reject, counter-offer), and "trade offers" for giving up to two resources (same or different kind) and receiving one. The resources available are Clay, Ore, Sheep, Wheat and Wood—thus the "trade offers" include all possible combinations of giving and receiving those resources. Some trade offers examples are: *Trade 1 Clay for 1 Wheat, Trade 2 Woods for 1 Clay, Trade 1 Ore and 1 Sheep for 1 Wood*. Overall, we have 72 actions $\in A^0$, 70 for trade offers, and 2

---

[4] In general, our action and game set up follows [4].

Table 1. State features retrieved from jSettlers. Number is the number of elements needed to describe this feature and Domain includes the feature's possible values.

Num	Feature	Domain	Description
1	Clay	$\{0, .., 10\}$	Player's number of Clay Units
1	Ore	$\{0, .., 10\}$	Player's number of Ore Units
1	Sheep	$\{0, .., 10\}$	Player's number of Sheep Units
1	Wheat	$\{0, .., 10\}$	Player's number of Wheat Units
1	Wood	$\{0, .., 10\}$	Player's number of Wood Units
49	Hexes	$\{0, .., 5\}$	Type of resource in a hex (unknown element = 0, clay = 1, ore = 2, sheep = 1, wheat = 4, wood = 5)
54	Nodes	$\{0, .., 4\}$	Builds ownership on each node (settlements and cities): (no builds = 0, opponents' builds = 1, 2, agents' builds = 3, 4)
80	Edges	$\{0, .., 2\}$	Roads ownership on each edge: (no road = 0, opponents' road = 1, agents' road = 2)
1	Robber	$\{0, .., 5\}$	Robber's location (unknown element = 0, clay = 1, ore = 2, sheep = 1, wheat = 4, wood = 5)
1	Turns	$\{0, .., 100\}$	Number of game turns
1	VP	$\{0, .., 10\}$	Number of player's victory points

for reply actions. The counter-offer reply action is not considered as a separate action in the actual implementation, since the agent directly makes a new offer instead of accepting or rejecting a trade.

4.2   DRRL in the SoC Domain

In SoC, the DRRL agent is responsible for the following during game-play:

1. Decide whether to accept, reject or make a counter-offer to a given trade proposal from another player.
2. At each round decide if a new trade offer should be made or not.
3. Select the preferred trade offer, if such an offer is to be made.

Given our discussion above, we consider a_t^i as the action $i = \{1, 2, 3, ...72\}$ selected at t, where $a_i \in \{a^1, a^2, a^3, ..., a^{72}\} \equiv A^o$. Furthermore, assuming action-dependent state features, we consider $Q^i(s_t, a_t^i; \theta^i)$ as a partition of the Q-function, for taking action a^i in state s at t, given the network parameters θ^i. Thus we assign a different $\theta^i \in \{\theta^1, \theta^2, ..., \theta^{72}\}$ for each possible action.

The function $r(s_t, a_t^i)$ gives the returned reward when executing action a_t^i in state s_t at time step t. In our setting, the only reward signal provided by the jSettlers at a state is VPs accumulated so far. So, we formulate the reward function as follows:

$$r(s_t, a_t^i) = \begin{cases} \overline{VP}(s_t, a_t^i) \cdot k & \text{if } \overline{VP}(s_t, a_t^i) > 0 \\ -VP \cdot k & otherwise \end{cases} \qquad (7)$$

where VP are the accumulated victory points gained in the game so far, and $\overline{VP}(s_t, a_t^i)$ are the victory points gained in t by acting with a^i—i.e., the immediate reward provided by the difference between VPs in the previous and current state (which actually can sometimes be a negative number). The k parameter was set to 0.01 after trials, since we wanted the reward to have an impact on the Q-values, without creating a huge deviation among them.

The DRRL algorithm is taking action every time the agent wants to decide whether to trade or reply to an offer. In each round, the network receives the state of the game as an input. To achieve this, the architecture of the neural network is formed accordingly. This means that the LSTM layer consists of 72 independent LSTM units, and also 72 different Q^i and sets of θ^i are produced.

4.3   Simulations and Results

The DRRL agent was implemented in Python using the Tensorflow library. To measure agent's performance we manipulated the jSettlers Java API, in order to support python-based clients, and specifically ones with deep neural network architecture. We tested our algorithm both in CPU (Intel(R) Core(TM) i3-2120 @ 3.30 GH) and GPU (NVIDIA GeForce GTX 900), and actually it was performing slightly faster in the CPU. This probably happens since learning is taking place in batches of one, thus the GPU's accelerating aspects are not used.

Usually the training of deep neural networks needs thousands of iterations. For instance, in [4], they trained their DRL agent for SoC with 500,000 learning experiences, until it converges to an effective policy. In our approach, we do not train the neural network, thus we had to find an efficient way to optimize the network parameters within the given game limitations. To this end, we experimented with different learning rates for Stochastic Gradient Descent (SGD). For very small learning rates (i.e. exponential), the algorithm appeared to be incapable of exploring the action space within the course of one game, and alternated among using 3 or 4 actions only. On the other hand, for higher learning rates, the algorithm gets stuck at local minima, and ended up exploring during the first game rounds, but then kept selecting the same preferred action. After some experimentation, we set the learning rate to 0.0023.

The θ^i model parameters were initialized with a truncated normal distribution. We also tested the network initialized with random, normal, and uniform distributions, but this did not help, because it introduced a bias towards specific actions in some cases. Finally, the $\hat{Q}$ (i.e. the Q-value of the previous chosen action) as well as the local Q^is were initialized with zeros.

We pit our agent against the "standard" jSettler agent, and also against the three MCTS implementations from [12]: BUCT, UCT, and VPI (Sect. 2.3). A SoC game involves four agents, thus in each game four of the aforementioned five agents face each other, chosen randomly and also in random order. We made sure that every agent faces every possible combination of others an equal number of times: in total, we ran five 4-agent tournaments of 20 games each, and each agent participated in four of those tournaments.

In every new game, all network parameters are initialized, as if the agent has never played before. Each game lasts about 15–26 rounds, and the agent gains at most about 70 *learning experiences* (since an experience is gained in every turn where she has to propose an offer, or respond to one).[5] The DRRL algorithm itself runs in about 56 s, and the whole game in approximately half to one hour and a half (depending on the number of rounds). As mentioned above, in total we ran 100 games, by combining the agents in five different pools of four agents (each corresponding to a "tournament"). Thus, every agent faces all different combinations of opponents, itself participating in 80 games in total. We used as our principal comparison metric, the agents' *win ratio*, calculated for each agent over the number of games that agent participates in.

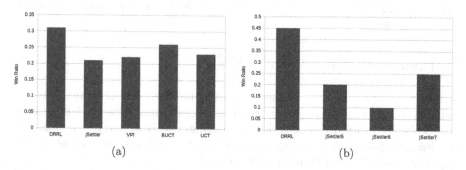

Fig. 3. (a) The agents' winning ratio over 80 games against all opponents. (b) The agents' winning ratio over 20 games against jSettlers.

Our evaluation results shown in summarized in Fig. 3a show that the DRRL agent is capable to outperform all others without pre-training, but by only training itself in real time while playing a game, and rapidly adapts to the game environment. Some instability is noticed from game to game, but this is normal due to the fact that each game is a first time experience for the agent, yet we plan to further investigate this in the future by learning parameters over a sequence of games. We can also report that the DRRL agent was consistently proposing significantly more offers (and counter-offers) than its opponents (specifically, 10–40 per game), and actually in a very early stage of the game, while other agents did not. This increased trading greatly benefited the agent. Yet, it was rarely accepting other players' trades offers—it would rather counter-offer instead. We also report that the MCTS agents perform much better when the DRRL agent is in the game, since it favors trading, while the jSettler does not.

---

[5] Compare this number to the $500,000$ learning experiences required by the DRL agent in [4].

We also ran experiments in a pool with only jSettlers as opponents. This allows us to make an indirect comparison of our DRRL agent against the "best" DRL agent built in [4], namely a DRL agent whose policy was pre-trained using about 500,000 learning experiences while facing jSettler opponents. Once trained, that DRL agent was able to achieve a win ratio of 53.36% when playing against three jSettler agents. We see in Fig. 3b that our DRRL agent achieves a win ratio of 45% against jSettlers, using at most 70 learning experiences; we note however that 45% is higher than the results achieved by most other DRL agents in [4]. We then increased our training horizon, allowing our DRRL agent to train itself across a series of 30 games; that is, the agent continued to update its network weights across 30 games ran sequentially, and not just one. This corresponds to ∼2,000 learning experiences. We can report that now the win rate of our agent rises to 56%, with DRRL now winning 17/30 games, and with the agent winning nine out of the fifteen last games played. Therefore our DRRL agent *(a)* matches (and slightly surpasses) the results of the DRL agent in [4]; and *(b)* we can also reasonably deduce that adding more training experience benefits the agent.

5   Conclusions and Future Work

In this paper we presented a novel approach for deep RL in strategic board games. We optimize the Q-value function through a neural network, but perform this locally for every different action, employing recurrent neural networks and Q-value decomposition. Our algorithm managed to outperform state-of-the-art opponents, and did so with minimal training.

Regarding future work, we intend to examine performance differences when using stacked LSTMs compared to the vanilla ones we used now, or even GRUs. Furthermore, we plan to extend the algorithm to larger or different action sets, and also create DRRL agents for other games. A final goal is to incorporate natural language processing, in order for the deep network to take into account the players' conversations during the game.

References

1. Afantenos, S., Kow, E., Asher, N., Perret, J.: Discourse parsing for multi-party chat dialogues. Proc. EMNLP **2015**, 928–937 (2015)
2. Anschel, O., Baram, N., Shimkin, N.: Deep reinforcement learning with averaged target DQN. CoRR abs/1611.01929 (2016)
3. Bellman, R.: Dynamic programming. Courier Corporation, Chelmsford (2013)
4. Cuayáhuitl, H., Keizer, S., Lemon, O.: Strategic dialogue management via deep reinforcement learning. In: Proceedings of the NIPS Deep Reinforcement Learning Workshop (NIPS 2015) (2015)
5. Dearden, R., Friedman, N., Russell, S.: Bayesian Q-learning. In: AAAI/IAAI, pp. 761–768 (1998)

6. Dobre, M.S., Lascarides, A.: Online learning and mining human play in complex games. In: 2015 IEEE Conference on Computational Intelligence and Games (CIG), pp. 60–67. IEEE (2015)
7. Finnman, P., Winberg, M.: Deep reinforcement learning compared with Q-table learning applied to backgammon (2016)
8. Guhe, M., Lascarides, A.: Game strategies for the Settlers of Catan. In: Computational Intelligence and Games (CIG), pp. 1–8. IEEE (2014)
9. van Hasselt, H., Guez, A., Silver, D.: Deep reinforcement learning with double Q-learning. CoRR abs/1509.06461 (2015)
10. Hausknecht, M., Stone, P.: Deep recurrent Q-learning for partially observable MDPs. CoRR, abs/1507.06527 **7**(1) (2015)
11. Hochreiter, S., Schmidhuber, J.: Long short-term memory. Neural Comput. **9**(8), 1735–1780 (1997)
12. Karamalegkos, E.: Monte Carlo tree search in the "Settlers of Catan" strategy game, Senior Undergraduate Diploma thesis, School of Electrical and Computer Engineering, Technical University of Crete (2014). https://goo.gl/rU9vG8
13. Keizer, S., et al.: Evaluating persuasion strategies and deep reinforcement learning methods for negotiation dialogue agents. In: Proceedings of the 15th Conference of the European Chapter of the Association for Computational Linguistics: Volume 2, Short Papers, vol. 2, pp. 480–484 (2017)
14. Kok, J.R., Vlassis, N.: Collaborative multiagent reinforcement learning by payoff propagation. J. Mach. Learn. Res. **7**(Sep), 1789–1828 (2006)
15. Lai, M.: Giraffe: using deep reinforcement learning to play Chess. arXiv preprint arXiv:1509.01549 (2015)
16. Lillicrap, T.P., et al.: Continuous control with deep reinforcement learning. CoRR abs/1509.02971 (2015)
17. Mnih, V., Kavukcuoglu, K., Silver, D., et al.: Human-level control through deep reinforcement learning. Nature **518**(7540), 529–533 (2015)
18. Oh, J., Guo, X., Lee, H., Lewis, R.L., Singh, S.: Action-conditional video prediction using deep networks in Atari games. In: Advances in Neural Information Processing Systems, pp. 2863–2871 (2015)
19. Osband, I., Blundell, C., Pritzel, A., Roy, B.V.: Deep exploration via bootstrapped DQN. CoRR abs/1602.04621 (2016)
20. Panousis, K.P.: Real-time planning and learning in the "Settlers of Catan" strategy game, Senior Undergraduate Diploma thesis, School of Electrical and Computer Engineering, Technical University of Crete (2014). https://goo.gl/4Hpx8w
21. Pfeiffer, M.: Reinforcement learning of strategies for Settlers of Catan. In: International Conference on Computer Games: Artificial Intelligence (2018)
22. Russell, S.J., Zimdars, A.: Q-decomposition for reinforcement learning agents. In: Proceedings of the 20th International Conference on Machine Learning (ICML-03), pp. 656–663 (2003)
23. Schmidhuber, J.: Deep learning in neural networks: an overview. Neural Netw. **61**, 85–117 (2015)
24. Silver, D., et al.: Mastering the game of go without human knowledge. Nature **550**, 354–359 (2017)
25. Stone, P., Veloso, M.: Team-partitioned, opaque-transition reinforcement learning. In: Proceedings of the Third Annual Conference on Autonomous Agents, pp. 206–212. ACM (1999)
26. Sutton, R.S., Barto, A.G.: Reinforcement Learning: An Introduction. MIT Press, Cambridge (1998)

27. Szita, I., Chaslot, G., Spronck, P.: Monte-Carlo tree search in Settlers of Catan. In: van den Herik, H.J., Spronck, P. (eds.) ACG 2009. LNCS, vol. 6048, pp. 21–32. Springer, Heidelberg (2010). https://doi.org/10.1007/978-3-642-12993-3_3

28. Thomas, R.S.: Real-time decision making for adversarial environments using a plan-based heuristic. Ph.D. thesis, Northwestern University (2003)

Counterfactually Fair Prediction Using Multiple Causal Models

Fabio Massimo Zennaro$^{(\boxtimes)}$ (ID) and Magdalena Ivanovska (ID)

Department of Informatics, University of Oslo,
PO Box 1080, Blindern, 0316 Oslo, Norway
{fabiomz,magdalei}@ifi.uio.no

Abstract. In this paper we study the problem of making predictions using multiple *structural causal models* defined by different agents, under the constraint that the prediction satisfies the criterion of *counterfactual fairness*. Relying on the frameworks of causality, fairness and opinion pooling, we build upon and extend previous work focusing on the qualitative aggregation of causal Bayesian networks and causal models. In order to complement previous qualitative results, we devise a method based on Monte Carlo simulations. This method enables a decision-maker to aggregate the outputs of the causal models provided by different agents while guaranteeing the counterfactual fairness of the result. We demonstrate our approach on a simple, yet illustrative, toy case study.

Keywords: Causality · Structural causal networks · Fairness ·
Counterfactual fairness · Opinion pooling · Judgement aggregation ·
Monte Carlo sampling

1 Introduction

In this paper we analyze the problem of integrating together the information provided by multiple agents in the form of potentially-unfair, predictive *structural causal models* in order to generate predictions that are *counterfactually fair*. We follow the tradition of social choice theory and assume that the opinions of independent, non-interacting agents or experts must be aggregated in a single final decision by a third party.

This work is rooted in two main fields of research: *causality* and *fairness*. Causality deals with the definition and the study of causal relationships; structural causal models, in particular, are versatile and theoretically-grounded models that allow us to express causal relations and to study these relationships via interventions and counterfactual reasoning [9]. Fairness is a research topic interested in evaluating if and how prediction systems deployed in sensitive scenarios may be guaranteed to support fair decisions; counterfactual fairness, in particular, is a concept of fairness developed in relation to causal models [7]. The use of causal models in societally-sensitive contexts has been advocated by several researchers on the ground that the additional structure of these models and

© Springer Nature Switzerland AG 2019
M. Slavkovik (Ed.): EUMAS 2018, LNAI 11450, pp. 249–266, 2019.
https://doi.org/10.1007/978-3-030-14174-5_17

the possibility of evaluating the effect of interventions would allow for deeper understanding and control in critical situations [2].

So far, little research has addressed the problem of aggregating multiple causal models. With no reference to fairness, [3] studied a method to aggregate causal Bayesian networks, while [1] introduced a notion of compatibility to analyze under which conditions causal models may be combined. Taking fairness into account, [11] proposed to tackle the problem of integrating multiple causal models as an optimization problem under the constraint of an ϵ-relaxation of fairness.

In this paper we offer a solution for the problem of generating counterfactually-fair predictions given a set of causal models. Differently from [3], we focus our study on structural causal models instead of causal Bayesian networks, as the latter ones can encode causal relationships but do not support counterfactual reasoning [9]; similarly to [3], though, we opt for a two-stage approach, made up of a qualitative stage, in which we work out the topology of an aggregated counterfactually-fair model, and a quantitative step, in which we use this topology to generate counterfactually-fair results out of individual causal models. The qualitative stage builds upon our previous work on this same topic in [13], and relies on the framework for judgment aggregation [6] and the work on pooling of causal Bayesian networks [3]. This paper reviews and rectifies this previous work and it defines the new quantitative stage which relies on Monte Carlo simulations [8] and, again, on opinion pooling [4].

The rest of the paper is organized as follows: Sect. 2 reviews basic concepts in the research areas considered; Sect. 3 provides the formalization of our problem and our contribution; Sect. 4 draws conclusions on this work and indicates future avenues of research.

2   Background

This section reviews basic notions used to define the problem of generating predictions out of multiple causal models under fairness: Sect. 2.1 recalls the primary definitions in the study of causality; Sect. 2.2 discusses the notion of fairness in machine learning; Sect. 2.3 offers a formalization of the problem of opinion pooling.

2.1   Causality

Following the formalism in [9], we provide the basic definitions for working with causality.

Causal Model. A *(structural) causal model* $\mathcal{M}$ is a triple $(\mathcal{U}, \mathcal{V}, \mathcal{F})$ where:

- $\mathcal{U}$ is a set of exogenous variables $\{U_1, U_2, \ldots, U_m\}$ representing background factors that are not affected by other variables in the model;
- $\mathcal{V}$ is a set of endogenous variables $\{V_1, V_2, \ldots, V_n\}$ representing factors that are determined by other exogenous or endogenous variables in the model;

- $\mathcal{F}$ is a set of functions $\{f_1, f_2, \ldots, f_n\}$, one for each variable V_i, such that the value v_i is determined by the *structural equation*:

$$v_i = f_i\left(v_{pa_i}, u_{pa_i}\right),$$

where v_{pa_i} are the values assumed by the variables in the set $\mathcal{V}_{pa_i} \subseteq \mathcal{V}\backslash\{V_i\}$ and u_{pa_i} are the values assumed by the variables in the set $\mathcal{U}_{pa_i} \subseteq \mathcal{U}$; that is, for each endogenous variable, there is a set of (parent) endogenous and a set of (parent) exogenous variables that determine its value through the corresponding structural equations.

Causal Diagram. The *causal diagram* $\mathcal{G}\left(\mathcal{M}\right)$ associated with the causal model $\mathcal{M}$ is the directed graph (V, E) where:

- V is the set of vertices representing the variables $\mathcal{U} \cup \mathcal{V}$ in $\mathcal{M}$;
- E is the set of edges determined by the structural equations in $\mathcal{M}$; edges are coming to each endogenous node V_i from each of its *parent nodes* $\mathcal{V}_{pa_i} \cup \mathcal{U}_{pa_i}$; we denote $V_j \rightarrow V_i$ the edge going from V_j to V_i.

Assuming the acyclicity of causality, we will take that a causal model $\mathcal{M}$ entails an acyclic causal diagram $\mathcal{G}\left(\mathcal{M}\right)$ represented as a *directed acyclic graph* (DAG).

Context. Given a causal model $\mathcal{M} = (\mathcal{U}, \mathcal{V}, \mathcal{F})$, we define a *context* $\overrightarrow{u} = (u_1, u_2, \ldots, u_m)$ as a specific instantiation of the exogenous variables, $U_1 = u_1, U_2 = u_2, \ldots, U_m = u_m$. Given an endogenous variable V_i, we will use the shorthand notation $V_i\left(\overrightarrow{u}\right)$ to denote the value of the variable V_i under the context $\overrightarrow{u}$. This value is obtained by propagating the context $\overrightarrow{u}$ through the causal diagram according to the structural equations.

Intervention. Given a causal model $\mathcal{M} = (\mathcal{U}, \mathcal{V}, \mathcal{F})$, we define *intervention* $do(V_i = \bar{v})$ as the substitution of the structural equation $v_i = f_i\left(v_{pa_i}, u_{pa_i}\right)$ in the model $\mathcal{M}$ with the equation $v_i = \bar{v}$. Given two endogenous variables X and Y, we will use the shorthand notation $Y_{X \leftarrow x}$ to denote the value of the variable Y under the intervention $do(X = x)$.

Notice that, from the point of view of the causal diagram, performing the intervention $do(X = x)$ is equivalent to setting the value of the variable X to x and removing all the incoming edges $\cdot \rightarrow X$ in X.

Counterfactual. Given a causal model $\mathcal{M} = (\mathcal{U}, \mathcal{V}, \mathcal{F})$, the context $\overrightarrow{u}$, two endogenous variables X and Y, and the intervention $do(X = x)$, a *counterfactual* is the value of the expression $Y_{X \leftarrow x}(\overrightarrow{u})$.

Note that, under the given context $\overrightarrow{u}$, the variable Y takes the value $Y(\overrightarrow{u})$. Instead, the counterfactual $Y_{X \leftarrow x}(\overrightarrow{u})$ represents the value that Y would have taken in the context $\overrightarrow{u}$ had the value of X been x.

Probabilistic Causal Model. A *probabilistic causal model* $\mathcal{M}$ is a tuple $(\mathcal{U}, \mathcal{V}, \mathcal{F}, P(U))$ where:

- $(\mathcal{U}, \mathcal{V}, \mathcal{F})$ is a causal model;
- $P(U)$ is a probability distribution over the exogenous variables. The probability distribution $P(U)$, combined with the dependence of each endogenous variable V_i on the exogenous variables, as specified in the structural equation for v_i, allows us to define a probability distribution $P(V)$ over the endogenous variables as: $P(V = v) = \sum_{\{\vec{u}|V=v\}} P(U = \vec{u})$.

Notice that we overload the notation $\mathcal{M}$ to denote both (generic) causal models and probabilistic causal models; the context will allow the reader to distinguish between them.

2.2   Fairness

Following the work of [7], we review the topic of fairness, with a particular emphasis on counterfactual fairness for predictive models.

Fairness and Learned Models. Black-box machine learning systems deployed in sensitive contexts (e.g.: police enforcement or educational grants allocation) and trained on historical real-world data have the potential of perpetuating, or even introducing [7], socially or morally unacceptable discriminatory biases (for a survey, see, for instance, [14]). The study of *fairness* is concerned with the definition of new metrics to assess and guarantee the social fairness of a predictive decision system.

Fairness of a Predictor. A predictive model can be represented as a (potentially probabilistic) function of the form $\hat{Y} = f(Z)$, where $\hat{Y}$ is a *predictor* and Z is a vector of covariates. An observational approach to fairness states that the set of covariates can be partitioned in a set of *protected attributes* $\mathcal{A}$, representing discriminatory elements of information, and a set of *features* $\mathcal{X}$, carrying no sensitive information. The predictive model can then be redefined as $\hat{Y} = f(A, X)$ and the fairness problem is expressed as the problem of learning a predictor $\hat{Y}$ that does not discriminate with respect to the protected attributes $\mathcal{A}$. Given the complexities of social reality and disagreement over what constitutes a fair policy, different measures of fairness may be adopted to rule out discrimination (e.g.: counterfactual fairness or fairness through unawareness); for a more thorough review of different types of fairness and their limitations, see [5] and [7].

Fairness Over Causal Models. Given a probabilistic causal model $(\mathcal{U}, \mathcal{V}, \mathcal{F}, P(U))$ fairness may be evaluated following an observational approach. Let us take $\hat{Y}$ to be an endogenous variable whose structural equation provides the predictive function $f_{\hat{Y}}$; let us also partition the remaining variables $\mathcal{U} \cup \mathcal{V} \backslash \{\hat{Y}\}$ into a set of protected attributes $\mathcal{A}$ and a set of features $\mathcal{X}$. Then we can evaluate the fairness of the predictor $\hat{Y}$ with respect to the discriminatory attributes $\mathcal{A}$.

Counterfactual Fairness. Given a probabilistic causal model $\mathcal{M} = (\mathcal{U}, \mathcal{V}, \mathcal{F}, P(U))$, a predictor $\hat{Y}$, and a partition of the variables $\mathcal{U} \cup \mathcal{V} \setminus \{\hat{Y}\}$ into $(\mathcal{A}, \mathcal{X})$, the predictor $f_{\hat{Y}}$ is *counterfactually fair* if, for every context $\overrightarrow{u}$,

$$P\left(\hat{Y}_{A \leftarrow a}(\overrightarrow{u}) | X = x, A = a\right) = P\left(\hat{Y}_{A \leftarrow a'}(\overrightarrow{u}) | X = x, A = a\right),$$

for all values y of the predictor, for all values a' in the domain of A, and for all x in the domain of X [7].

In other words, the predictor $\hat{Y}$ is counterfactually fair if, under all the contexts, the prediction on $\hat{Y}$ given the observation of the protected attributes $A = a$ and the features $X = x$ would not change if we were to intervene $do(A = a')$ to force the value of the protected attributes A to a', for all the possible values that the protected attributes can assume.

Denoting $Desc_{\mathcal{M}}(\mathcal{A})$ the descendants of the nodes in $\mathcal{A}$ in the model $\mathcal{M}$, an immediate property follows from the definition of counterfactual fairness:

Lemma 1 *(Lemma 1 in [7]). Given a probabilistic causal model $(\mathcal{U}, \mathcal{V}, \mathcal{F}, P(U))$, a predictor $\hat{Y}$ and a partition of the variables into $(\mathcal{A}, \mathcal{X})$, the predictor $\hat{Y}$ is counterfactually fair if $f_{\hat{Y}}$ is a function depending only on variables that are not in $Desc_{\mathcal{M}}(\mathcal{A})$.*

2.3 Opinion Pooling

Following the study of [4], we introduce the framework for opinion pooling.

Opinion Pooling. Assume there are N experts, each one expressing his/her opinion o_i, $1 \leq i \leq N$. The problem of *pooling* (or *aggregating*) the opinions o_i consists in finding a single pooled opinion o^* representing the collective opinion that best represents the individual opinions in the given context.

Probabilistic Opinion Pooling. Given opinions in the form of probability distributions $p_i(x)$, $1 \leq i \leq N$, defined over the same domain, *probabilistic opinion pooling* is concerned with finding a single pooled probability distribution $p^*(x) = F(p_1, \ldots, p_n)(x)$, where F is a functional mapping a tuple of pdfs to a single probability distribution [4].

Now, given a set of probabilistic opinions $p_i(x)$, different functionals F may be chosen to perform opinion pooling, either using a principled approach such as an axiomatic approach based on the definition of a set of desired properties [4], or using standard statistical operators, such as arithmetic averaging or geometric averaging.

Judgment Aggregation. Given opinions in the form of a set of Boolean functions $j_i(x)$, *judgment aggregation* is concerned with finding a single Boolean function $j^*(x) = F(j_1, \ldots, j_n)(x)$, where F is a functional mapping a tuple of Boolean functions to a single Boolean function [6].

As in the case of probabilistic opinion pooling, given a set of judgments $j_i(x)$, different functions F may be chosen to perform judgment aggregation, such as majority voting or intersection [3].

Aggregation of Causal Bayesian Networks. Given opinions expressed in the form of causal Bayesian networks[1] (CBN) $\mathcal{B}_i$, aggregation of CBNs is concerned with defining a single pooled CBN $\mathcal{B}^*$.

A seminal study in aggregating CBNs is offered by [3]. They suggest a *two-stage approach* to the problem of aggregating N causal Bayesian networks $\mathcal{B}_i$. In the first qualitative stage, they determine the graph of the pooled CBN reducing the problem to a judgment aggregation over the edges in the individual CBNs; namely, for every two variables X and Y, the presence of an edge from node X to node Y in the model $\mathcal{B}_i$ is represented as the i-th expert casting the judgment $j_i(X \to Y) = 1$, and the absence of it as the judgment $j_i(X \to Y) = 0$; the problem of defining the pooled graph is then reduced to a judgment aggregation problem over the judgments $j_i()$. In the second quantitative step, they derive the conditional probability distributions for the pooled graph applying probabilistic opinion pooling to the corresponding conditional probability distributions in the individual CBNs. A critical result in the study of [3] is the translation of the classical impossibility theorem from judgment aggregation [6] into an impossibility theorem for the qualitative aggregation of CBNs:

Theorem 1 *(Theorem 1 in [3]). Given a set of CBNs defined over at least three variables, there is no judgment aggregation rule f that satisfies all the following properties:*

- Universal Domain: *the domain of f includes all logically possible acyclic causal relations;*
- Acyclicity: *the pooled graph produced by f is guaranteed to be acyclic over the set of nodes;*
- Unbiasedness: *given two variables X and Y, the causal dependence of X on Y in the pooled graph rests only on whether X is causally dependent on Y in the individual graphs, and the aggregation rule is not biased towards a positive or negative outcome;*
- Non-dictatorship: *the pooled graph produced by f is not trivially equal to the graph provided by one of the experts.*

As a consequence of this theorem, no unique aggregation rule satisfying the above properties can be chosen for the pooling of causal judgments in the first step of the two-stage approach. A relaxation of these properties must be decided depending on the scenario at hand.

3   Aggregation of Causal Models Under Fairness

This section analyzes how probabilistic causal models can be aggregated under fairness: Sect. 3.1 provides a formalization of our problem; Sect. 3.2 discusses how

---

[1] A Bayesian network (BN) [10] is a structured representations of the joint probability distribution of a set of variables in the form of a directed acyclic graph with associated conditional probability distributions. A causal BN is a BN where all the edges represent causal relations between variables.

to define the topology of a counterfactually-fair causal graph by pooling together the graphs of different probabilistic structural causal models; Sect. 3.3 explains how to evaluate a counterfactually-fair prediction out of the individual models using the pooled graph; finally, Sect. 3.4 offers an illustration of the use of our method on a toy case study.

3.1   Problem Formalization

Let us consider the case in which N experts are required to provide a probabilistic causal model $\mathcal{M}_i = (\mathcal{U}, \mathcal{V}, \mathcal{F}_i, P_i(U))$ representing a potentially socially-sensitive scenario. For simplicity, we assume that the experts are provided with a fixed set of variables $(\mathcal{U}, \mathcal{V})$. The task of the experts can be summarized in two modeling phases: (i) a qualitative phase, in which they define the causal topology of the graph $\mathcal{G}(\mathcal{M}_i)$ (which variables are causally influencing which other variables); and, (ii) a quantitative phase, specifying the probability distribution functions $P_i(U)$ (how the stochastic behavior of the exogenous variables is modeled) and the structural equations $\mathcal{F}_i$ (how each endogenous variable is causally influenced by its parents variables).

Critically, we are not requesting the experts to provide fair models. Individual experts may not be aware of specific discrimination issues, they may have different understandings of fairness or, simply, they may not have the technical competence to formally evaluate or guarantee fairness. The task of defining which form of fairness is relevant, and enforcing it, is up to the final decision-maker only.

The (potentially unfair) models $\mathcal{M}_i$ defined by the experts are then provided to a decision-maker, who wants to exploit them to compute a single counterfactually-fair predictive output $\hat{Y}^*$. We assume the decision-maker to be knowledgeable of fairness implications and to be responsible for partitioning the exogenous and endogenous variables $\mathcal{U} \cup \mathcal{V}$ into sensitive attributes $\mathcal{A}$ and non-sensitive attributes $\mathcal{X}$.

In summary, our problem may be expressed as follows: given N (potentially counterfactually-unfair) probabilistic causal models $\mathcal{M}_i$ defined on the same variables $(\mathcal{U}, \mathcal{V})$, and a partition of the variables into $(\mathcal{A}, \mathcal{X})$, can we define a pooling algorithm that allows us to construct an aggregated counterfactually-fair causal model $\mathcal{M}^* = f(\mathcal{M}_i)$ and compute a counterfactually-fair predictive output $\hat{Y}^*$?

Mirroring the modeling approach of the experts, we propose to solve this problem by adopting a two-stage approach. We reduce the task of the final decision-maker to the two following phases: (i) a qualitative phase, in which we compute the topology of a counterfactually-fair pooled graph $\mathcal{G}(\mathcal{M}^*)$ (which nodes and causal links from the individual models can be retained under a requirement of counterfactual fairness); and, (ii) a quantitative phase, in which we provide a method to evaluate a probability distribution over the final predictor $\hat{Y}^*$ (how is the final counterfactually-fair predictor computed).

3.2 Qualitative Aggregation over the Graph

In the qualitative phase of our approach, we consider the different expert models $\mathcal{M}_i$ and we focus on the problem of defining the topology of an aggregated graph $\mathcal{G}(\mathcal{M}^*)$ that guarantees counterfactual fairness.

We tackle this challenge following the solution proposed in [3] to perform a qualitative aggregation of causal Bayesian networks (see Sect. 2.3): in each causal model $\mathcal{M}_i$, we convert each edge in a binary judgment and we then perform judgment aggregation according to a chosen aggregation rule JARule (). However, this solution presents two shortcomings: (i) it does not guarantee that the predictor $\hat{Y}^*$ in the aggregated probabilistic causal model $\mathcal{M}^*$ will be counterfactually fair; and (ii) because of Theorem 1, we cannot apply this procedure without first choosing one of the properties in the theorem statement to sacrifice. We solve the first problem by relying and enforcing the condition specified in Lemma 1; practically, we introduce in our algorithm a *removal step*, in which we remove all the protected attributes and their descendants from the aggregated model $\mathcal{M}^*$. This procedure satisfies by construction the condition in Lemma 1, and thus guarantees counterfactual fairness. We address the second problem by arguing that the structure of the causal graph immediately suggests the possibility of dropping the property of unbiasedness, which requires that the presence of an edge in the pooled graph depends only on the presence of the same edge in each individual graph; we can relax this property, by making the presence of an edge in the final graph dependent also on an ordering of the edges. In our specific case, we can easily introduce an ordering of the edges with respect to the predictor $\hat{Y}$, as a starting point. The ordering produced in this way may not be total, and we may still have to introduce another rule to break potential ties (e.g., random selection or an alphabetical criterion). We formalize this procedure in an additional algorithmic step, *pooling step*, in which we order edges in relation to their distance from the predictor and then we perform judgment aggregation using the rule JARule () so that, if the edge is selected for insertion in the pooled model $\mathcal{G}(\mathcal{M}^*)$, then it is added as long as acyclicity is not violated [3].

The two algorithmic steps defined above may be interchangeably combined. This gives rise to two alternative algorithms: a *removal-pooling* algorithm (Algorithm 1), and a *pooling-removal* algorithm (Algorithm 2). We provide the pseudo code for the two functions, Removal() and Pooling(), in Algorithms A.1 and A.2 in the appendix; also, for a detailed description and analysis of these functions and their outputs we refer the reader to [13].

At the end, both algorithms return the skeleton of a pooled causal model $\mathcal{M}^*$ that is guaranteed to be counterfactually fair. However, the result, so far, contains only a qualitative description of the causal model: we determined the topology of the graph $\mathcal{G}(\mathcal{M}^*)$, but the structural equations of the pooled causal model $\mathcal{M}^*$ are left undefined.

Algorithm 1. Removal-Pooling Algorithm

1: **Input:** N graph models $\mathcal{G}(\mathcal{M}_i)$ over the variables $\{\mathcal{U}, \mathcal{V}\}$, a predictor $\hat{Y} \in \mathcal{V}$, a partitioning of the variables $\{\mathcal{U}, \mathcal{V}\} \backslash \{\hat{Y}\}$ into protected attributes $\mathcal{A}$ and $\mathcal{X}$, a judgment aggregation rule JARule ()

2:

3: $\{\mathcal{M}'_i\}_{i=1}^{N} = $ Removal $\left(\{\mathcal{M}_i\}_{i=1}^{N}, \mathcal{A}\right)$

4: $\mathcal{M}^* = $ Pooling $\left(\{\mathcal{M}'_i\}_{i=1}^{N}, \text{JARule}()\right)$

5: **return** $\mathcal{M}^*$

Algorithm 2. Pooling-Removal Algorithm

1: **Input:** N graph models $\mathcal{G}(\mathcal{M}_i)$ over the variables $\{\mathcal{U}, \mathcal{V}\}$, a predictor $\hat{Y} \in \mathcal{V}$, a partitioning of the variables $\{\mathcal{U}, \mathcal{V}\} \backslash \{\hat{Y}\}$ into protected attributes $\mathcal{A}$ and $\mathcal{X}$, a judgment aggregation rule JARule ()

2:

3: $\mathcal{M}' = $ Pooling $\left(\{\mathcal{M}_i\}_{i=1}^{N}, \text{JARule}()\right)$

4: $\mathcal{M}^* = $ Removal $(\mathcal{M}', \mathcal{A})$

5: **return** $\mathcal{M}^*$

3.3   Quantitative Aggregation over the Distribution of the Predictor

In the quantitative phase of our approach, we study how we can use the topology of the pooled and counterfactually-fair causal graph $\mathcal{G}(\mathcal{M}^*)$ that we have generated in the first step to produce quantitative and counterfactually-fair outputs.

After the phase of qualitative aggregation, the individual models provided by the experts $\mathcal{M}_i$ have been aggregated only with respect to their nodes and edges; the pooled graph $\mathcal{M}^*$ encodes the topology of a counterfactually-fair causal model, but it lacks the definition of probability distributions over the exogenous nodes and structural equations over the endogenous nodes in order to be complete and usable.

Defining the probability distributions and the structural equations in the aggregated model $\mathcal{M}^*$ by pooling together individual functions in each expert model $\mathcal{M}_i$ is a particularly challenging task: different experts may provide substantially different functions, functions may be defined on different domains (since the same node may have different incoming edges in different expert graphs), and domains may have been changed in the aggregated model (since nodes may have been dropped in order to guarantee counterfactual fairness). Therefore, instead of finding an explicit form for the probability distributions and the structural equations in the aggregated model $\mathcal{M}^*$, we suggest to compute the distribution of the predictor $\hat{Y}_i$ in each expert model $\mathcal{M}_i$ while *integrating out* all the components that do not belong to the counterfactually-fair graph $\mathcal{G}(\mathcal{M}^*)$, and finally aggregate them to obtain the final counterfactually-fair predictor $\hat{Y}^*$.

More formally, let $\mathcal{Z} \subseteq \mathcal{U} \cup \mathcal{V}$ be the set of *fair features* corresponding to nodes that are present in $\mathcal{G}(\mathcal{M}^*)$, and let $\bar{\mathcal{Z}} \subseteq \mathcal{U} \cup \mathcal{V}$ be the set of *unfair features* corresponding to nodes that are not present in $\mathcal{G}(\mathcal{M}^*)$. Now, if we are given an instance of fair features $Z = z$, we can compute the probability distribution of the predictor $\hat{Y}_i$ in each model $\mathcal{M}_i$ by integrating out the unfair features:

$$P(\hat{Y}_i|Z = z) = \int P(\hat{Y}_i|Z = z, \bar{Z})d\bar{Z}.$$

In other words, we use the aggregated model $\mathcal{G}(\mathcal{M}^*)$ to identify in each expert model a counterfactually fair sub-graph and to integrate out the contributions of the rest of the graph. Practically, this operation of integration and estimation of the distribution of $P(\hat{Y}_i|Z = z)$ can be efficiently carried out using Monte Carlo sampling [8].

The final result will then consist of a set of N individual pdfs $P(\hat{Y}_i|Z = z)$. In order to take a decision, these pdfs can be simply merged together using simple statistical operators (e.g.: by taking the mean of the expected values) or relying on standard opinion pooling operators [4].

3.4   Illustration

Here we give a simple illustration of the problem of causal model aggregation under counterfactual fairness, which we recover from [13].

Setup. In this example, we imagine that the head of a Computer Science department asked two professors, Alice and Bob, to design a predictive system to manage PhD selections. In particular, we imagine that Bob and Alice were required to define a causal model over the endogenous variables *age* (Age), *gender* (Gnd), *MSc university department* (Dpt), *MSc final mark* (Mrk), *experience in the job market* (Job), *quality of the cover letter* (Cvr), the relative exogenous variables (U.) and a predictor ($\hat{Y}$).

The graphs of the two models provided by Alice $\mathcal{G}(\mathcal{M}_A)$ and Bob $\mathcal{G}(\mathcal{M}_B)$ are given in Figs. 1 and 2, respectively.

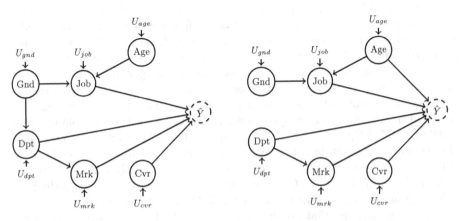

Fig. 1. Graph $\mathcal{G}(\mathcal{M}_A)$.   **Fig. 2.** Graph $\mathcal{G}(\mathcal{M}_B)$.

Moreover, Alice and Bob came up with the following probability distributions over the exogenous nodes $\mathcal{U}$:

$$U_{age\_A} \sim \text{Poisson}(\lambda = 3) \qquad U_{age\_B} \sim \text{Poisson}(\lambda = 4)$$
$$U_{job\_A} \sim \text{Bernoulli}(p = 0.3) \qquad U_{job\_B} \sim \text{Bernoulli}(p = 0.2)$$
$$U_{gnd\_A} \sim \text{Bernoulli}(p = 0.5) \qquad U_{gnd\_B} \sim \text{Bernoulli}(p = 0.5)$$
$$U_{dpt\_A} \sim \text{Categorical}([0.7, 0.2, 0.1]) \quad U_{dpt\_B} \sim \text{Categorical}([0.7, 0.15, 0.1, 0.05])$$
$$U_{mrk\_A} \sim \text{Beta}(\alpha = 2, \beta = 2) \qquad U_{mrk\_B} \sim \text{Beta}(\alpha = 2, \beta = 2)$$
$$U_{cvr\_A} \sim \text{Beta}(\alpha = 2, \beta = 5) \qquad U_{cvr\_B} \sim \text{Beta}(\alpha = 2, \beta = 5)$$

and they defined the structural equations for the endogenous nodes $\mathcal{V}$:

$$V_{age\_A} = 20 + U_{age\_A}$$
$$V_{gnd\_A} = U_{gnd\_A}$$
$$V_{job\_A} = U_{job\_A} + \frac{V_{gnd\_A}}{2} + \frac{V_{age\_A}}{100}$$
$$V_{dpt\_A} = \text{if } (V_{gnd\_A} = 1) \text{ then } 0 \text{ else } \frac{U_{dpt\_A}}{10}$$
$$V_{mrk\_A} = \text{if } (V_{dpt\_A} = 0) \text{ then } U_{mrk\_A} + 0.1 \text{ else } U_{mrk\_A} - 0.1$$
$$V_{cvr\_A} = U_{cvr\_A} + 0.2$$
$$\hat{Y}_A = V_{job\_A} + V_{dpt\_A} + V_{mrk\_A} + V_{cvr\_A}$$

$$V_{age\_B} = 19 + U_{age\_B}$$
$$V_{gnd\_B} = U_{gnd\_B}$$
$$V_{job\_B} = U_{job\_B} + \frac{V_{age\_B}}{100} + [\text{if } (V_{gnd\_B} = 1) \text{ then } 0.5 \text{ else } 0]$$
$$V_{dpt\_B} = \frac{U_{dpt\_B}}{10}$$
$$V_{mrk\_B} = \text{if } (V_{dpt\_B} = 0) \text{ then } U_{mrk\_B} + 0.1 \text{ else } U_{mrk\_B}$$
$$V_{cvr\_B} = U_{cvr\_B} + 0.1$$
$$\hat{Y}_A = \frac{V_{job\_A}}{100} + V_{job\_A} + V_{dpt\_A} + V_{mrk\_A} + V_{cvr\_A}$$

Notice that these probability distributions and structural equations are pure examples, and do not have any deep meaningful relation with the scenario at hand; they were chosen mainly to illustrate the use of a variety of distributions and functions, and to output as a predictor $\hat{Y}$ a score that can be used for decision-making. In reality, such functions would be determined via machine learning methods or carefully defined by a modeler.

Notice that the models and the structural equations were defined by Alice and Bob with no explicit concern about any form of fairness. Now, however, the head of the department wants to aggregate these models in a way that guarantees counterfactual fairness with respect to the gender of the PhD candidate.

Qualitative Aggregation. As a first step, the head of the department assumes all the exogenous variables to be non-sensitive and partitions the endogenous ones into protected attributes $\mathcal{A} = \{\text{Gnd}\}$ and features $\mathcal{X} = \{\text{Age}, \text{Dpt}, \text{Mrk}, \text{Job}, \text{Cvr}\}$, and then chooses as a judgment aggregation rule `JARule ()` the *strict majority rule*.

She then decides to apply the *pooling-removal* algorithm to the models $\mathcal{M}_A$ and $\mathcal{M}_B$. A detailed explanation of the application of this algorithm is available in [13]. The resulting pooled counterfactually-fair model $\mathcal{M}^*$ is illustrated in Fig. 3.

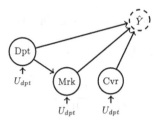

Fig. 3. Graph of the pooled counterfactually-fair model $\mathcal{G}(\mathcal{M}^*)$ after applying the *pooling-removal* algorithm.

Quantitative Aggregation. At this point, the head of the department can use the individual expert models $\mathcal{M}_A$ and $\mathcal{M}_B$, and the aggregated counterfactually-fair model $\mathcal{G}(\mathcal{M}^*)$ to compute predictive scores for the PhD applicants.

Suppose, for instance, that the following candidates were to submit their application:

App$_1$ = {Age = 22; Gnd = F; Dpt = Computer Science; Mrk = 0.8; Job = True; Cvr = 0.4}
App$_2$ = {Age = 22; Gnd = M; Dpt = Computer Science; Mrk = 0.8; Job = True; Cvr = 0.4}

From a formal point of view, we may imagine the second candidate as the result of the intervention $do(\text{Gnd} = M)$ on the first candidate, thus forcing the gender to male.

Now, for the sake of illustration, we implemented the models and the candidates using Edward [12], a Python library for probabilistic modeling, and we made the code available online[2]. Whenever using Monte Carlo sampling, we collected 10^5 samples.

The models provided by Alice and Bob obviously define two different predictors $\hat{Y}_A$ and $\hat{Y}_B$ with two dissimilar probability distributions (see Fig. 4 in the appendix for an estimation of these pdfs). If the head of the department were to feed the data about the candidates to the two models, she would receive different and unfair results:

$$\hat{Y}_A(\text{App}_1) = 4.720 \quad \hat{Y}_B(\text{App}_1) = 3.340$$
$$\hat{Y}_A(\text{App}_2) = 2.720 \quad \hat{Y}_B(\text{App}_2) = 2.840$$

---

[2] https://github.com/FMZennaro/Fair-Pooling-Causal-Models.

Indeed, these results show both a legitimate disagreement between Alice and Bob on how they score individual candidates, but they also show a troubling internal disagreement in that both experts assign different scores to the two applicants when the only difference between them is their gender. Since the head of the department considers gender a protected attribute, this result is deemed unfair. More formally, if the second candidate were to be seen as an intervention we would have:

$$P\left(\hat{Y}_{Gnd \leftarrow F}(\overrightarrow{u}) | X = x, A = a\right) \neq P\left(\hat{Y}_{Gnd \leftarrow M}(\overrightarrow{u}) | X = x, A = a\right),$$

thus denying counterfactual fairness.

To tackle the problem, the head of the department decides to evaluate the predictive score for the candidates by computing the distribution of the predictor $\hat{Y}_i$ given only the fair features Z from the pooled counterfactually-fair model $\mathcal{M}^*$:

$$P(\hat{Y}_A | Z = z) = \int_{Age, Gnd, Job} P(\hat{Y}_A | Dpt = CS, Mrk = 0.8, Cvr = 0.4)$$

$$P(\hat{Y}_B | Z = z) = \int_{Age, Gnd, Job} P(\hat{Y}_B | Dpt = CS, Mrk = 0.8, Cvr = 0.4)$$

This step does not provide a scalar output as in the previous evaluation, but it defines two probability distributions $P(\hat{Y}_A | Z = z)$ and $P(\hat{Y}_B | Z = z)$ (see Fig. 5 in the appendix for an estimation of these pdfs). These two pdfs can now be pooled together for final decision making. After deciding to compute the average of the expected value of the pdfs[3], the head of the department obtains the following fair results:

$$E\left[P\left(\hat{Y}_A(App_1) | Z\right)\right] = 3.022 \quad E\left[P\left(\hat{Y}_B(App_1) | Z\right)\right] = 2.312$$
$$E\left[P\left(\hat{Y}_A(App_2) | Z\right)\right] = 3.030 \quad E\left[P\left(\hat{Y}_B(App_2) | Z\right)\right] = 2.313$$

These results are more comforting in that, while they still allow room for disagreement between Alice and Bob over the evaluation of individual candidates, they guarantee that the two applicants, who differ only on a protected attribute, receive identical predictive scores (within the numerical precision of a Monte Carlo simulation[4]). Again, formally, if we were to see the second candidate as an intervention on the first, we would have:

$$P\left(\hat{Y}_{Gnd \leftarrow F}(\overrightarrow{u}) | X = x, A = a\right) = P\left(\hat{Y}_{Gnd \leftarrow M}(\overrightarrow{u}) | X = x, A = a\right),$$

---

[3] Notice that the decision of considering just the expected value of the pdfs may not be ideal in this case, given the multimodality of these pdfs, as shown in Fig. 5 in the appendix.

[4] This precision can be increased by incrementing the number of Monte Carlo samples collected.

thus satisfying counterfactual fairness. Therefore, these counterfactually-fair scores can now be safely averaged into a final fair predictor $\hat{Y}^*$ by the head of the department and used for decision-making.

4   Conclusion

This paper offers a complete approach to the problem of computing aggregated predictive outcomes from a collection of causal models while respecting a principle of counterfactual fairness. Our solution comprises two phases: (i) a qualitative step, in which we use judgment aggregation to determine a counterfactually-fair pooled model; and, (ii) a quantitative step, in which we use Monte Carlo sampling to evaluate the predictive output of each model by integrating out unfair components, and then we perform opinion pooling to aggregate these outputs. The entire approach was illustrated on the toy-case of PhD admissions, showing that it does indeed provide counterfactually-fair results.

However, this work represents just a first attempt at solving the problem of aggregating multiple causal models in order to provide counterfactually-fair predictions. Some avenues for future development that we are investigating include:

- our method presupposes causal models defined over the same set of exogenous and endogenous variables; however, our solution is quite flexible and, with little formal work, it may be extended to produce fair outcomes from the aggregation of causal graphs defined on different sets of exogenous and endogenous variables;
- from a formal point of view, it may be interesting to investigate extreme cases (e.g.: scenarios in which qualitative aggregation provides no fair model), examine what are the conditions for a fair model to exist, and evaluate how these conditions may be relaxed to allow the *most fair possible* aggregation of causal models;
- more importantly, it may be worth to study how a purely observational approach to fairness may be integrated by a pro-active affirmative approach. In this last more realistic approach, the aim is not only to guarantee unbiased outcomes with respect to the available historical data (which may itself be biased), but purposefully and actively compensate existing bias through policies and interventions.

Appendix A: Algorithms

Algorithm A.1. Removal Function

1: **Input:** N graph models $\mathcal{G}(\mathcal{M}_j) = (\mathsf{V}_j, \mathsf{E}_j)$, where the vertex set V_j is defined over the exogenous and endogenous variables $\mathcal{U} \cup \mathcal{V}$; a partitioning of the variables $\mathcal{U} \cup \mathcal{V} \setminus \{\hat{Y}\}$ into protected attributes $\mathcal{A}$ and $\mathcal{X}$.

2:

3: Initialize $\mathcal{W}_{fair} := \mathcal{U} \cup \mathcal{V}$

4: **for** $j = 1$ **to** N **do**

5:     $\mathcal{W}_{\neg} := \{W \,|\, (W \in \mathcal{A}) \vee (W \in Desc_{\mathcal{M}_j}(\mathcal{A}))\}$

6:     $\mathcal{W}_{fair} := \mathcal{W}_{fair} \setminus \mathcal{W}_{\neg}$

7: **end for**

8: **for** $j = 1$ **to** N **do**

9:     Remove from the edge set E_j of $\mathcal{G}(\mathcal{M}_j)$ all edges $V_x \to V_y \,|\, (V_x \notin \mathcal{W}_{fair} \vee V_y \notin \mathcal{W}_{fair})$

10: **end for**

11:

12: **return** $\mathcal{M}_j$

Algorithm A.2. Pooling Function

1: **Input:** N graphs models $\mathcal{G}(\mathcal{M}_j) = (\mathsf{V}_j, \mathsf{E}_j)$, where the vertex set V_j is defined over the exogenous and endogenous variables $\mathcal{U} \cup \mathcal{V}$; a judgment aggregation rule $\mathtt{JARule}\,()$.

2:

3: Initialize D to the length of the longest path in the models $\mathcal{M}_j$

4: Initialize $\mathcal{M}^*$ by setting up the graph $\mathcal{G}(\mathcal{M}^*)$ in which $\mathsf{V}^* = \mathcal{U} \cup \mathcal{V}$ and $\mathsf{E}^* = \emptyset$

5: **for** $j = 1$ **to** N **do**

6:     Initialize the vertex set $\mathsf{V}_{j,0} := \hat{Y}$

7:     Initialize the edge set $\mathsf{E}_{j,0} := \emptyset$

8: **end for**

9: **for** $j = 1$ **to** N **do**

10:     **for** $d = 1$ **to** D **do**

11:         $\mathsf{E}_{j,d} := \{(V_x \to V_y) \,|\, (V_x \to V_y) \in \mathsf{E}_j \wedge (V_x \in \mathsf{V}_{j,d-1} \vee V_y \in \mathsf{V}_{j,d-1})\}$

12:         $\mathsf{V}_{j,d} := \{V_x \,|\, (V_x \to \cdot) \in \mathsf{E}_{j,d} \vee (\cdot \to V_x) \in \mathsf{E}_{j,d}\}$

13:     **end for**

14: **end for**

15: **for** $j = 1$ **to** N **do**

16:     **for** $d = 1$ **to** D **do**

17:         $\forall (V_x \to V_y) \in \mathsf{E}_{j,d}$, if $(\mathtt{JARule}\,(V_x \to V_y) = 1) \vee (\mathsf{E}^* \cup \{V_x \to V_y\}$ is acyclic) then $\mathsf{E}^* := \mathsf{E}^* \cup \{V_x \to V_y\}$

18:     **end for**

19: **end for**

20: **return** $\mathcal{M}^*$

Appendix B: Figures

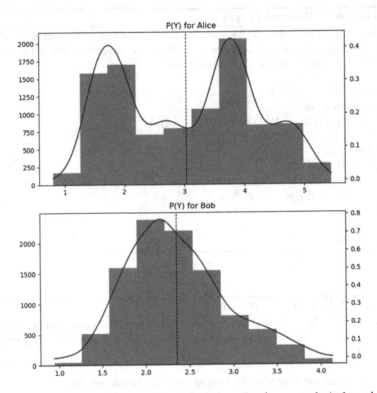

Fig. 4. Histogram and probability distribution function (computed via kernel density estimation) of $P(\hat{Y})$ in the model provided by Alice and Bob. The x-axis reports the domain of the outcome of the predictor $\hat{Y}$; the left y-axis reports the number of samples used to compute the histogram, while the right y-axis reports the normalized values used to compute the pdf.

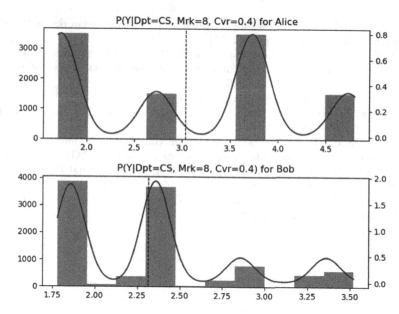

Fig. 5. Histogram and probability distribution function (computed via kernel density estimation) of $P(\hat{Y}|Dpt = \text{CS}, Mrk = 0.8, Cvr = 0.4)$ in the model provided by Alice and Bob. The x-axis reports the domain of the outcome of the predictor $\hat{Y}$; the left y-axis reports the number of samples used to compute the histogram, while the right y-axis reports the normalized values used to compute the pdf.

References

1. Alrajeh, D., Chockler, H., Halpern, J.Y.: Combining experts' causal judgments (2018)
2. Barabas, C., Dinakar, K., Virza, J.I., Zittrain, J., et al.: Interventions over predictions: reframing the ethical debate for actuarial risk assessment. arXiv preprint arXiv:1712.08238 (2017)
3. Bradley, R., Dietrich, F., List, C.: Aggregating causal judgments. Philos. Sci. **81**(4), 491–515 (2014)
4. Dietrich, F., List, C., Hájek, A., Hitchcock, C.: Probabilistic Opinion Pooling: Oxford Handbook of Probability and Philosophy. Oxford University Press, Oxford (2016)
5. Gajane, P.: On formalizing fairness in prediction with machine learning. arXiv preprint arXiv:1710.03184 (2017)
6. Grossi, D., Pigozzi, G.: Judgment aggregation: a primer. Synth. Lect. Artif. Intell. Mach. Learn. **8**(2), 1–151 (2014)
7. Kusner, M.J., Loftus, J., Russell, C., Silva, R.: Counterfactual fairness. In: Advances in Neural Information Processing Systems, pp. 4069–4079 (2017)
8. MacKay, D.J.: Information Theory, Inference and Learning Algorithms. Cambridge University Press, Cambridge (2003)
9. Pearl, J.: Causality. Cambridge University Press, Cambridge (2009)

10. Pearl, J.: Probabilistic Reasoning in Intelligent Systems: Networks of Plausible Inference. Elsevier, Amsterdam (2014)
11. Russell, C., Kusner, M.J., Loftus, J., Silva, R.: When worlds collide: integrating different counterfactual assumptions in fairness. In: Advances in Neural Information Processing Systems, pp. 6417–6426 (2017)
12. Tran, D., Kucukelbir, A., Dieng, A.B., Rudolph, M., Liang, D., Blei, D.M.: Edward: a library for probabilistic modeling, inference, and criticism. arXiv preprint arXiv:1610.09787 (2016)
13. Zennaro, F.M., Ivanovska, M.: Pooling of causal models under counterfactual fairness via causal judgement aggregation. ICML/IJCAI/AAMAS Workshop on Machine Learning for Causal Inference, Counterfactual Prediction, and Autonomous Action. arXiv preprint arXiv:1805.09866 (2018)
14. Zliobaite, I.: A survey on measuring indirect discrimination in machine learning. arXiv preprint arXiv:1511.00148 (2015)

Author Index

Printed in the United States
By Bookmasters